T0178169

Physical Chemistry from a Different Angle Workbook

Georg Job · Regina Rüffler

Physical Chemistry from a Different Angle Workbook

Exercises and Solutions

Springer

Georg Job
Job Foundation
Hamburg, Germany

Regina Rüffler
Job Foundation
Hamburg, Germany

ISBN 978-3-030-28490-9 ISBN 978-3-030-28491-6 (eBook)
https://doi.org/10.1007/978-3-030-28491-6

This Springer imprint is published by the registered company Springer Nature Switzerland AG
The registered company address is: Gewerbestrasse 11, 6330 Cham, Switzerland

Preface

As a companion to the textbook "Physical Chemistry from a Different Angle," this workbook offers an excellent opportunity to deepen the understanding of the concepts presented in the textbook by addressing specific problems. The workbook is divided into two parts: a first part with exercises and a second part with the corresponding completely worked-out solutions.

The first part of the book comprises almost 200 exercises, which are thematically linked to the textbook. Exercises with a higher degree of difficulty are marked with *. The exercises marked with † are based on examples provided by Prof. Friedrich Herrmann.

The following procedure is recommended when dealing with exercises where a numerical value is required: First, the general formula is specified. In order to avoid calculation errors, the physical quantities are given in SI units (with the corresponding sign), meaning the volume is not given in liters, but in m^3, the mass not in g, but in kg, etc. Eventually, the final result is calculated. For intermediate calculations, it is useful to express the unit prefixes as powers of ten (except k for kg) and to only use the prefixes again in the final result.

In the second part of the book, the solutions are explained in detail step by step, thus enabling students to learn independently. Short intermediate calculations (designated with dotted-dashed underlines) are inserted into the main calculation of the quantity in question, longer intermediate calculations are placed in front of the main calculation.

All references, including equation numbers, relate to the first edition of the textbook "Physical Chemistry from a Different Angle" published in 2016 by Springer.

We gratefully acknowledge the constant support and patience of the board of the Job Foundation. Our special thanks go to Eduard J. Job†, who founded the Job Foundation in 2001, and to his brother Norbert Job, who has been financing the Foundation since 2017. Additionally, we would like to thank Dr. Steffen Pauly, Beate Siek and Dr. Charlotte Hollingworth at Springer for their excellent cooperation and Prof. Friedrich Herrmann for allowing us to use some of his exercise examples.

Hamburg, Germany
February 2019

Georg Job
Regina Rüffler

The original version of the book was revised: The correction to the book is available at https://doi.org/10.1007/978-3-030-28491-6_3

About the Authors

Georg Job studied chemistry at the University of Hamburg, where he received in 1968 his Ph.D. degree supervised by A. Knappwost. He was a lecturer at the Institute of Physical Chemistry of the University of Hamburg from 1970 to 2001. Two guest lectureships brought him to the Institute for Didactics of Physics at the University of Karlsruhe (1979–80) and to the Tongji University in Shanghai (1983).

Already as a student, G. Job was looking for ways to make the abstract conceptual structure of thermodynamics easier to understand and thus easier to use. This finally led to the publication of the book *A New Concept of Thermodynamics* in 1972. In the following, the new teaching concept was consistently developed and extended in its application, so that it eventually included most parts of physical chemistry. G. Job presented the new concept in numerous articles and lectures at national and international conferences. In cooperation with R. Rüffler, the textbook *Physical Chemistry from a Different Angle* was published. As a companion to the textbook, this workbook with numerous exercises and the corresponding detailed solutions was written.

Regina Rüffler studied chemistry at the Saarland University, where she received in 1991 her Ph.D. degree supervised by U. Gonser. She was a lecturer at the Institute of Physical Chemistry of the University of Hamburg from 1989 to 2002, interrupted by a two-year stay as a guest researcher at the Saarland University. During her lectureship she gave numerous lectures and supervised practical courses and numerical exercises at the bachelor's as well as the master's level.

Her passion for teaching led her to join the Eduard Job Foundation in 2002. In addition to writing the textbook as well as the workbook in cooperation with G. Job, she creates instructions for the more than one hundred demonstration experiments integrated into the textbook and produces corresponding educational videos for which she has won several prizes (https://job-stiftung.de/index.php?demonstration-experiments). She has also presented the new teaching concept in all its facets at numerous national and international conferences and has implemented it in the experimental lecture "Thermodynamics" for students of wood science at the University of Hamburg.

Contents

1 Exercises

1.1 Introduction and First Basic Concepts

1.1.1 Molar concentration

A sample of 500 cm^3 of an aqueous solution (S) of glucose (substance B; $C_6H_{12}O_6$) contains 45 g of this sugar. What is the molar concentration c_B of the solution?

Hint: The molar masses M of the considered substances are usually not given because they can be easily calculated with the help of the content formula: They correspond to the sums of the molar masses (whose numerical values are equal to the relative atomic masses) of the elements that form the compound multiplied by the corresponding content numbers. Glucose, for example, is composed of the elements carbon, hydrogen and oxygen with the molar masses of 12.0×10^{-3} kg mol^{-1}, 1.0×10^{-3} kg mol^{-1} and 16.0×10^{-3} kg mol^{-1}. Finally, the molar mass M_B of glucose results in $(6 \times 12.0 + 12 \times 1.0 + 6 \times 16.0) \times 10^{-3}$ kg mol^{-1} = 180.0×10^{-3} kg mol^{-1}.

1.1.2 Mass fraction and mole fraction

A volume of 100 mL of a saline solution (S) [consisting of sodium chloride (substance B) and water (substance A)] contains 15.4 g of dissolved sodium chloride at 25 °C. Calculate the mass fraction w_B and the mole fraction x_B, if the density of the saline solution is ρ_S = 1.099 g mL^{-1}.

1.1.3 Mass fraction

For simplicity, hard liquor can be considered as a homogeneous mixture of ethanol (substance B) and water (substance A). We would like to make a "hard liquor" with a mass fraction w_B of 33.5 % of alcohol, which corresponds to 40 percent alcohol by volume (vol%), the unit common in liquor trade. (Many varieties of vodka, for example, have such an alcoholic content.) For this purpose, we have 100 g of ethanol at hand. How many grams of water must be added to obtain the desired alcoholic content?

1.1.4 Composition of a gas mixture

What are the molar concentration and mass concentration, c_B and β_B, as well as the mole fraction and mass fraction, x_B and w_B, of oxygen (substance B) in air at standard conditions? [Air is roughly a mixture of O_2 and N_2 (substance A) where 21 % of the molecules are O_2.]

G. Job and R. Rüffler, *Physical Chemistry from a Different Angle Workbook*,
https://doi.org/10.1007/978-3-030-28491-6_1

Hint: For the sake of simplicity, assume an amount of air of one mole and take into account that one mole of any gas, be it pure or mixed, has a volume V of about 24.8 L at standard conditions [298 K (25 °C), 100 kPa (1 bar)].

1.1.5* Converting measures of composition

In Table 1.1 in the textbook "Physical Chemistry from a Different Angle", the mathematical relations for converting the most common measures of composition for binary mixtures of two components A and B are given. Prove that the expression

$$\frac{M_A c_B}{\rho - c_B(M_B - M_A)}$$

in the second column of the first line corresponds to x_B.

1.1.6 Description of a reaction process

The ammonia synthesis is the main industrial process for the conversion of the hardly reactive atmospheric nitrogen into usable nitrogenous compounds (such as fertilizers). We consider the formation of ammonia in a continous-flow reactor under stationary conditions for a short while.

	Symbol B	B′	D
Conversion formula:	$N_2 + 3\ H_2 \rightarrow 2\ NH_3$.		

Extent of conversion: $\xi(10^h10^m) = 13$ mol, $\xi(10^h40^m) = 19$ mol.

To simplify matters, one uses here and in the following appropriate symbols instead of the full names or formulas of the substances.

a) What are the conversion numbers ν_B, $\nu_{B'}$ and ν_D?

b) Calculate the conversion $\Delta\xi$ during the observation period.

c) Calculate the changes in amount and mass of all three substances in this period.

1.1.7 Application of the basic stoichiometric equation in titration

After the addition of a few drops of a phenolphthalein solution to 250 mL of a sulfuric acid solution (B) of unknown concentration $c_{B,0}$, this solution is titrated with a sodium hydroxide solution (B′) (concentration $c_{B',0} = 0.1\ \text{kmol m}^{-3}$) until the color of the solution changes from colorless to pink. 24.40 mL of base are required.

a) Formulate the conversion formula of the underlying acid-base reaction as well as the corresponding basic stoichiometric equation.

b) What is the molar concentration $c_{B,0}$ of the sulfuric acid solution?

c) Calculate the mass $m_{B,0}$ of H_2SO_4 (in mg) contained in the initial solution.

1.1.8 Application of the basic stoichiometric equation in precipitation analysis

A volumetric pipette (*2*) is used to transfer a certain volume of a Ba^{2+} solution (such as a solution of barium nitrate) from a volumetric flask (*1*) into a beaker. An excess of diluted sulfuric acid is added whereby a precipitate of slightly soluble $BaSO_4$ forms. This precipitate is filtered off, washed, dried and weighed; its final mass in the filter crucible is 467 mg.

	Symbol	B	B′	D
Conversion formula:		$Ba^{2+}\|w$ +	$SO_4^{2-}\|w$ →	$BaSO_4\|s$

Basic stoichiometric equation: $\Delta\xi = \dfrac{\Delta n_B}{\nu_B} = \dfrac{\Delta n_{B'}}{\nu_{B'}} = \dfrac{\Delta n_D}{\nu_D}$

Again, appropriate symbols are used instead of the formulas of the substances.

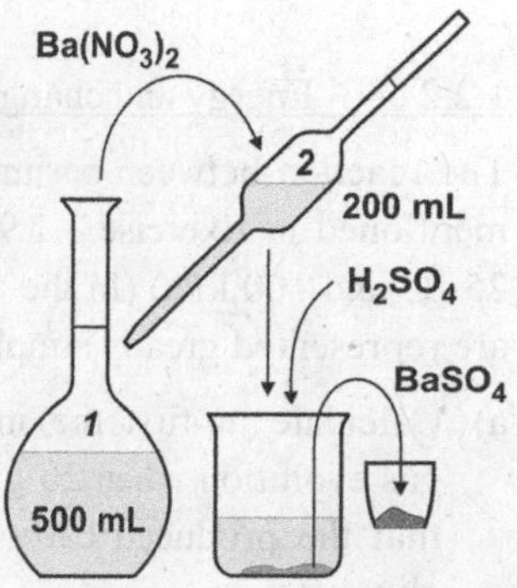

a) Calculate the conversion $\Delta\xi$ in case of complete precipitation.

b) What value results for the change Δn_B of the amount of Ba^{2+}?

c) What is the Ba^{2+} concentration $c_{B,1}$ of the barium nitrate solution in the volumetric flask?

d) What was the amount $n_{B,1}$ of Ba^{2+} in the full volumetric flask, i.e. before taking the sample?

1.1.9 Basic stoichiometric equation with participation of gases

When $BaCO_3$ is dissolved in diluted nitric acid (to prepare, for example, the barium nitrate solution from Exercise 1.1.8) CO_2 is released.

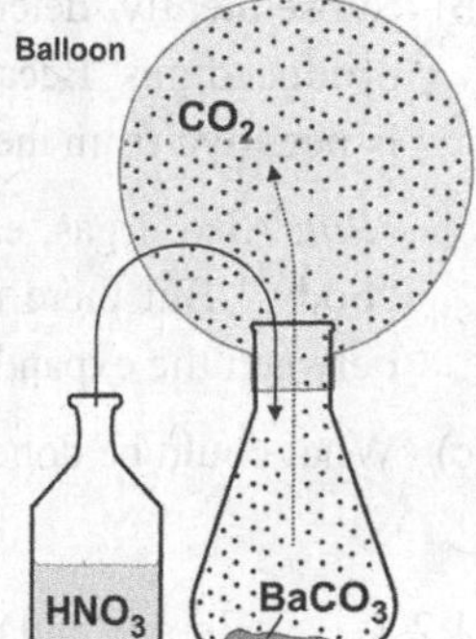

a) Formulate the conversion formula of this process and also the basic stoichiometric equation.

b) What is the minimum volume of nitric acid (c = 2 kmol m^{-3}) required to dissolve 1 g of $BaCO_3$?

c) What volume of carbon dioxide is produced during the process (under standard conditions)? (In the adjoining figure, the gas molecules are represented greatly simplifying by dots.)

Hint: Remember that one mole of any gas has a volume of about 24.8 L under standard conditions.

1.2 Energy

1.2.1 Energy expenditure for stretching a spring

A steel spring, such as the one on the right, has a spring stiffness D of the order of 10^5 N m^{-1}.

a) How much energy W_1 is needed to stretch the spring a distance of 10 cm from its unstretched length l_0?

b) How much energy W_2 is needed to stretch the spring from 50 cm to 60 cm?

c) A person with a mass of 50 kg hangs on the spring attached to the ceiling. Calculate the extension Δl of the spring.

1.2.2 Energy and change of volume (I)

The reaction between barium carbonate and diluted nitric acid mentioned in Exercise 1.1.9 takes place in an open beaker (at 25 °C and 100 kPa) (In the adjoining figure, the gas molecules are represented greatly simplifying by dots.).

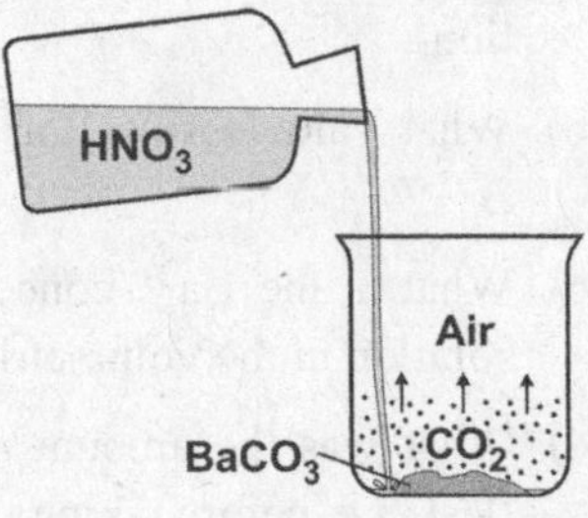

a) Calculate at first the increase ΔV in volume due to the gas evolution when 20 g of carbonate are used. We imagine that the produced carbon dioxide gas drives back the air above it.

   Hint: Remember that one mole of any gas, be it pure or mixed, has a volume V of about 24.8 L at standard conditions (25 °C, 100 kPa). Therefore, the mixing of the carbon dioxide with the air above it practically does not change the value of ΔV.

b) Subsequently, determine the expenditure of energy ΔW required to make room for the produced gas. Because the stored energy decreases when the system expends energy, ΔW is negative from the point of view of the system.

   Hint: Also a gas, e.g. in a cylinder with movable piston, can be considered as an elastic "body." But there is even no need for a real piston: we can imagine that the boundary between the expanding gas and the surrounding air is acting as a kind of piston.

c) What could be done to make the change in volume ΔV visible?

1.2.3 Energy and change of volume (II)

Octane (C_8H_{18}), a constituent of motor fuel, is combusted with air supply and the exhaust gases are cooled down to room temperature. Calculate the energy exchanged between system and surroundings under standard conditions (298 K, 100 kPa) due to the increase ΔV in volume. One liter of liquid octane (density $\rho = 0.70$ g cm^{-3}) should be used. At first, insert the missing conversion numbers in the following conversion formula:

Symbol	B		B′		D		D′

$\square\ C_8H_{18}|l + \square\ O_2|g \rightarrow \square\ CO_2|g + \square\ H_2O|g\,.$

Comment: To prevent the water vapor in the exhaust gas from condensation, the fivefold amount of (dry) air compared to that necessary for complete combustion is needed. Thereby, the volume V of the system in question changes remarkably, but not ΔV.

1.2.4 Energy of a body in motion

Starting from rest, a car with a mass of 1.5 t (1 t = 1000 kg) is accelerated on a straight road to a speed of 50 km h^{-1}.

a) Calculate its final kinetic energy W_{kin}.

b) How high could the car with an initial speed of 50 km h^{-1} coast up (engine disengaged) a ramp if friction would be negligible?

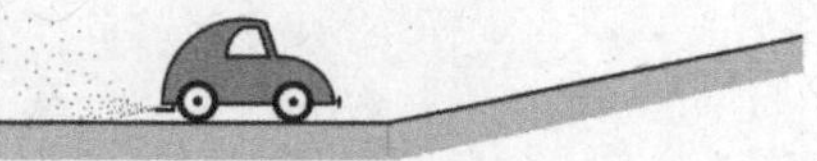

1.2.5 Falling without friction †

An elephant (E; m_E = 2000 kg) and a mouse (M; m_M = 20 g) jump off a bridge into the water.

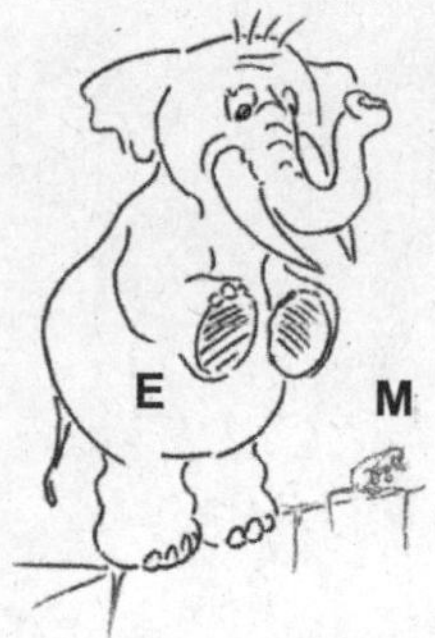

a) The duration Δt_E of the fall of the elephant is 2 s. What is its speed v_E and how much momentum p_E does it have when it reaches the water?

b) What is the duration Δt_M of the fall of the mouse and its speed v_M and how much momentum p_M does it have when it reaches the water?

c) Calculate the height of the bridge (above the surface of the water).

1.2.6 Energy of a raised body †

A pumped-storage plant has an upper reservoir at a height h_u of 800 m and a lower reservoir at a height h_l of 400 m.

a) The usable volume of water is 8 million cubic meters. How much energy can be stored? (The density of water is supposed to be 1 g cm^{-3}.)

b) The generators supply a power $P = W/\Delta t$ of 800 MW [1 Watt (W) = 1 J s^{-1}] at the maximum. How long will the stored energy last?

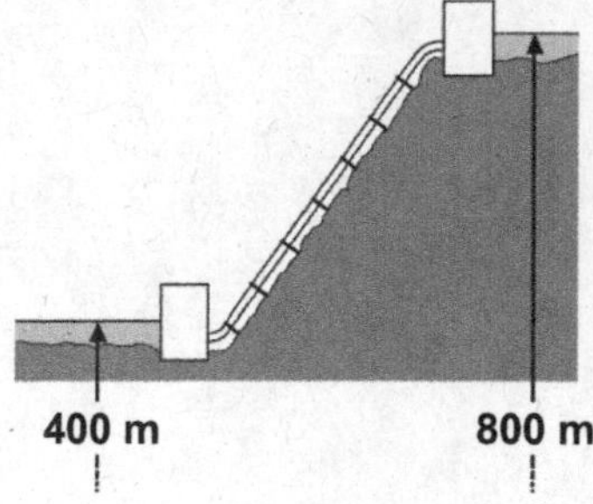

1.2.7 Energy expenditure for climbing

A tourist (T) with a mass of 80 kg climbs up a dune with a height of 100 m at the sea in order to have a better view.

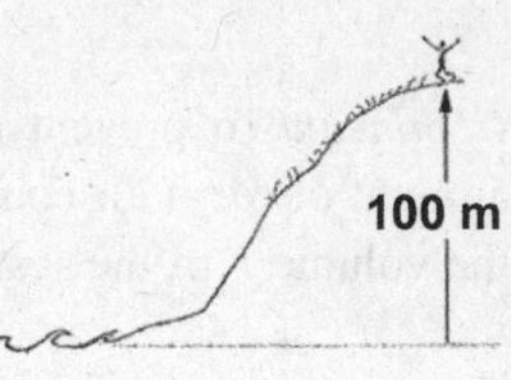

a) How much energy has to be expended against gravity?

b) How high could an astronaut (A) with a mass of 200 kg (including the necessary equipment) climb from the surface of the moon ($g_{Moon} = 1.60\ \mathrm{m\,s^{-2}}$) by expending the same energy?

1.3 Entropy and Temperature

1.3.1 Ice calorimeter

a) 20 g of a mixture of iron powder and sulfur powder with a mole ratio of $n_{Fe}:n_S = 1:1$ are placed into the test tube in the ice calorimeter from Experiment 3.5 in the textbook "Physical Chemistry from a Different Angle." The mixture is ignited and after the end of the reaction the produced water is collected in a graduated cylinder. The measured volume of water V is 63 mL. (Keep in mind: A volume V_I of 0.82 mL of melt water corresponds to an amount of entropy S_I of 1 Ct). What amount of entropy S' would be released during the reaction of $n' = 1$ mole of iron with $n' = 1$ mole of sulfur?

b) Using the ice calorimeter, the melting temperature of ice was determined to be $T =$ 273 K (Experiment 3.6). For this purpose, the immersion heater [power $P = 1000$ W (= 1000 J s^{-1})] was switched on for a time period of duration Δt of 27 s. What volume V of melt water was produced?

1.3.2 Measuring the entropy content

The diagram shows the heating curve for a block of copper with a mass of 1 kg.

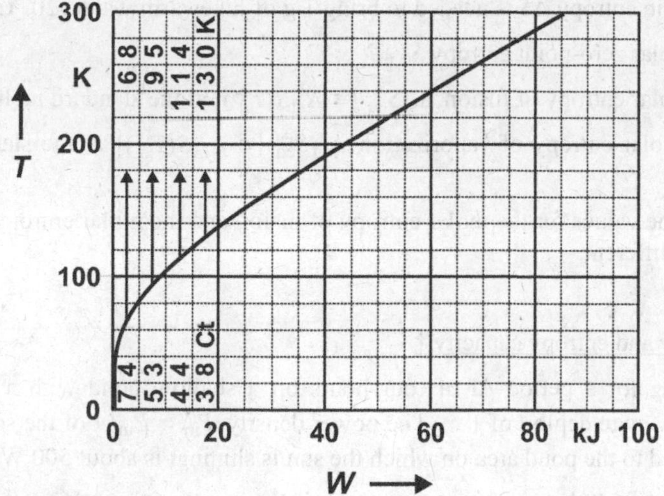

a) Write, as indicated, temperatures and entropy changes ΔS_i for every 5 kJ of energy added vertically into the columns.

b) How much entropy does the block contain at 300 K in total and how much related to the amount of substance of copper?

1.3.3 Entropy of water

The following diagram shows the entropy of 1 g of water as a function of temperature (at constant pressure).

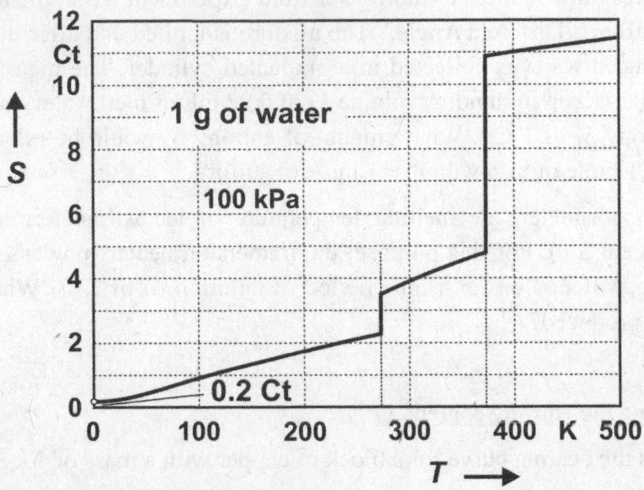

a) What increase in entropy ΔS is needed to bring 1 g of lukewarm water (20 °C) to a boil?

b) What is the molar zero-point entropy $S_{0,m}$?

c) What is the molar entropy of fusion $\Delta_{sl}S_{eq}^{\ominus}$ [$\equiv \Delta_{sl}S(T_{sl}^{\ominus})$] at the standard melting point?

d) What is the molar entropy of vaporization $\Delta_{lg}S_{eq}^{\ominus}$ [$\equiv \Delta_{lg}S(T_{lg}^{\ominus})$] at the standard boiling point?

e) Explain why the values for the molar entropy of fusion and the molar entropy of vaporization are quite different.

1.3.4* Entropy and entropy capacity †

The sun is shining for a period Δt of four hours on a shallow pond with a surface A of 6000 m^2 and an average depth l of 1 m. The power density $P''_{in} = P_{in}/A$ of the solar radiation, in this case referred to the pond area on which the sun is shining, is about 500 W m^{-2}.

a) How much energy W is transferred into the pond water if only 80 % of the radiation is absorbed (efficiency $\eta_{abs} = 0.80$), while the rest is reflected?

b) What is the increase in entropy ΔS of the pond water when its temperature is about 298 K?

c) By how many degrees does the temperature of the water rise? [molar entropy capacity and density of water at 298 K (and 100 kPa): $\mathcal{C}_m = 0.253$ Ct mol^{-1}K^{-1} and $\rho = 997$ kg m^{-3}]

1.3.5 Heat pump †

A paddling pool is heated by a heat pump, which pumps entropy from the cool ambient air ($\vartheta_1 = 17$ °C) into the warm pool water ($\vartheta_2 = 27$ °C). The power consumed by the heat pump is 3 kW. What is the amount of entropy S_t transferred by the heat pump in one hour from the air into the water in the ideal case?

1.3.6 Heat pump versus electric heating †

a) An (ideal) heat pump is used to heat a house. The outdoor temperature is $\vartheta_1 = 0$ °C and the indoor temperature $\vartheta_2 = 25$ °C. The heat pump transfers an amount of entropy of 30 Ct per second from outside into the house (in order to compensate the loss of entropy of the house through the walls). What is the transferred entropy S_t and what is the energy consumption W_t of the pump in one day?

b) The same house is heated under otherwise identical conditions by common electric (resistance) heating, i. e. the amount of entropy of 30 Ct per second necessary for compensation of entropy loss is not transferred from outside but generated in the house. The temperature in the house is supposed to be again 25 °C. How much entropy S_g has to be generated in one day and what is the corresponding consumption of energy W_b?

1.3.7* Thermal power plant (I)

In the common steam power plants, the energy W_{use} ($= -W_t$) is used, which can be gained during the transfer of entropy from the steam boiler ($T_1 \approx 800$ K) into the cooling tower ($T_2 \approx 300$ K). The entropy S_g itself is generated in the boiler by consumption of energy W_1.

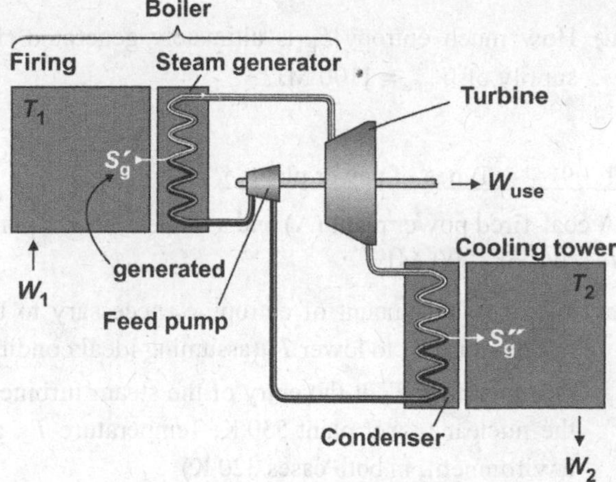

a) How large would the values for the energies W_1 and W_{use} be in the case of $S_g = 1$ Ct?

b) What efficiency η_{ideal} could be achieved in the ideal case ($\eta_{ideal} = W_{use}/W_1$)?

c) The actual value is $\eta_{real} \approx 40$ %. How much entropy S'_g is initially generated in the firing for each kJ of gained energy? What amount of entropy S''_g is finally released to the environment?

d) Where do the energy losses of 60 % occur? What contribution does alone the firing make, and what contribution does the rest of the plant make?

1.3.8* Thermal power plant (II)

A coal-fired power plant (ambient temperature $T_a \approx 300$ K) delivers a usable energy of 1100 MJ. That is roughly the electric energy released by a large block-unit power station to the grid in one second.

a) If 1 kg of hard coal (m_0) reacts with atmospheric oxygen, one can gain theoretically (meaning at an efficiency of $\eta_{ideal} = 100$ %) a usable energy $W_{use,0}$ of about 35 MJ. What is the mass m_1 of coal that one has to burn in order to provide the usable energy $W_{use,1}$ of 1100 MJ mentioned above?

b) The actual efficieny η_{real} of modern thermal power plants is about 40 % and the coal consumption is correspondingly higher. Calculate the mass m_2 of coal that has to be used in this case.

c) Hence, 60 % of the usable energy gets already lost in the power plant under generation of entropy. Please mark the parts with a cross where the energy is mainly lost.

boiler □, turbine □, generator □, condenser + cooling tower □.

At what point does the generated entropy mainly leave the power plant?

boiler □, turbine □, generator □, condenser + cooling tower □.

What are the parts through which the entropy is transported without noticeable increase in its amount?

boiler □, turbine □, generator □, condenser + cooling tower □.

d) How much entropy S_g is ultimately generated (in all parts together) in the case of the supply of $W_{use} = 1100$ MJ?

1.3.9* Types of power plants †

A coal-fired power plant (A) and a nuclear power plant (B) each release a usable energy W_{use} ($= -W_t$) of 1200 MJ.

a) What is the amount of entropy S_t necessary to be transferred in each case from higher temperature T_1 to lower T_2 (assuming ideal conditions)?

(Temperature T_1 at the entry of the steam turbine in the coal-fired power plant 800 K, in the nuclear power plant 550 K; Temperature T_2, at which the entropy is released into the environment, in both cases 320 K)

b) How much energy W_1 would be necessary to generate this entropy in each case?

c) Which efficiency $\eta_{ideal} = W_{use}/W_1$ could ideally be achieved in a coal-fired power plant and which one in a nuclear power plant?

d) What is the situation regarding generated entropy and efficiency in a hydroelectric power plant (C)?

1.3.10 Entropy generation during entropy conduction †

The heating wire of a cooking plate with a power P of 1000 W has a temperature T_1 of 1000 K.

a) How much entropy S'_g is generated in the heating wire in one second?

b) A pot with water is placed on the cooking plate, the water temperature is $\vartheta_2 = 100$ °C. How much entropy S_g is generated along the path from the heating wire to the water?

c) How much entropy in total does flow into the water per second (assuming that no entropy is released into the environment)?

1.4 Chemical Potential

The required data can be found in Table A2.1 in the Appendix of the textbook "Physical Chemistry from a Different Angle."

1.4.1 Stability of states

Use the chemical potentials to decide which state of a substance is stable at standard conditions.

a) Graphite — Diamond

b) Rhombic sulfur — Monoclinic sulfur

c) Solid iodine — Liquid iodine — Iodine vapor

d) Water ice — Water — Water vapor

e) Ethanol — Ethanol vapor

1.4.2 Predicting Reactions

Using chemical potentials, check whether the chemical drive of the following reactions is positive and if—after removal of possible inhibitions—a reaction can be expected (at standard conditions). As a first step, write down the particular conversion formula.

a) Binding of carbon dioxide by quicklime (CaO)

b) Combustion of ethanol vapor (while water vapor is produced)

c) Decomposition of silver oxide into its elements

d) Reduction of hematite (Fe_2O_3) with carbon (graphite) to iron (thereby releasing carbon monoxide gas)

1.4.3 Dissolving Behavior

Using the chemical potentials for pure and dissolved states, estimate (in zero-order approximation) whether the following substances dissolve easily in water or not. More precise data can be obtained by considering mass action (Chapter 6 in the textbook "Physical Chemistry from a Different Angle").

a) Cane sugar (sucrose)

b) Table salt (sodium chloride)

c) Limestone (calcite)

d) Oxygen

e) Carbon dioxide

f) Ammonia

1.5 Influence of Temperature and Pressure on Transformations

The required data can be found in Table A2.1 in the Appendix of the textbook "Physical Chemistry from a Different Angle."

1.5.1 Temperature dependence of chemical potential and transition temperature

a) Calculate the change $\Delta\mu$ in chemical potential when ethanol is heated from 25 °C to 50 °C at 100 kPa.

b) What value is obtained in first approximation for the standard boiling temperature $T_{lg}^{\ominus}$ of ethanol?

1.5.2 Decomposition and reaction temperatures

Calculate the decomposition temperature T_D or the reaction temperature T_R (i.e. the minimum temperature needed for the reaction) of the following processes from drive $\mathcal{A}^{\ominus}$ and its temperature coefficient α. Formulate first the conversion formula of the corresponding process.

a) Decomposition of limestone (calcite, $CaCO_3$) (resulting in the formation of quicklime)

b) Reduction of magnetite (Fe_3O_4) with carbon (graphite) to iron (thereby releasing carbon dioxide gas)

1.5.3 Pressure dependence of chemical potential

a) How does the chemical potential μ_l [= $\mu(H_2O|l)$] of (liquid) water (β_l = 18.1 µG Pa^{-1}) change when the pressure is increased from 100 kPa to 200 kPa (at 298 K).

b) How does the chemical potential μ_g [= $\mu(H_2O|g)$] of water vapor change when the pressure is increased under the same conditions as under a). Bear in mind that water vapor has to be treated as (ideal) gas.

1.5.4 Behavior of gases at pressure change

Carbon dioxide is the primary greenhouse gas emitted through human activities. Calculate the chemical potential μ_B [= $\mu(CO_2|g)$] of this gas at 25 °C when its pressure is reduced from 1.0 bar to 0.00039 bar, the partial pressure of CO_2 in the air.

1.5.5 "Boiling pressure"

The experiments 5.3 and 5.4 in the textbook demonstrate that the boiling temperature of water depends on pressure. More precisely, the lower the pressure, the lower the boiling temperature. At what pressure $p_{lg}^{\ominus}$ will water already boil at room temperature (25 °C)?

1.5.6 Temperature and pressure dependence of a fermentation process

Cane sugar (sucrose) can be fermented in aqueous solution into alcohol (ethanol) by means of suitable yeasts (after enzymatic cleavage) in order to finally produce beverages such as rum. The conversion formula for the fermentation process in total is (insert the conversion numbers into the blank boxes in the formula):

Symbol	B		B′		D		D′
	$\square\ C_{12}H_{22}O_{11}\|w$	+	$\square\ H_2O\|l$	$\rightarrow$	$\square\ C_2H_6O\|w$	+	$\square\ CO_2\|g$.

For the sake of simplicity, one uses again appropriate symbols instead of the full names or formulas of the substances.

a) Calculate first the temperature coefficient α of drive $\mathcal{A}^{\ominus}$ and then the change $\Delta\mathcal{A}$ in drive when the temperature is increased from 25 °C to 50 °C. The temperature coefficient α of ethanol in aqueous solution is $-148\ \mathrm{G\,K^{-1}}$. The other values can be found in Table A2.1 mentioned above.

b) Discuss the result from the point of view of matter dynamics.

c*) If the sugar is fermented in a closed container, the pressure p will increase because of the escape of CO_2 gas from the solution. What is the effect of an increase in pressure from 1 to 10 bar at standard temperature on the drive $\mathcal{A}$ of the fermentation process?

d) Discuss also this result from the point of view of matter dynamics.

1.5.7 Transition temperature and transition pressure

At 298 K and 100 kPa, the chemical drive for the transition from solid rhombic sulfur (rhom) to solid monoclinic sulfur (mono) is –75.3 G.

a) Which of the two modifications is stable under the mentioned conditions?

b) The temperature coefficient α of the chemical potential of rhombic sulfur is $-32.07\ \mathrm{G\,K^{-1}}$, that of the chemical potential of monoclinic sulfur $-33.03\ \mathrm{G\,K^{-1}}$. Can an increase in temperature be expected to make the other modification more stable? If so, at what temperature will the phase transition occur at 100 kPa?

c) The pressure coefficient β of the chemical potential of rhombic sulfur is $15.49\ \mathrm{\mu G\,Pa^{-1}}$, that of the chemical potential of monoclinic sulfur $16.38\ \mathrm{\mu G\,Pa^{-1}}$. Can an increase in pressure be expected to make the other modification more stable? If so, at what pressure will the phase transition occur at 298 K?

1.5.8* Pressure dependence of chemical potential and shift of freezing point

a) Calculate the change in chemical potential of (liquid) water and ice (both at 0 °C) when the pressure is increased from 0.1 MPa to 5 MPa. Usually, not the pressure coefficients of the substances under various conditions are tabulated, but their densities. The densities of water and ice at the standard melting point are $1.000\ \mathrm{g\,cm^{-3}}$ and $0.917\ \mathrm{g\,cm^{-3}}$, respectively. Use the memory aid "$\beta = V_m = V/n$" mentioned in Section 5.3 of the textbook "Physical Chemistry from a Different Angle" for the calculation.

b) Discuss the result from the point of view of matter dynamics.

c) Calculate the melting point of ice under a pressure of 5 MPa. The temperature coefficient α of the chemical drive for the melting process is +22.0 G K^{-1}.

1.5.9 Decomposition pressure at elevated temperature

Silver oxide (Ag_2O) is a blackish brown powder and stable under standard conditions (298 K, 100 kPa). It only decomposes completely when it is heated above the minimum temperature T_D in an *open* system (Experiment 5.2). In a *closed* (previously evacuated) glass flask, however, equilibrium is established and a specific decomposition pressure of silver oxide meaning a specific oxygen pressure $p(O_2)$ can be observed. Calculate this oxygen pressure at 400 K.

1.6 Mass Action and Concentration Dependence of Chemical Potential

1.6.1 Concentration dependence of chemical potential

A volume of 500 cm^3 of an aqueous solution of glucose (Glc) ($C_6H_{12}O_6$) contains 10 g of this sugar. Calculate the chemical potential μ_B (= μ(Glc|w) of glucose in the solution at 25 °C.

1.6.2* Concentration dependence of a fermentation process

We again consider the fermentation of cane sugar (sucrose) in aqueous solution by means of suitable yeasts to alcohol (ethanol). The conversion formula for the process in total is:

$$\begin{array}{llcll} \text{Symbol} & \text{B} & \text{B}' & \text{D} & \text{D}' \\ & C_{12}H_{22}O_{11}|w & + \ H_2O|l \ \rightleftarrows & 4\ C_2H_6O|w & +\ 4\ CO_2|g. \end{array}$$

a) Calculate the chemical drive $\mathcal{A}^\ominus$ for the fermentation process under standard conditions. The standard value of the chemical potential of ethanol in aqueous solution is −181 kG. The remaining values can again be found in Table A2.1 in the Appendix of the textbook "Physical Chemistry from a Different Angle."

b) What sugar concentration $c_{B,1}$ is necessary in the initial solution in order to achieve an alcohol concentration $c_{D,1} \approx 1\ \text{mol}\,\text{L}^{-1}$ (like in regular beers) at the end?

c) What are the concentrations of sugar and alcohol, $c_{B,2}$ and $c_{D,2}$, when about half of the initial sugar ($c_{B,1}$) is fermented, and what is the drive $\mathcal{A}$ in this case at standard temperature (only the substances dissolved in water have to be considered)?

1.6.3 Dependence of the chemical drive of ammonia synthesis on the gas composition

In Exercise 1.1.6 we already briefly dealt with the industrially important ammonia synthesis. Now it should be investigated how the drive of this process changes with the gas composition. Under standard conditions, the drive $\mathcal{A}^\ominus$ of the reaction represented by the following conversion formula,

$$\begin{array}{llll} \text{Symbol} & \text{B} & \text{B}' & \text{D} \\ & N_2|g & +\ 3\ H_2|g \ \rightleftarrows & 2\ NH_3|g, \end{array}$$

is +32.9 kG. What is the drive at 298 K when the partial pressures of nitrogen, hydrogen, and ammonia are 25 kPa, 52 kPa, and 75 kPa, respectively? In which direction does the reaction take place spontaneously under these conditions?

1.6.4 Mass action law (I)

Sn^{2+} ions in aqueous solution (containing hydrochloric acid) react after addition of iodine solution to Sn^{4+} ions:

$$Sn^{2+}|w + I_2|w \rightleftarrows Sn^{4+}|w + 2\ I^-|w.$$

The standard chemical potential $\mu^\ominus$ of Sn^{4+} ions in the acidic solution is +2.5 kG. Calculate the equilibrium number $\mathcal{K}_c^\ominus$ and the conventional equilibrium constant $K_c^\ominus$ of the reaction at 25 °C.

1.6.5 Mass action law (II)

A mixture of the gases bromine and chlorine is heated to 1000 K, whereby some bromine monochloride gas is formed:

$$\begin{array}{lccc} \text{Symbol} & \text{B} & \text{B}' & \text{D} \\ & Br_2|g + & Cl_2|g \rightleftarrows & 2\,BrCl|g. \end{array}$$

The equilibrium number $\mathring{\mathcal{K}}_c$ for the reaction at 1000 K is 0.2.

a) Calculate the chemical drive $\mathring{\mathcal{A}}$ for the process.

b) What is the concentration c_D of BrCl in an equilibrium mixture with the concentrations $c_B = 1.45\ \text{mmol L}^{-1}$ of Br_2 and $c_{B'} = 2.41\ \text{mmol L}^{-1}$ of Cl_2?

1.6.6 Composition of an equilibrium mixture (I)

The monosaccharide D-mannose (D-Man; $C_6H_{12}O_6$) is a chemical building block of many plant polysaccharides (mannans). Like D-glucose, it exists in two stereoisomeric forms, α-D-mannose and β-D-mannose. If, for example, pure α-D-mannose is dissolved in water at a concentration c_0 of $0.1\ \text{kmol m}^{-3}$, it partly transforms into β-D-mannose:

$$\begin{array}{lcc} \text{Symbol} & \alpha & \beta \\ & \alpha\text{-D-Man}|w \rightleftarrows & \beta\text{-D-Man}|w. \end{array}$$

The standard drive of the reaction is $\mathcal{A}^\ominus = -1.7$ kG. Calculate the composition of a solution in which α-D-mannose (c_α) and β-D-mannose (c_β) are in equilibrium at 25 °C.

1.6.7* Composition of an equilibrium mixture (II)

Gaseous hydrogen iodide (HI) decomposes at 700 K partially into hydrogen and iodine according to the conversion formula

$$\begin{array}{lccc} \text{Symbol} & \text{B} & \text{D} & \text{D}' \\ & 2\,HI|g \rightleftarrows & H_2|g + & I_2|g. \end{array}$$

The conventional equilibrium constant $\mathring{K}_c$ at this temperature is 0.0185. At the beginning of the reaction, only pure hydrogen iodide with a concentration $c_{B,0}$ of $10\ \text{mol m}^{-3}$ should be present in the flask. Calculate the composition of the equilibrium mixture, i.e. specify the concentrations c_B, c_D and $c_{D'}$.

Hint: For better overview, it is advisable to create first a kind of table as shown in Section 6.3 in the textbook "Physical Chemistry from a Different Angle" (Table 6.1).

1.6.8 Decomposition of silver oxide

We have already dealt with silver oxide, Ag_2O, and its decomposition into the elements according to the conversion formula

$$2\,Ag_2O|s \rightleftarrows 4\,Ag|s + O_2|g$$

from different angles. Here, the decomposition pressure of Ag_2O at standard conditions should be determined. Calculate first the equilibrium number $\mathcal{K}_p^{\ominus}$ from the chemical potentials and then the conventional equilibrium constant $K_p^{\ominus}$. Subsequently, formulate the mass action law.

1.6.9 Solubility of silver chloride

Silver chloride is hardly soluble in water. Calculate, based on the conversion formula

$$AgCl|s \rightleftarrows Ag^+|w + Cl^-|w\,,$$

the saturation concentration c_{sd} of the salt in aqueous solution at 25 °C. Proceed as in the previous exercise, i.e. first determine the equilibrium number $\mathcal{K}_{sd}^{\ominus}$ from the chemical potentials and then the conventional equilibrium constant $K_{sd}^{\ominus}$.

1.6.10* Solubility product (I)

Another very slightly soluble salt is calcium fluoride (CaF_2). In nature, it occurs in large amounts as mineral fluorite (also called fluorspar). The "solubility product" $\mathcal{K}_{sd}^{\ominus}$ of calcium fluoride is 3.45×10^{-11}.

a) What is the saturation concentration c_{sd} of calcium fluoride in water at 25 °C?

b) Estimate the concentration of the Ca^{2+} ions when calcium fluoride is dissolved in an aqueous NaF solution (0.010 kmol m^{-3}) (also at 25 °C)?

1.6.11* Solubility product (II)

Calcium phosphates are very important biominerals, because they play a major role in building up the bones and teeth of vertebrates. In the following, we would like to take a closer look at the nearly insoluble tricalcium phosphate (TCP), $Ca_3(PO_4)_2$. The equilibrium number for the dissolution process is $\mathcal{K}_{sd}^{\ominus} = 2.07 \cdot 10^{-33}$.

a) Determine how many milligrams of phosphate will theoretically dissolve in 500 mL of water at 25 °C.

b) Some anions such as PO_4^{3-}, but also CO_3^{2-} and S^{2-}, tend to react with water ("hydrolysis"). Describe qualitatively how the solubility is affected by this reaction.

1.6.12 Oxygen content in water

In a garden pond, a fountain ensures that the water is well-aerated. What is the mass concentration of oxygen (substance B) in air-saturated water, $\beta(\mathrm{B}|\mathrm{w})$, in $\mathrm{mg\,L^{-1}}$ at an ambient pressure of 100 kPa? HENRY's law constant $\overset{\ominus}{K_{\mathrm{H}}}$ (= $\overset{\ominus}{K_{\mathrm{gd}}}$) for the solubility of oxygen in water is $1.3 \times 10^{-5}\ \mathrm{mol\,m^{-3}\,Pa^{-1}}$.

1.6.13 Solubility of CO_2

a) What molar concentration of carbon dioxide (substance B), $c_1(\mathrm{B}|\mathrm{w})$, results in a mineral water at 25 °C when CO_2 at normal pressure [$p_1(\mathrm{B}|\mathrm{g}) = 1$ bar] is above the liquid? First calculate HENRY's law constant $\overset{\ominus}{K_{\mathrm{gd}}}$ by means of chemical potentials. Use the conversion formula $CO_2|\mathrm{g} \rightleftarrows CO_2|\mathrm{w}$ as the basis for the calculation.

b) What volume $V_1(\mathrm{B}|\mathrm{g})$ of gas corresponds to the amount of CO_2 dissolved in 1 L?

c*) What volume $V_{\mathrm{O}}(\mathrm{B}|\mathrm{g})$ of carbon dioxide bubbles out of an open 1-L bottle over time when the mineral water was bottled at a CO_2 pressure $p_2(\mathrm{B}|\mathrm{g})$ of 3 bar?

1.6.14* Absorption of CO_2 in lime water

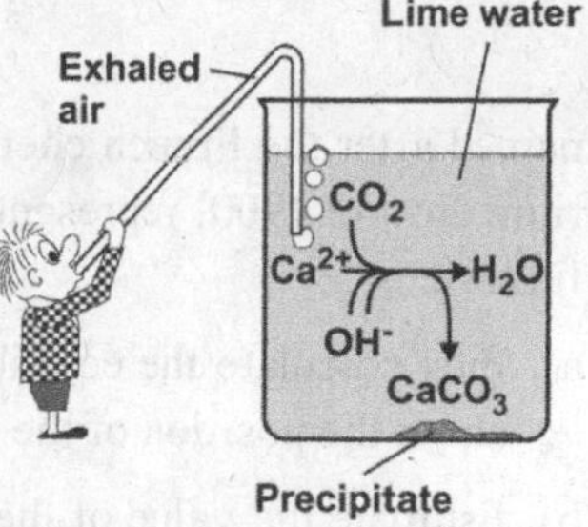

If one blows exhaled air (which contains carbon dioxide) into lime water (an aqueous calcium hydroxide solution) the liquid will turn milky due to the precipitation of calcium carbonate:

Symbol	B	B′	B″	D	D′

$$Ca^{2+}|\mathrm{w} + 2\ OH^{-}|\mathrm{w} + CO_2|\mathrm{g} \rightleftarrows CaCO_3|\mathrm{s} + H_2O|\mathrm{l}.$$

The partial pressure of CO_2 in the air escaping from the lime water should be determined. The concentration of the lime water is about $20\ \mathrm{mol\,m^{-3}}$. First calculate the drive $\overset{\ominus}{\mathcal{A}}$ of the precipitation reaction as well as the equilibrium number $\overset{\ominus}{\mathcal{K}_{pc}}$ and the conventional equilibrium constant $\overset{\ominus}{K_{pc}}$. We speak of a so-called "mixed" equilibrium number or constant, since both concentrations and (partial) pressures are included in the mass action law.

1.6.15 Distribution of iodine

A separation technique commonly used in the laboratory is "extraction by shaking" (extracting a substance B from its solution by using another solvent which is practically immiscible with the first one and in which the substance dissolves much better). As demonstrated in Experiment 4.3, iodine (substance B) should be extracted from an aqueous solution; but chloroform (trichloromethane, $CHCl_3$, short chl) is used instead of ether:

$$I_2|\mathrm{w} \rightleftarrows I_2|\mathrm{Chl}$$

$$\overset{\ominus}{\mu}/\mathrm{kG} \qquad \ldots.. \qquad 4.2$$

a) What fraction a_W of iodine ($a_W = n(B|w)/n_{total}$) remains in the aqueous phase if the extraction is performed with an equal volume of chloroform compared to that of the aqueous solution ($V_{S,w} = V_{S,Chl}$)? (Here, the water layer floats on top of the specifically heavier chloroform layer.)

b*) How does the result change if one first extracts the iodine with half the volume of chloroform, then releases the iodine-containing lower phase and repeats the process with the second half of the chloroform?

1.6.16* Distribution of iodine for advanced students

We can also use carbon disulfide, CS_2, instead of ether or chloroform for the extraction of iodine (substance B) from an aqueous solution. 500 mL of the solution should contain 500 mg of iodine. What mass m_x of iodine remains in the aqueous phase after extraction by 50 mL of carbon disulfide? The distribution coefficient $K^{\ominus}_{dd}$ of iodine between carbon disulfide and water is 588.

1.6.17 BOUDOUARD reaction

The BOUDOUARD reaction between solid carbon, carbon dioxide and carbon monoxide,

$$\begin{array}{llll} \text{Symbol} & \text{B} & \text{B}' & \text{D} \\ & \text{C|Graphite} + & CO_2|\text{g} \rightleftarrows & 2\ CO|\text{g}, \end{array}$$

named after the French chemist Octave Leopold BOUDOUARD, who investigated this equilibrium around 1900, represents an important part of the smelting process of iron ore in a blast furnace.

a) First calculate the equilibrium constant $K^{\ominus}_p$ under standard conditions. What can you say about the position of the equilibrium?

b) Estimate the value of the equilibrium constant $\mathring{K}_p$ at a temperature of 800 °C. Thus, how would the system react to the temperature increase?

c) The partial pressure of the carbon dioxide in the equilibrium mixture at 800 °C is supposed to be 30 kPa. What is then the partial pressure of the carbon monoxide?

d) What would happen if the produced carbon monoxide were continously removed from the system?

1.6.18* BOUDOUARD reaction for advanced students

Pure carbon dioxide gas with an initial pressure of 1 bar is subjected to glowing carbon in a closed container at a temperature of 700 °C. Calculate the partial pressures of carbon dioxide and carbon monoxide that will be established in equilibrium as well as the total pressure. The conventional equilibrium constant $\mathring{K}_p$ at 700 °C is 0.81 bar.

Hint: For better overview, it is again advisable to create first a kind of table as shown in Section 6.3 in the textbook (Table 6.1). In addition, it is necessary to solve a quadratic equation.

1.7 Consequences of Mass Action: Acid-Base Reactions

1.7.1 Proton potential of strong acid-base pairs (I)

In a cabinet in the laboratory, there is a reagent bottle (*1*) containing diluted hydrochloric acid with a concentration of 0.50 kmol m^{-3}.

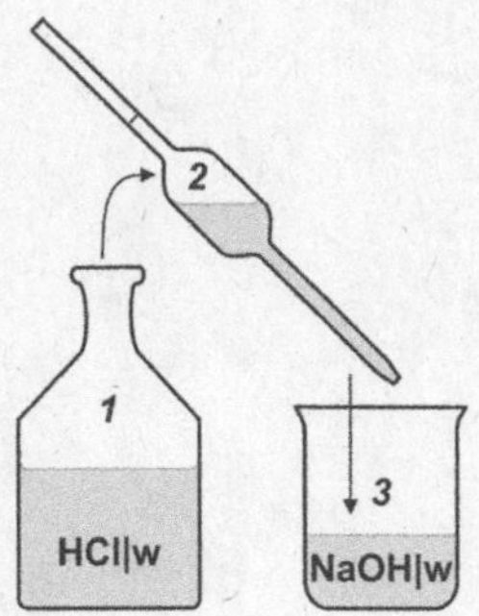

a) Calculate the proton potential of the hydrochloric acid at 25 °C.

b) 50 mL of hydrochloric acid is taken with a volumetric pipette (*2*) from the bottle and poured into a beaker (*3*), which already contains 50 mL of sodium hydroxide solution with a concentration of 0.20 kmol m^{-3}. Calculate the actual concentration of oxonium ions H_3O^+ and the proton potential.

1.7.2 Proton potential of strong acid-base pairs (II)

12.0 g of sodium hydroxide pellets were weighed out in a beaker, dissolved in a little bit of water and the solution was transferred to a 500 mL volumetric flask (*1*). Subsequently, the flask was made up to the mark with water and the solution was mixed by shaking.

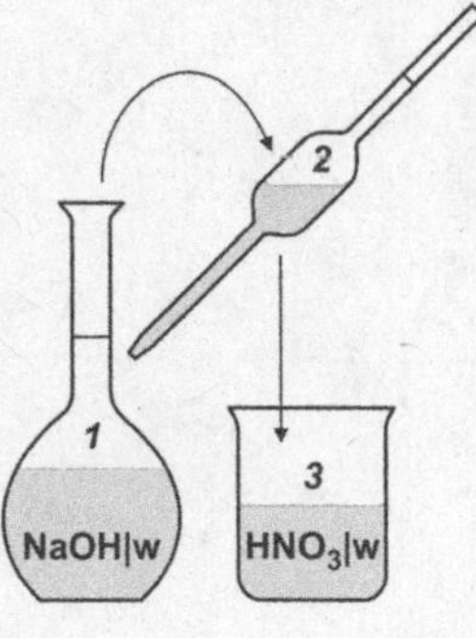

a) Calculate the concentration of hydroxide ions and the proton potential in the solution at 25 °C.

b) 50 cm^3 of the alkaline solution is added with the help of a volumetric pipette (*2*) to 150 cm^3 of diluted nitric acid with a concentration of 0.10 kmol m^{-3} in a beaker (*3*). Calculate the actual concentration of hydroxide ions and the proton potential.

1.7.3 Proton potential of weak acid-base pairs (I)

a) What is the proton potential in a lactic acid solution with a concentration of $c_{\mathrm{HLac}} =$ 0.30 kmol m^{-3} at 25 °C? (The abbreviation Lac is used for the C_2H_4OHCOO group.)

b) Calculate the degree of protonation Θ. The equation, which had to be used in part a), was derived by assuming that the acid of a weak acid-base pair is only dissociated to a very small extent in aqueous solution. Decide on the basis of the Θ value whether this assumption was justified in the case of the present example or not.

1.7.4 Proton potential of weak acid-base pairs (II)

The proton potential of an aqueous solution containing 245 mg of sodium cyanide (NaCN) in 100 mL of solution (V_S) is −62.6 kG at 25 °C. Determine the standard value $\mu_p^\ominus(\mathrm{HCN/CN^-})$ of the proton potential of the weak acid-base pair.

1.7.5 Titration of a weak acid

A volume (V_0) of 100 mL of a benzoic acid solution (HBenz; C_6H_5COOH) with a concentration of 0.100 kmol m^{-3} is filled in a titration flask and subsequently titrated with sodium hydroxide solution with a concentration of 2.000 kmol m^{-3}. The standard proton potential of the acid-base pair $C_6H_5COOH/C_6H_5COO^-$ is –23.9 kG.

a) What is the proton potential of the starting solution at 298 K?

b) Calculate the proton potential of the solution after adding 2.00 mL of the titrator ($V_{T,1}$).

c) What is the proton potential of the solution after adding another milliliter of the titrator (in total, the volume $V_{T,2}$ was added to the benzoic acid solution)?

d) What volume of hydroxide solution is required to reach the equivalence point ($V_{T,3}$)?

e) Calculate the proton potential of the solution at the equivalence point.

1.7.6 Buffer action

500 mL of a buffer solution containing equal amounts of acetic acid (HAc) and acetate (Ac^-) are prepared by using an aqueous acetic acid solution with a concentration of 0.20 kmol m^{-3} and an aqueous sodium acetate solution of the same concentration.

a) What is the proton potential in the buffer solution (at 298 K)?

b) What is the proton potential after adding 500 μL of hydrochloric acid with a concentration of 2.00 kmol m^{-3} to the buffer solution with the help of a microliter pipette?

c) What is the proton potential when the same amount of hydrochloric acid is added to 500 mL of pure water?

1.7.7 Buffer capacity

A buffer solution consists of ammonium chloride in a concentration of 0.060 kmol m^{-3} and ammonia in a concentration of 0.040 kmol m^{-3}.

a) Calculate the proton potential in the buffer solution (at 25 °C).

b) Determine how many milliliters of a sodium hydroxide solution with a concentration of 0.1 kmol m^{-3} can be added to 50 cm^3 of the buffer solution without its buffer action being depleted?

   As a rule of thumb, a buffer will be depleted as soon as the ratio of acid to corresponding base exceeds the value of 1:10 (or 10:1, respectively). Therefore, the proton potential of a buffer can fluctuate by about ± 6 kG (this corresponds to the decapotential μ_d) upon addition of acids and bases before it is depleted.

1.7.8 Indicator

The standard value $\mu_p^{\ominus}$ of the proton potential of the acid-base indicator phenol red is −45.1 kG. The indicator acid (HInd) is yellow, while the corresponding indicator base (Ind^-) is red. A color change of the indicator to yellow is recognizable if the concentration of the yellow form is about 30 times greater than the concentration of the red form. The color change to red is recognizable if the concentration ratio of the red to the yellow form is about 2:1. Determine the limiting proton potentials of the transition range of the indicator.

1.8 Side Effects of Transformations of Substances

1.8.1 Application of partial molar volumes

50.0 mL of water (W) and 50.0 mL of ethanol (E) were filled into a mixing cylinder and subsequently mixed by shaking upside down (based on Experiment 8.2 in the textbook "Physical Chemistry from a Different Angle"). Afterwards, a decrease in volume of the mixture can be observed. Estimate the total volume of the mixture using the diagram below, which illustrates the dependence of the molar volume of water (W) and ethanol (E) in water-ethanol mixtures from the mole fraction x_E of ethanol at 298 K [based on the data from Benson G C, Kiyohara O (1980) Thermodynamics of Aqueous Mixtures of Non-electrolytes. I. Excess Volumes of Water – n-Alcohol Mixtures at Several Temperatures. J Solution Chem 9:791–803]. The densities of the pure liquids at this temperature are $\rho_W = 0.997\ g\ cm^{-3}$ and $\rho_E = 0.789\ g\ cm^{-3}$, respectively. First calculate the amounts of water and ethanol in the mixture in question and then the mole fraction of ethanol.

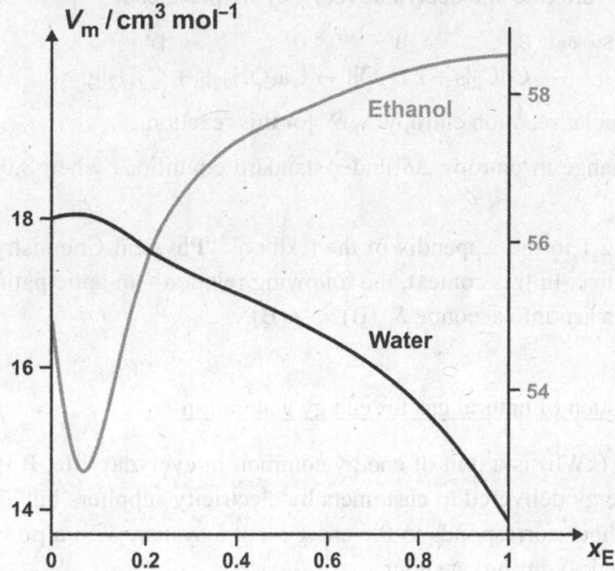

1.8.2 Volume demand of dissolved ions

How does the volume V_W of 10 L of water change under standard conditions,

a) if 1.00 cm³ of solid calcium hydroxide $Ca(OH)_2$ (≙ 0.030 mol) is dissolved in it?

Symbol B D D′

$$Ca(OH)_2|s \rightarrow Ca^{2+}|w + 2\ OH^-|w\,.$$

First calculate the molar reaction volume $\Delta_R V^\ominus$ and then the final volume V_{final} of the solution (under standard conditions).

b) if the water dissociates from an ion-free state until equilibrium is established?

Symbol B D D′

$$H_2O|l \rightarrow H^+|w + OH^-|w\,,$$

Again, calculate the molar reaction volume $\Delta_R V^\ominus$ and then the change in volume ΔV (under standard conditions).

Necessary data:

	$Ca^{2+}|w$	$H^+|w$	$OH^-|w$
$V_m^\ominus$ / cm^3 mol^{-1}	−17.7	0.2	−5.2

as well as Table 8.1 in the textbook "Physical Chemistry from a Different Angle."

1.8.3 Reaction entropy

The demonstration experiment 4.10 deals with the reaction of calcium carbide with water whereby calcium hydroxide and acetylene (ethyne) are produced:

Symbol B B′ D D′

$$CaC_2|s + 2\,H_2O|l \rightarrow Ca(OH)_2|s + C_2H_2|g.$$

a) Calculate the molar reaction entropy $\Delta_R S^\ominus$ for this reaction.

b) What is the change in entropy ΔS under standard conditions when 8.0 g of calcium carbide are used?

Hint: Use Table A2.1 in the Appendix of the textbook "Physical Chemistry from a Different Angle" as data source. In this context, the following relation—in anticipation of its later derivation—has to be taken into account: $S_m(B) = -\alpha(B)$.

1.8.4 Combustion of natural gas for energy generation

The kilowatt hour (kWh) is a unit of energy common in everyday life. It is mainly used as a billing unit for energy delivered to customers by electricity suppliers but also by heat suppliers. One kilowatt hour corresponds to the energy that a system with a power of one kilowatt consumes (or provides) during one hour.

By combustion of natural gas,

Symbol B B′ D D′

$$\square\ CH_4|g + \square\ O_2|w \rightarrow \square\ CO_2|w + \square\ H_2O|l$$

(in the conversion formula simplistically represented by methane, CH_4), a freely usable energy W^*_{use} ($= -W_{\rightarrow\xi}$) of 1 kWh is supposed to be released (under standard conditions). (The index * indicates the environment.)

a) What is the amount of natural gas required for the process? First insert the conversion numbers in the blank boxes in the conversion formula above.

b) What is the change in volume of the natural gas during the reaction?

1.8.5 Entropy balance of a reaction

In the following, we will consider the reaction of hydrogen with oxygen,

$$\begin{array}{lccc} \text{Symbol} & \text{B} & \text{B}' & \text{D} \\ & 2\,H_2|g & +\ O_2|g \rightarrow & 2\,H_2O|l, \end{array}$$

the so-called oxyhydrogen reaction, from different angles. On the one hand, the hydrogen can be combusted in a gas burner. Thereby, a very hot flame is produced that can be used, for example, for welding. The energy released during the reaction remains unused, meaning it is completely used up to generate entropy (case 1, $\eta = 0$). Alternatively, the energy can also be utilized electrically in a hydrogen oxygen fuel cell (to drive a motor, for example). In this context, we consider a common fuel cell (case 2, $\eta \approx 70$ %) as well as a lossless fuel cell (case 3, $\eta = 100$ %). In all three cases, 1.0 kg of water should be produced at room temperature (298 K) (i.e. in the case of the burner after the combustion gases have cooled down).

a) First calculate the molar reaction entropy $\Delta_R S^\ominus$ for the oxyhydrogen reaction and then the change in entropy ΔS (under standard conditions) when the desired mass of water is produced.

b) What is the chemical drive $\mathcal{A}^\ominus$ of the reaction and the maximum value of useful energy $W^*_{\text{use,max}}$ (i.e. for $\eta = 1$)?

c) Determine how much entropy is generated (S_g) and how much is exchanged with the environment (S_e) in the three different cases. The energy $T \times S_e$ is called "heat" by thermodynamicists (related to the conversion, $T \times S_e / \Delta\xi$, also "molar heat of reaction"). Calculate this energy, too.

1.8.6 Energy and entropy balance of water evaporation

We study the evaporation of 1.0 cm³ of water at a relative humidity of 70 % and room temperature (25 °C):

$$\begin{array}{lcc} \text{Symbol} & \text{B} & \text{D} \\ & H_2O|l \rightarrow & H_2O|g,\text{air}\,. \end{array}$$

The chemical potential of water vapor in ambient air is −238.0 kG under the specified conditions.

a) What is the chemical drive $\mathcal{A}$ for the evaporation process and the maximum usable energy $W^*_{\text{use,max}}$?

b) Calculate the height to which a weight of 1 kg can be lifted by complete use of the energy.

c) How much entropy S_g is generated when the energy remains completely unused?

d*) The value μ_D [$= \mu(H_2O|g,air)$] given above should be deduced by means of the mass action equation. Keep in mind that the drive $\mathcal{A}$ disappears at a relative humidity of 100 %.

1.8.7 "Working" duck

The "drinking duck" known as a toy slowly swings back and forth, finally dips its beak into the water and comes back up nodding. After a number of oscillations, the process starts anew. The movement of the duck can be used to lift a weight by means of an appropriate equipment (Experiment 8.5). The functional principle of the duck is based on the evaporation of water (see Exercise 1.8.6). Let us assume that 1 mmol of water evaporates at a relative humidity of 50 %.

a) Calculate the chemical potential of water vapor μ_D [$= \mu(H_2O|g,air)$] using the mass action equation.

b) What is the maximum usable energy $W^*_{use,max}$?

c) What would be the efficiency η if the duck lifted a small block of 10 g by 10 cm?

1.8.8 "Muscular energy" by oxidation of glucose

A chimney sweeper, weighing 70 kg, has to climb stairs, gaining 24 m in height in order to reach his workplace on the roof of the house. The necessary energy is supplied by oxidation of glucose ($C_6H_{12}O_6$) to carbon dioxide and water in the muscular tissue. First, insert the conversion numbers into the blank boxes in the conversion formula below:

Symbol	B		B′		D		D′
	$\square\ C_6H_{12}O_6\vert w$	+	$\square\ O_2\vert w$	→	$\square\ CO_2\vert w$	+	$\square\ H_2O\vert l$.

a) Calculate the chemical drive $\mathcal{A}^\ominus$ for the oxidation process (under standard conditions).

b) What is the minimum amount of energy W_{min} required for the man to reach his workplace?

c) Glucose tablets, which are popular as a fast energy supplier, weigh approximately 6 g. How much energy W^*_{use} usable for the climbing is supplied by such a tablet if the estimated efficiency η is 20 %?

d) How many grams of glucose does the chimney sweeper eventually consume for the climbing or, asked differently, what is the change in mass Δm^*_B of his glucose reservoir as a result of the climbing? How many glucose tablets does he have to eat at the minimum to get up to the roof?

1.8.9 Fuel cell

We imagine a fuel cell in which propane reacts with oxygen:

Symbol	B		B′		D		D′
	□ $C_3H_8|g$	+	□ $O_2|g$	→	□ $CO_2|g$	+	□ $H_2O|g$.

Such fuel cells are used, for example, to generate electricity in camper vans. First, insert the missing conversion numbers into the blank boxes of the conversion formula above.

a) Calculate the chemical drive $\mathcal{A}^\ominus$ for the oxidation process (under standard conditions).

b) What is the molar reaction entropy $\Delta_R S^\ominus$ in this case?

c) Determine how much usable energy W^*_{use} can be gained at standard temperature with a conversion of $\Delta\xi = 2$ mol and an efficiency of $\eta = 80$ %. The fuel cell is supposed to have a maximum power output of 250 W. How long does it run under these conditions?

d) How much entropy S_g is generated in the cell under the conditions in c) and how much entropy S_e is exchanged with the environment?

e) How much entropy S'_g is generated and how much entropy S'_e is exchanged with the environment when the fuel cell is short-circuited?

1.8.10 Calorimetric measurement of chemical drive

To determine the chemical drive for the combustion of ethanol, which is meanwhile also added to gasoline, 0.001 moles of the substance were combusted in the reaction vessel of a calorimeter in an oxygen atmosphere at a temperature of around 25 °C and a pressure of 1 bar, which corresponds more or less to standard conditions. The combustion resulted in an increase in temperature of 2.49 K. In order to calibrate the calorimeter, its heating element was connected for a period of 150 s to a voltage source, which delivered a voltage of 12 V and a current of 1.5 A.

a) What is the entropy $-S_e$ released by the sample during the reaction if the electric heating of the calorimeter resulted in an increase in temperature of 4.92 K?

b) Calculate the chemical drive of the reaction. The molar reaction entropy $\Delta_R S^\ominus$ for the combustion of ethanol is -138.8 Ct mol^{-1} (under standard conditions). What would be the percentage error if the latent entropy is not taken into consideration?

1.9 Coupling

1.9.1 Flip rule

Apply the flip rule to deduce the relation $\beta = V_m$. In this context, β is the pressure coefficient of the chemical potential μ of a substance and V_m its molar volume. Comment each step.

1.9.2 Increase in temperature by compression

An object that expands when entropy is added to it—and this is valid for almost all—will, inversely, become warmer when compressed.

a) The expansion with increasing entropy can be quantified by the coefficient $(\partial V/\partial S)_p > 0$. By "flipping," one obtains a new coefficient. What coefficient is this and why is the sentence mentioned at the beginning a consequence of it?

b) How does ice water (water below 4 °C) behave in this regard?

1.9.3 Change in volume of a concrete wall

What is the average change in volume of a concrete wall with a length l of 10 m, a height h of 3 m and a width w of 20 cm between winter ($\vartheta_1 = -15$ °C) and summer ($\vartheta_2 = +35$ °C) in cubic centimeters? The volumetric thermal expansion coefficient γ for concrete is supposed to be about 36×10^{-6} K^{-1}.

1.9.4* Gasoline barrel

A steel barrel (S) has a volume of 216 L at 20 °C. What is the maximum volume of gasoline (G) that may be filled into the barrel? For safety reasons, one has to calculate with a final temperature of 50 °C and there should be no losses due to overflow. The volumetric thermal expansion coefficient γ of steel is 35×10^{-6} K^{-1}, that of gasoline 950×10^{-6} K^{-1}.

1.9.5 Compressibility of hydraulic oil

A hydraulic oil in a cylinder has a volume of 1 L under an external pressure of 10 bar. Subsequently, the pressure is increased to 20 bar. Thereby, the volume is reduced to 0.9997 L. Estimate the compressibility of the oil. (The temperature is supposed to be approximately constant.)

1.9.6* Density of water in the depth of the ocean

By what percentage does the density of water increase at a depth of 200 m compared to its density at the water surface? The hydrostatic pressure of sea water at a depth of 200 m is 2 MPa; for the compressibility χ of water a value of 4.6×10^{-10} Pa^{-1} is assumed. Additionally, the water temperature is supposed to have approximately a constant value independent from the depth.

1.9.7 Increase in pressure in a liquid-in-glass thermometer

We discuss a glass thermometer filled with ethanol. While the volume of the ethanol changes noticeably with an increase in temperature or pressure, the glass tube can be regarded as thermally and mechanically rigid.

Imagine that by accident the end of the measuring range of the thermometer, which is at about 60 °C, had been reached, so that the entire volume of the glass would have been filled with ethanol. Estimate the expected increase in pressure Δp inside if the temperature had been increased by further 5 °C. The volumetric thermal expansion coefficient γ of ethanol is $1.1 \times 10^{-3}\ \mathrm{K}^{-1}$ and its compressibility χ is $1.5\ \mathrm{GPa}^{-1}$ at 60 °C. The corresponding values for glass, $\gamma \approx 0.02 \times 10^{-3}\ \mathrm{K}^{-1}$ and $\chi \approx 0.01\ \mathrm{GPa}^{-1}$, are very much smaller.

1.9.8 Isochoric molar entropy capacity

At 20 °C, ethanol has an isobaric molar entropy capacity $\mathcal{C}_{\mathrm{m},p}$ of $0.370\ \mathrm{Ct\,mol^{-1}\,K^{-1}}$. Its isochoric molar entropy capacity $\mathcal{C}_{\mathrm{m},V}$ is to be calculated. For this purpose, the following data are available: volumetric thermal expansion coefficient $\gamma = 1.40 \times 10^{-3}\ \mathrm{K}^{-1}$; compressibility $\chi = 11.2 \times 10^{-10}\ \mathrm{Pa}^{-1}$; density $\rho = 0.789\ \mathrm{g\,cm^{-3}}$.

1.9.9* Isentropic compressibility

The speed of sound c in liquids or gases obeys the simple equation

$$c = \sqrt{\frac{1}{\chi_S \cdot \rho}},$$

in which ρ denotes the density and χ_S the compressibility, not the isothermal $\chi = \chi_T = -V^{-1}(\partial V/\partial p)_T$, but the isentropic $\chi_S = -V^{-1}(\partial V/\partial p)_S$. The reason for this is that the entropy propagates only slowly compared to the sound. The entropy contained in a small section comprising the amount Δn remains constant, even if the pressure and the temperature there fluctuate noticeably when a sound wave passes. The quantities χ_S and χ_T can be converted into each other by the following relationship:

$$\chi_S = \chi_T - \frac{M\gamma^2}{\rho \mathcal{C}_\mathrm{m}}$$

[M molar mass, γ volumetric thermal expansion coefficient, ρ density, $\mathcal{C}_\mathrm{m}$ (isobaric) molar entropy capacity].

a) Derive the above relationship between isothermal and isentropic compressibility by means of the calculation rules for differential quotients presented in Section 9.4 of the textbook "Physical Chemistry from a Different Angle."

b) What is the value of the isentropic compressibility of ethanol using the data given in Exercise 1.9.8?

c) What is the speed of sound in ethanol?

1.9.10* Compression of ice water

How does the temperature of water change when it is (isentropically) compressed at 0 °C with an overpressure of 1000 bar? Let us assume that the volumetric thermal expansion coefficient $\gamma = -70 \times 10^{-6}\ \mathrm{K}^{-1}$ and the (isobaric) molar entropy capacity $\mathcal{C}_\mathrm{m} = 0.28\ \mathrm{Ct\,K^{-1}\,mol^{-1}}$ are approximately constant.

1.10 Molecular-Kinetic View of Dilute Gases

1.10.1 Gas cylinder

Nitrogen is stored in a steel cylinder with a volume V_0 of 50 L at a pressure p_0 of 5 MPa and a temperature ϑ_0 of 20 °C. Under these conditions, the gas can still be treated as ideal.

a) What is the amount n and the mass m of nitrogen stored in the cylinder?

b) In the external warehouse, the gas cylinder was exposed to direct sunlight for a longer period of time. As a result, it and its content heated up to $\vartheta_1 = 45$ °C. What is now the pressure p_1 in the cylinder?

1.10.2* Weather balloon

A weather balloon is filled with hydrogen at sea level ($p_0 = 100$ kPa) at a temperature ϑ_0 of 25 °C until its radius r_0 is 1.5 m. Subsequently, the balloon is launched. What is the volume V_1 and the radius r_1 of the balloon at a height of 10 km when the pressure p_1 is there 30 kPa and the temperature ϑ_1 −50 °C? Hydrogen can be considered ideal in good approximation under these conditions. Additionally, we assume that the ballon is spherical and that the inside and outside pressure is the same.

1.10.3 Potassium superoxide as lifesaver

Potassium superoxide (KO_2) has the remarkable property of binding carbon dioxide and releasing simultaneously oxygen:

Symbol	B	B′	D	D′

$$\square\, KO_2|s + \square\, CO_2|g \rightarrow \square\, K_2CO_3|s + \square\, O_2|g.$$

Therefore, it can be used in canisters for rebreathers (e.g. for fire fighting or mine rescue work). What mass of KO_2 is necessary if a volume of 40 dm^3 of CO_2 is supposed to be bound at 10 °C and 100 kPa?

1.10.4* Divers' disease

A diver with a mass of 80 kg prepares on a boat for his dive.

a) Determine the amount $n(B|Bl)_0$ of nitrogen (substance B) in his blood (Bl). A person has a blood volume of about 70 mL per kilogram of body weight. For simplicity, we can use HENRY's law constant for the solubility of nitrogen in water instead of that for the solubility of nitrogen in blood, meaning $\overset{\circ}{K}_H (= \overset{\circ}{K}_{gd}) = 5.45\times10^{-6}$ mol m^{-3} Pa^{-1} at 37 °C, a person's normal body temperature. Ambient air consists of 78 % nitrogen; the atmospheric pressure $p_{total,0}$ at sea level is supposed to be 100 kPa.

b) The diver reaches a depth of 20 m during his dive. There, he breathes air with a pressure of $p_{\text{total},1} = 300$ kPa delivered by the diving cylinder. Determine the amount $n(\text{B}|\text{Bl})_1$ of nitrogen, which is now dissolved in the same blood volume.

c) Subsequently, the diver ascends very fast to the water surface. What volume $\Delta V(\text{B}|\text{g})_2$ of N_2 gas is released as bubbles in the diver's bloodstream? Arterial gas embolism caused by gas bubbles in the pulmonary capillaries is the reason for the dreaded "divers' disease."

1.10.5* Volumetric determination of gases

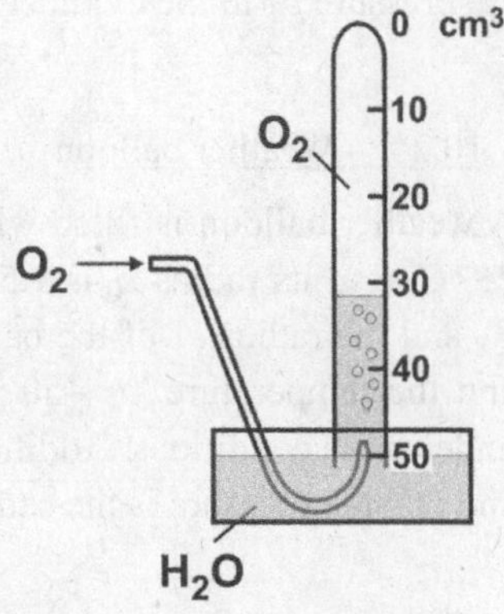

In order to volumetrically measure a gas released during a reaction (see also Experiment 16.8 in the textbook), a eudiometer (graduated glass tube closed at the top) can be used. The gas bubbles through the water into the eudiometer so that its amount can be determined by means of the displaced volume of water (see figure on the right). On the one hand, the gas dissolves partly in water; on the other hand, water vapor is mixed with the gas. The first process reduces the volume of trapped gas, the second one increases it. Both influences are to be estimated for oxygen (substance B), which is collected in a eudiometer under standard conditions (100 kPa, 298 K).

a) First, calculate the solubility of O_2 in water meaning the saturation concentration $c(\text{B}|\text{w})$. HENRY's law constant $\overset{\ominus}{K_{\text{H}}}$ $(= \overset{\ominus}{K_{\text{gd}}})$ is 1.3×10^{-5} mol m^{-3} Pa^{-1}.

b) Then calculate the saturation vapor pressure $p_{\text{lg,D}}$ of water (substance D) by means of the chemical potentials.

c) How much O_2 dissolves when $V_{\text{D}} = 50$ cm^3 of water is saturated with it? What is the gas volume $V(\text{B}|\text{g})$ corresponding to this amount?

d) What is the total pressure $p_{\text{total,Eu}}$ and the O_2 partial pressure $p(\text{B}|\text{g})_{\text{Eu}}$ in the gas-filled part of the eudiometer when the gas volume is 50 cm^3?

e) What contribution $V(\text{D}|\text{g})$ does the water vapor make to the total volume of 50 cm^3?

1.10.6* Inflating a bicycle tire

The tires of common bicycles have very roughly a constant volume V_0 of 2 L. What is the minimum amount of energy W required to inflate the tire from 1 to 5 bar?

Hint: For simplicity's sake, let's consider a pump with a cylinder capacity of $4V_0$, so that the tire can be brought from the initial pressure to the final pressure in one single push. The process is supposed to be, however, so slowly that the temperature does not increase remarkably.

1.10.7 Cathode ray tube

In the past, cathode ray tubes were used to display images in television sets and computer monitors. In order for the electron beam to be able to produce an undisturbed image on the screen, as few gas molecules as possible should be "in the way" of the electrons. How many gas molecules are present in every cm^3 of a cathode ray tube when the pressure in the tube is about 0.1 mPa and the temperature is 20 °C?

1.10.8 Speed and kinetic energy of translation of gas molecules

Calculate the root mean square speed and the average molar kinetic energy of

a) hydrogen molecules at $T_1 = 298$ K and $T_2 = 800$ K as well as

b) oxygen molecules at the same temperatures.

1.10.9 Speed of air molecules

How does the root mean square speed of air molecules change if one compares a hot summer day ($\vartheta_1 = 35$ °C) with a cold winter day ($\vartheta_2 = -10$ °C)?

1.10.10* MAXWELL's speed distribution

Use MAXWELL's speed distribution to estimate the fraction of molecules with a speed between 299.5 m s^{-1} and 300.5 m s^{-1} in nitrogen gas at a temperature of 25 °C.

Hint: Unfortunately, the factor 2 has been lost in the exponential term in Equation (10.54) for MAXWELL's speed distribution in the textbook "Physical Chemistry from a Different Angle." The correct term is $\exp(-mv^2/2k_BT)$.

1.11 Substances with Higher Density

1.11.1 VAN DER WAALS equation

A container with a volume of 5.00 L is filled with 352 g of carbon dioxide.

a) Calculate the pressure at a temperature of 27 °C using the VAN DER WAALS equation.

b) What would be the pressure if one assumed an ideal behavior of the gas?

1.11.2 Phase diagram of water

a) Calculate values for the melting, boiling and sublimation pressure curves of water using the equations presented in the textbook "Physical Chemistry from a Different Angle" and insert them into the tables. For this purpose, select first the necessary data.

Necessary data:

Use as data source for the chemical potentials $\mu^\ominus$ and their temperature coefficients α Table A2.1 in the Appendix and for the pressure coefficients β Table 5.2.

	$H_2O\vert s$	$H_2O\vert l$	$H_2O\vert g$
$\mu^\ominus / \text{kG}$			
$\alpha / \text{G K}^{-1}$			
$\beta / \mu\text{G Pa}^{-1}$			

Melting:

The slope of the melting pressure curve is very steep, almost vertical. Therefore, it is recommendable to solve the general equation for calculating this curve for T and to insert subsequently the initial and final values of the pressure range under consideration (in this case 0 and 50 kPa).

Corresponding table:

p / kPa	0	50
T / K		

Boiling:

Corresponding table:

T / K	240	260	280	300	320	340	360
p / kPa							

Sublimation:

Corresponding table:

T/K	240	260	280	300	320	340
p/kPa						

Use the calculated points to plot the melting, boiling and sublimation pressure curves of water into the grid table.

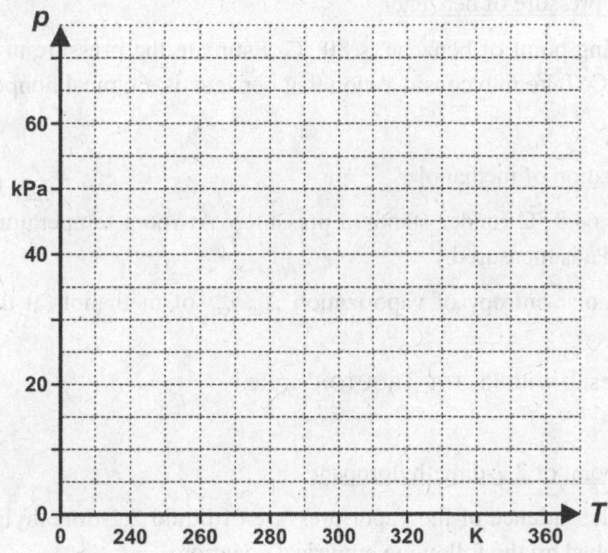

b) Using the boiling pressure curve (vapor pressure curve), estimate the boiling temperature of water on the summit of Mount Everest (height of 8848 m). The average air pressure at this altitude is about 32.6 kPa.

1.11.3 Evaporation rate

How much higher, roughly estimated, is the evaporation rate ω_1 of water on a hot summer day (ϑ_1 = 34 °C) compared to that (ω_2) on a cool autumn day (ϑ_2 = 10 °C)? The conditions are such that the (temperature-dependent) vapor pressure p_{lg} of water determines the evaporation rate quite significantly.

1.11.4 Atmosphere in the bathroom

A person takes an extensive hot bath in a small bathroom without ventilation with a floor area of 4 m^2 und a height of 2.5 m. What is the amount and mass of water in the air of the bathroom when the air is saturated with water vapor at 38 °C?

1.11.5 Maximum Allowable Concentration (MAC)

An open bottle containing acetone stands against orders for a long time in a laboratory. How many times is the maximum allowable concentration in workplace air of $1200\,\text{mg}\,\text{m}^{-3}$ exceeded at a temperature of 20 °C if we assume that the air in the room is saturated with acetone? The standard boiling point $T_{\text{lg}}^{\ominus}$ of acetone is 329 K, its molar entropy of vaporization (at the standard boiling point) $\Delta_{\text{lg}}S_{\text{eq}}^{\ominus}$ [$\equiv \Delta_{\text{lg}}S(T_{\text{lg}}^{\ominus})$] is $88.5\,\text{Ct}\,\text{mol}^{-1}$.

1.11.6 Boiling pressure of benzene

The standard boiling point of benzene is 80 °C. Estimate the pressure at which benzene already boils at 60 °C. Take into consideration that benzene is a typical nonpolar compound.

1.11.7 Vaporization of methanol

Methanol boils at 64.3 °C (under standard pressure). At room temperature (25 °C), a vapor pressure of 16.8 kPa is measured.

a) Estimate the molar entropy of vaporization $\Delta_{\text{lg}}S_{\text{eq}}^{\ominus}$ of methanol (at the standard boiling point).

b) Compare the result with PICTET-TROUTON's rule.

1.11.8* Triple point of 2,2-dimethylpropane

The temperature dependence of the vapor pressure of liquid 2,2-dimethylpropane can be approximately described by the following empirical equation,

$$\ln(p_{\text{lg}}/p^{\ominus}) = -\frac{2877.56\ \text{K}}{T} + 10.1945\,,$$

that of solid 2,2-dimethylpropane by the equation

$$\ln(p_{\text{sg}}/p^{\ominus}) = -\frac{3574.36\ \text{K}}{T} + 12.9086\,.$$

a) Estimate the molar entropy of vaporization $\Delta_{\text{lg}}S_{\text{eq}}^{\ominus}$ at the standard boiling point as well as the standard boiling point $T_{\text{lg}}^{\ominus}$ of 2,2-dimethylpropane by comparing the coefficients with Equation (11.11). Repeat the calculation process in an analogous manner for the molar entropy of sublimation $\Delta_{\text{sg}}S_{\text{eq}}^{\ominus}$ [$\equiv \Delta_{\text{sg}}S(T_{\text{sg}}^{\ominus})$] at the standard sublimation point as well as the standard sublimation point $T_{\text{sg}}^{\ominus}$.

b) Estimate the temperature $T_{\text{slg.}}$ and the vapor pressure p_{slg} at the triple point of 2,2-dimethylpropane.

1.11.9* Pressure dependence of the transition temperature

As shown in Chapter 5, transition temperatures depend on pressure, which is particularly noticeable in the case of boiling temperatures (see, for example, Experiments 5.3 and 5.4). This pressure dependence can generally be expressed by the coefficient $(\partial T/\partial p)_{\mathcal{A},\xi}$.

a) Let us consider the boiling process as an example:

$$\mathrm{B|l} \rightleftarrows \mathrm{B|g}\,.$$

Transform the above coefficient so that it is related to the molar volume of vaporization (change in molar volume that occurs on vaporization) and the molar entropy of vaporization.

b) The temperature unit was previously defined by the fact that the temperature difference between the freezing and boiling temperatures of water at 1 atm (101325 Pa) should correspond to 100 degrees (Celsius). Today, the unit is defined by the agreement that the absolute temperature of the triple point T_{slg} of water should be exactly 273.16 K. At the same time, the IUPAC (International Union of Pure and Applied Chemistry) has recommended to use 100 kPa (= 1 bar) instead of 1 atm as standard pressure for thermodynamic data. Calculate how this affects the boiling temperature of water. The molar entropy of vaporization $\Delta_{\mathrm{lg}}S_{\mathrm{eq}}$ (at the boiling point) is 109.0 Ct mol^{-1} and can be regarded as constant in the pressure range under consideration.

1.12 Spreading of Substances

1.12.1 Lowering of vapor pressure

Demonstration experiment 12.9 shows qualitatively that the vapor pressure of diethyl ether ($C_4H_{10}O$) is lowered by addition of oleic acid ($C_{18}H_{34}O_2$).

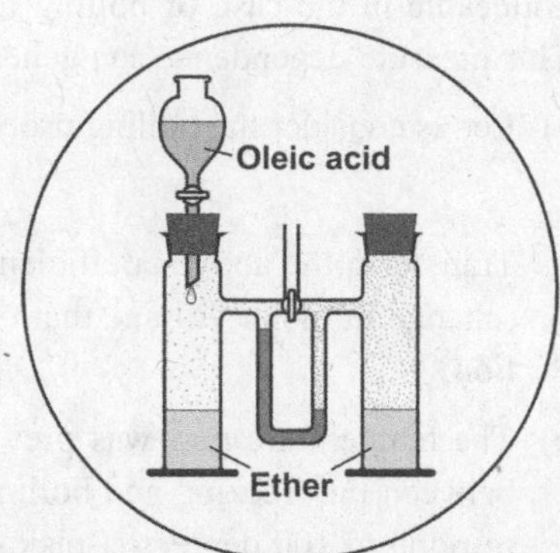

a) Calculate the lowering of vapor pressure when 11.3 g of oleic acid (substance B) are dissolved at 20 °C in 100.0 g of diethyl ether (substance A). The vapor pressure of the ether at this temperature is 586 hPa.

b) What is the vapor pressure of the solution?

c) What would be the difference in height between the liquid columns in the two legs of the manometer when water ($\rho \approx 998\ \text{kg m}^{-3}$ at 20 °C) is used as a barrier fluid?

1.12.2 "Household chemistry"

a) Three cubes of sugar (sucrose, $C_{12}H_{22}O_{11}$; substance B) are dissolved in 250 mL of water (substance A) at room temperature. The mass of one sugar cube should be about 3 g; the density of water is about 1 g cm^{-3} (at 25 °C). Estimate at what temperature the sugar solution would freeze. For this purpose, use the data in Table 12.1 of the textbook "Physical Chemistry from a Different Angle."

b) On a package of wheat noodles, the following instruction is found: "Place the noodles in salted water that boils and give the pasta occasionally a stir until it is "al dente"." A typical "noodle water" contains 10 grams of table salt (NaCl; substance B) per kilogram of water. Estimate at what temperature the saline solution boils (at 100 kPa). Keep in mind that the colligative properties depend on the number of dissolved particles.

1.12.3 Osmotic pressure

An aqueous urea solution has an osmotic pressure of 99 kPa at 298 K.

a) What is the amount n_B of urea in 1 L of the solution?

b) In order to estimate the much larger amount n_A of water in 1 L of the solution, the presence of urea in the thin solution can be ignored; volume, mass, density of the solution are supposed to correspond to that of pure water. What is the value for m_A and n_A? (The density of water at 298 K is about 1000 kg m^{-3}).

c) What is the mole fraction x_B of urea in the solution?

d) How does the chemical potential of the solvent water change by the addition of the foreign substance urea?

e) The freezing point of water is lowered by adding the foreign substance. Calculate this change in temperature. The molar entropy of fusion $\Delta_{sl}S^{\bullet}_{eq,A}$ of water (at the standard freezing point) is 22.00 Ct mol^{-1}.

f) What amount $n_{B'}$ of magnesium chloride ($MgCl_2$) ought to be used per liter to make a salt solution that has the same osmotic pressure as the urea solution in part a)?

1.12.4 Sea water

Sea water can be roughly considered as an aqueous saline solution with a mass fraction w_B of the salt of 3.5 %. The lowering of potential due to the salt content, for example, influences the vapor pressure, boiling point and freezing point of water. However, the relationships based on mass action can only be applied under reservation because the salt content of seawater is relatively high. But at least estimated values can be determined.

a) What are the amounts n_B of sodium chloride and n_A of the solvent water in 1000 g of sea water and what is therefore the mole fraction x_F of foreign substance?

b) Estimate the change in potential $\Delta\mu_A$ of the solvent water (at 298 K).

c) Sea water (S) evaporates more slowly than river water (R) (fresh water, $p^{\bullet}_{lg,A}$) because of its lower vapor pressure p_{lg}. Estimate roughly how much slower it evaporates (at 298 K).

d) How much lower is the freezing point of sea water (compared to that of fresh water)? The molar entropy of fusion $\Delta_{sl}S^{\bullet}_{eq,A}$ of water (at the standard melting point) is 22.00 Ct mol^{-1}.

e) How much higher is its boiling point (at standard pressure)? The molar entropy of vaporization $\Delta_{lg}S^{\bullet}_{eq,A}$ of water (at the standard boiling point) is 109.0 Ct mol^{-1}.

f) What is the minimum overpressure needed at 298 K to force sea water (solution S) through a membrane permeable only to H_2O molecules in order to desalinate it (reverse osmosis)? The density ρ_S of a saline solution with a mass fraction of 3.5 % is 1022 kg m^{-3} at the given temperature.

1.12.5 "Frost protection" in the animal world

The hemolymph (in everyday language also called "insect blood") of a parasitic wasp, *Bracon cephi*, contains in winter time a mass fraction w_B of about 0.3 of glycerol ($C_3H_8O_3$) in order to withstand prolonged exposure to low temperatures.

a) For the sake of simplicity, the hemolymph of the wasp can be considered to be an aqueous glycerol solution. What are the amounts n_B of glycerol and n_A of water in 100 g of the solution (S) and what is the mole fraction x_F of foreign substance?

b) What protection against freezing would be offered by the glycerol in the hemolymph, meaning estimate the lowering of freezing point [molar entropy of fusion of water (at the freezing point): $\Delta_{sl}S^{\bullet}_{eq,A} = 22.0 \text{ Ct mol}^{-1}$].

c) What is the osmotic concentration c_F (formerly known as osmolarity) of the hemolymph at 20 °C? Take into consideration that a glycerol solution with a mass fraction w of 0.3 of glycerol has a density ρ_S of about 1.065 g mL^{-1} at the given temperature.

d) What is the osmotic pressure of the hemolymph at 20 °C?

1.12.6 "Osmotic power plant"

The chemical potential μ_A of water in sea water (S) is lower than in river water (R) (fresh water, meaning almost pure water) because of the salt content of sea water: $\mu_{A,R} > \mu_{A,S}$. This potential gradient (it corresponds energetically to a waterfall of around 250 m) is sought to be exploited for generation of energy. A pilot plant was constructed at the Oslofjord in Norway. The data required in the following, the standard potential $\mu^{\ominus}$ as well as the corresponding temperature and pressure coefficients α and β, are listed in the following table:

Substance	Formula	$\mu^{\ominus}/\text{kG}$	α/GK^{-1}	β/GPa^{-1}
Water	$H_2O|l$	–237.14	–70	18×10^{-6}

a) Determine $\mu_{A,R}$ and $\mu_{A,S}$ at 100 kPa and a water temperature of 10 °C. Sea water can be regarded as an aqueous saline solution with a mole fraction $x_B = 0.011$ of sodium chloride (do not round the results in this case).

b) The lowering of the water potential to $\mu_{A,S}$ caused by the salt content can be compensated by increasing the pressure p on the sea water to such an extent that its chemical potential again corresponds to that of fresh water. The pressure change Δp to be calculated arises in an osmotic cell filled with sea water and immersed into fresh water by itself due to the influx of water. In an "osmotic power plant", sea water and fresh water are pumped into large tanks separated by a semipermeable membrane. The pressure generated by the migration of water through the membrane into the salt-containing solution is used to generate electricity by means of a water turbine.

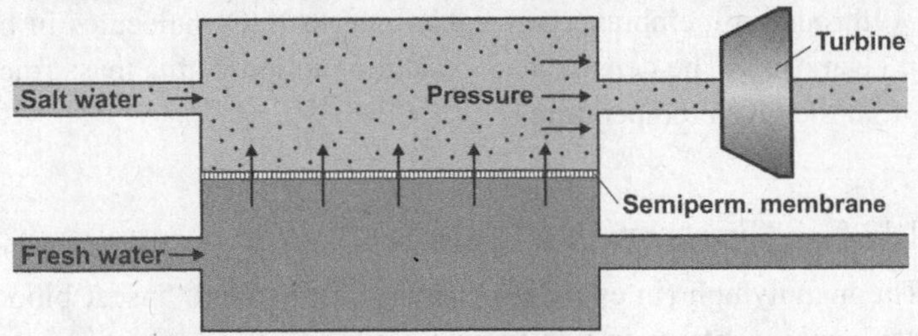

c) The migration of H_2O from fresh water to sea water, $H_2O|R \rightarrow H_2O|S$, can be interpreted as a chemical reaction. What is the chemical drive $\mathcal{A}$ under the conditions mentioned in part a)? The consumption of 1 m^3 of fresh water corresponds to a conversion $\Delta\xi$ of 55500 mol.

How much usable energy W^*_{use} can be gained in this way (efficiency on full load $\eta \approx$ 60 %)?

1.12.7 Isotonic saline solution

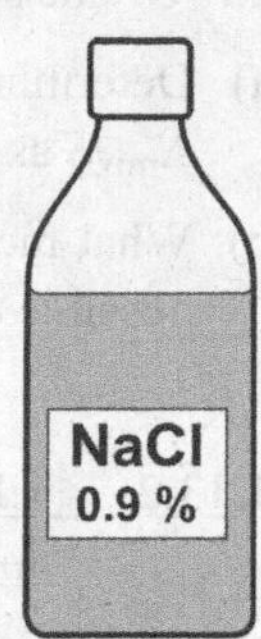

A saline solution used for intravenous infusion has to show at body temperature (37 °C) the same osmotic pressure as blood plasma ($p_{osm} \approx 7.38$ bar) in order not to damage the patient's red blood cells (so-called isotonic saline solution).

a) How many grams of sodium chloride are required to prepare 500 mL of an isotonic saline solution (based on the approximation for thin solutions)?

b) The isotonic saline solution used in medicine for intravenous therapy contains 9.0 g of salt per 1000 mL solution. Estimate the correction factor f [for explanation, see Solution 2.12.4.f)].

1.12.8 Determining molar masses by cryoscopy

From a historical viewpoint, cryoscopy was an important method in order to determine molar masses at a time when more precise methods did not exist. Camphor (substance A) was often used as solvent at higher temperatures because of its high cryoscopic constant. After addition of 40 mg of an unknown organic substance (substance B) to 10.0 g of liquid camphor, the freezing point of the mixture is lowered by 0.92 K.

a) What is the molar mass of the solute?

b) If the empirical formula of the unknown substance is C_5H_6O, what is its molecular formula? Which compound (used, for example, in dentistry) could it be?

1.12.9 Determining molar masses by osmometry

Due to its sensitivity, osmometry is particularly suitable for the investigation of macromolecular substances such as synthetic polymers, proteins or enzymes. An aqueous solution of 10.0 g of the enzyme catalase (substance B) (which catalyzes the decomposition of the cytotoxic agent hydrogen peroxide into the elements in aerobic organisms such as also human beings) has a volume of 1.00 L. The osmotic pressure of the solution at 27 °C is 104 Pa. Estimate the molar mass of the enzyme.

1.13 Homogeneous and Heterogeneous Mixtures

1.13.1 Ideal liquid mixture

0.8 mole of benzene (component A) and 1.2 mole of toluene (component B) are mixed at 25 °C. The solution should be considered ideal.

a) Determine the chemical drive $\mathcal{A}_{mix}$ for the mixing process, the molar entropy of mixing $\Delta_{mix}S$ as well as the molar volume of mixing $\Delta_{mix}V$.

b) What mole ratio of benzene to toluene has to be chosen in order to obtain the maximum possible entropy of mixing, referred to 1 mole of the mixture?

1.13.2 Chemical potential of a homogeneous mixture

a) Plot the chemical potentials of nitrogen-oxygen mixtures (under standard conditions) for x_B = 0 ... 1 according to the equation

$$\mu_M = x_A \times (\mathring{\mu}_A + RT \ln x_A) + x_B \times (\mathring{\mu}_B + RT \ln x_B)$$

into the grid on the right side. The index A stands for oxygen, the index B for nitrogen. Keep in mind that $x_A = 1 - x_B$ is valid.

b) What is the μ value of air?

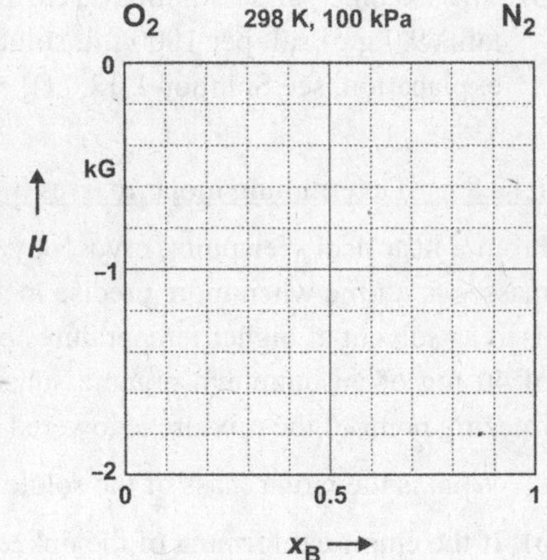

1.13.3 Mixing of ideal gases

A container with a volume of 10 L is separated into two equal parts by a partition wall. One half contains hydrogen gas (substance A) under a pressure of 100 kPa and the other half contains nitrogen gas (substance B) under the same pressure. The temperature of both gases is 15 °C. One assumes that the gases show ideal behavior. The partition wall is then removed so that the gases can mix spontaneously.

a) Calculate the chemical drive $\mathcal{A}_{mix}$ for the mixing process and from this the energy W_f $(= -W_{\rightarrow\xi})$, which is released during the mixing process.

b*) In a second experiment, one half still contains hydrogen gas under a pressure of 100 kPa, the other half, however, contains nitrogen gas under a pressure of 300 kPa. The temperature of both gases is again 15 °C. What energy W_f is released when the partition wall is removed?

1.13.4 Real mixture

In order to take into account the interactions occuring in real mixtures, the equation for the chemical potential μ_M of an ideal homogeneous mixture is supplemented by the extra potential $\overset{+}{\mu}_M$. Accordingly, this contribution describes the difference between the experimentally observed chemical potential and the chemical potential of the ideal homogeneous mixture. The dependence of this extra potential on the composition of a real mixture of two substances A and B should be given by the following empirical formula:

$$\overset{+}{\mu}_M(x_B) = 0.49RT \times x_B \times (1-x_B).$$

a) Are the components A and B of the mixture highly compatible, lowly compatible or incompatible?

b) Calculate the chemical drive $\mathcal{A}_{mix}$ for the mixing process as well as the molar entropy of mixing $\Delta_{mix}S$ when the components A and B are mixed at 30 °C in a mole ratio of 1:4.

c) How much energy W_f is released when 1 mole of component A is mixed with 4 moles of component B?

1.13.5 Boiling equilibrium in the butane-pentane system

The adjoining figure shows the $\mu(x_B)$ curve for a butane-pentane mixture as liquid (l) and as vapor (g) under standard conditions (index A butane, index B pentane). Which phases exist in equilibrium for the compositions characterized by a, b, c, d, e and f? How are the phases composed (x_B^l, x_B^g) and what are their particular fractions (x^l, x^g)?

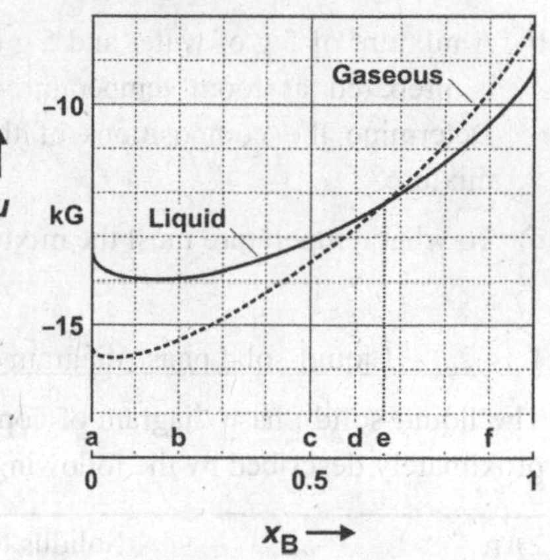

	liquid: x^l, x_B^l	vapor: x^g, x_B^g
a)	.... %, ... %	.... %, ... %
b)	.... %, ... %	.... %, ... %
c)	.... %, ... %	.... %, ... %
d)	.... %, ... %	.... %, ... %
e)	.... %, ... %	.... %, ... %
f)	.... %, ... %	.... %, ... %

1.14 Binary Systems

1.14.1 Miscibility diagram

Demonstration experiment 14.1 in the textbook "Physical Chemistry from a Different Angle" has shown that a mixture of phenol, C_6H_5OH, and water in a ratio of about 1:1 consists of two phases at room temperature. In 1937, A. N. CAMPBELL and A. J. R. CAMPBELL experimentally determined the adjacent miscibility diagram. In the diagram, temperature T was plotted against mass fraction w of phenol (substance B).

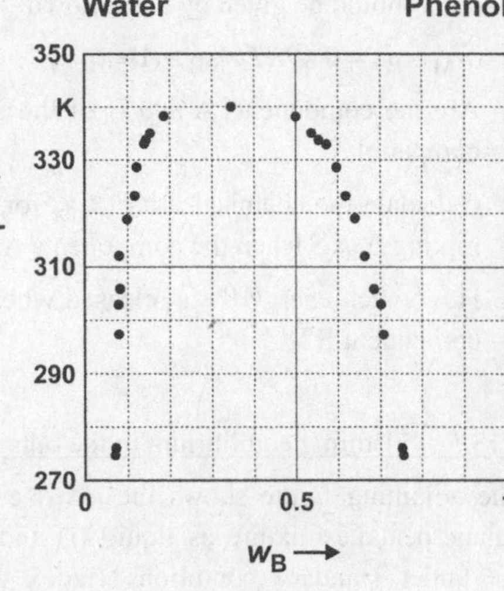

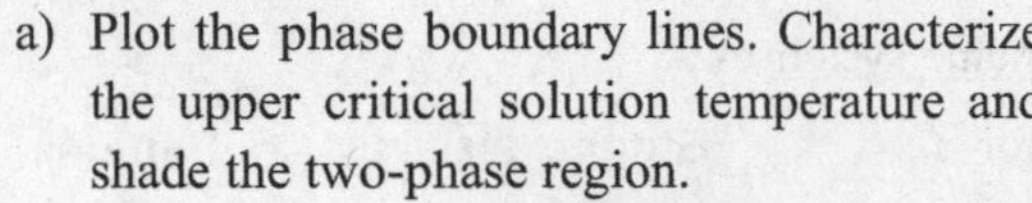

a) Plot the phase boundary lines. Characterize the upper critical solution temperature and shade the two-phase region.

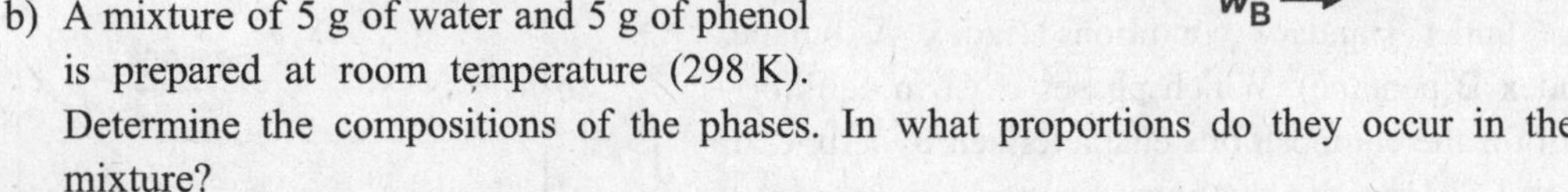

b) A mixture of 5 g of water and 5 g of phenol is prepared at room temperature (298 K). Determine the compositions of the phases. In what proportions do they occur in the mixture?

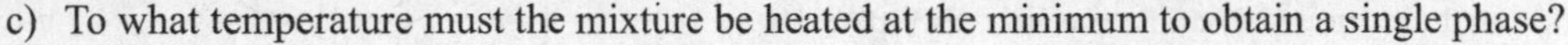

c) To what temperature must the mixture be heated at the minimum to obtain a single phase?

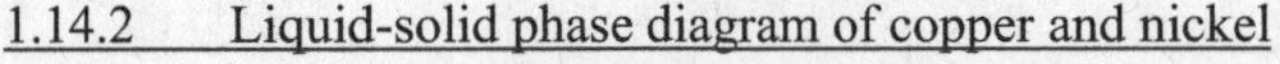

1.14.2 Liquid-solid phase diagram of copper and nickel

The liquid-solid phase diagram of copper (substance A) and nickel (substance B) can be approximately described by the following data:

w_B	Solidus temperature / °C	Liquidus temperature / °C
0	1085	1085
0.1	1105	1138
0.2	1129	1190
0.3	1157	1233
0.4	1190	1276
0.5	1223	1310
0.6	1262	1343
0.7	1300	1371
0.8	1343	1400
0.9	1400	1424
1.0	1452	1452

a) Plot the liquid-solid phase diagram into the coordinate grid below. Here, w_B represents the mass fraction of nickel. Label the diagram completely [that is, mark the solidus curve (melting curve) and the liquidus curve (freezing curve), respectively and indicate which phases are present in the different regions of the diagram]. Shade possible two-phase regions.

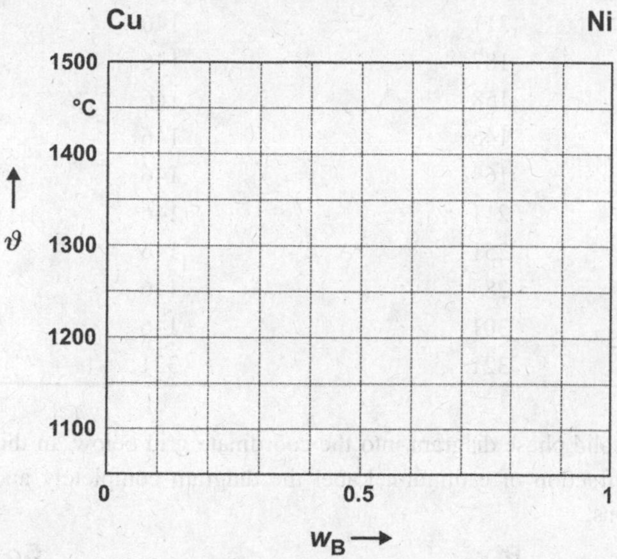

b) Decide whether the components of the alloy, here copper and nickel, behave highly compatible, indifferent, lowly compatible, incompatible, or completely incompatible.

c) Consider 10 kg of a melt of a Cu-Ni alloy with a mass fraction of nickel of $w_B = 0.25$. Such an alloy is, for example, commonly used in silver-colored modern-circulated coins. This melt is extremely slowly cooled down from a temperature of 1300 °C to a temperature of 1175 °C so that equilibrium conditions prevail. Which phases are present at 1175 °C and what are their compositions?

d) Determine the relative amount (in terms of mass fraction) of each phase at 1175 °C.

e) What is the total mass of mixed crystals at 1175 °C?

1.14.3 Liquid-solid phase diagram of bismuth and cadmium

The liquid-solid phase diagram of bismuth (substance A) and cadmium (substance B) can be approximately described by the following data [from: Moser Z, Dutkiewicz J, Zabdyr L, Salawa J (1988) The Bi-Cd (Bismuth-Cadmium) System. Bull Alloy Phase Diagr 9:445-448]:

x_B	Solidus temperature / °C	Liquidus temperature / °C
0	271	271
0.05	261	146
0.1	250	146
0.2	231	146
0.3	211	146
0.4	187	146
0.5	158	146
0.55	146	146
0.6	164	146
0.7	211	146
0.8	251	146
0.9	284	146
0.95	301	146
1.0	321	321

a) Plot the liquid-solid phase diagram into the coordinate grid below. In this case, x_B represents the mole fraction of cadmium. Label the diagram completely and shade possible two-phase regions.

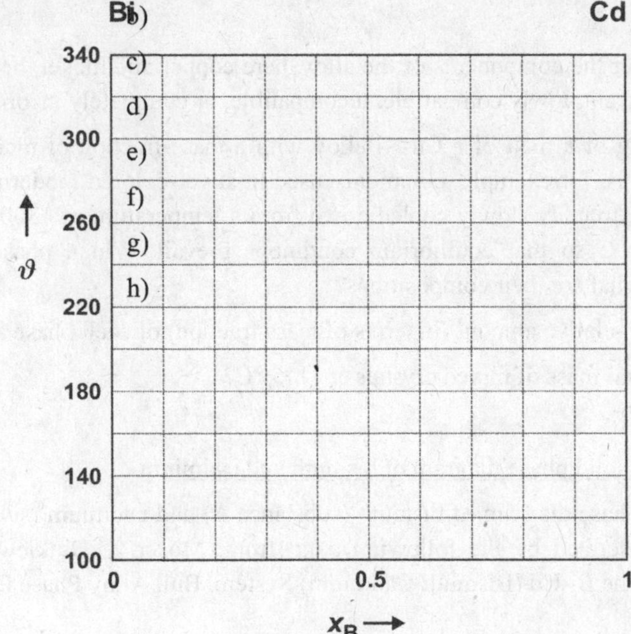

b) Decide whether the components of the alloy, here bismuth and cadmium, behave highly compatible, indifferent, lowly compatible, incompatible, or completely incompatible.

In the following, we consider a partially molten Bi-Cd alloy whose melt shows a mole fraction x_B of bismuth of 0.3. The total amount of liquid and crystalline phase is supposed to be 20 moles.

c) What is the temperature and the composition of the solid phase at this temperature?

d) At the temperature determined in part c), 10 moles of crystallites have already formed. Specify the composition of the starting mixture.

e) What is the amount of bismuth crystallites when the alloy was cooled down until it became completely solid?

1.14.4* Liquid-solid phase diagram of bismuth and lead

The following figure schematically shows the liquid-solid phase diagram of bismuth and lead. x_B represents the amount of lead (substance B).

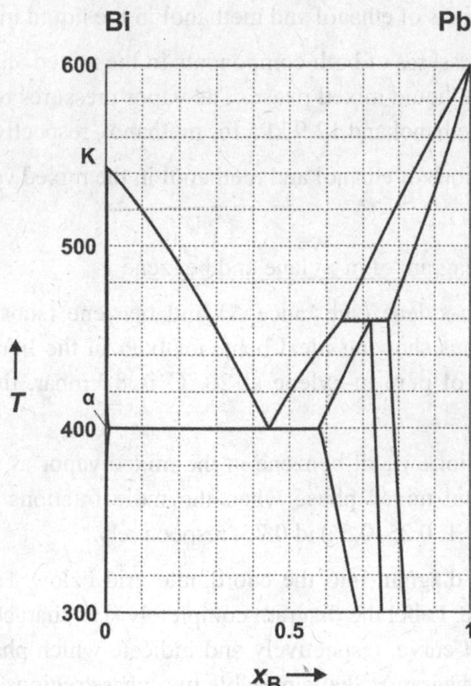

a) Trace the liquidus and solidus curves in different colors and shade the two-phase regions.

b) In total, there are four different "phases" (single-phase regions). Label them in the diagram with "l" if they are liquid, and proceeding from left to right, with Greek letters (α, β, γ...) if they are solid. Also mark the eutectic point.

c) Which of these phases occur when a melt containing equal amounts of bismuth and lead is cooled down? Determine in each case the mole fraction x of the phase in question and its lead content x_B.

T/K	Phase "l"	Phase "α"	Phase "β"	Phase "γ"
500	.... %, ... %	.... %, ... %	.... %, ... %	.... %, ... %
420	.... %, ... %	.... %, ... %	.... %, ... %	.... %, ... %
401	.... %, ... %	.... %, ... %	.... %, ... %	.... %, ... %
300	.... %, ... %	.... %, ... %	.... %, ... %	.... %, ... %

1.14.5 Liquid mixed phase and corresponding mixed vapor

Consider a mixture of 50 g of ethanol (substance A) and 50 g of methanol (substance B) at 20 °C, meaning a homogeneous mixture of two volatile liquids that are (almost) indifferent to each other.

a) Calculate the mole fractions of ethanol and methanol in the liquid mixed phase.

b) Determine the partial pressures of both components in the mixed vapor as well as the total vapor pressure above the liquid mixed phase. The vapor pressures of the pure components at 20 °C are 5.8 kPa for ethanol and 12.9 kPa for methanol, respectively.

c) Calculate the mole fractions of ethanol and methanol in the mixed vapor.

1.14.6 Vapor pressure diagram of m-xylene and benzene

The liquid hydrocarbons m-xylene (substance A) and benzene (substance B) are (almost) indifferent to each other, thus showing ideal behavior both in the liquid and in the gaseous phase. The vapor pressure of pure m-xylene at 20 °C is 8.3 mbar, that of pure benzene is 100 mbar.

a) Calculate the mole fractions x_B^g of benzene in the mixed vapor as well as the total vapor pressure above the liquid mixed phase when the mole fractions x_B^l of benzene in the liquid mixed phase are 0.1, 0.25, 0.5 and 0.75, respectively.

b) Plot the vapor pressure diagram into the coordinate grid below. Here, x_B represents the mole fraction of benzene. Label the diagram completely (i.e. characterize the boiling point curve and the dew point curve, respectively and indicate which phases are present in the different regions of the diagram). Shade possible two-phase regions.

c) Determine from the vapor pressure diagram at which pressure a liquid mixture of 2 moles of benzene and 1 mole of m-xylene begins to boil.

d) Calculate the composition of the corresponding mixed vapor.

e) Determine the composition and vapor pressure of the last remaining drops of liquid when almost all of the liquid mixed phase has been vaporized.

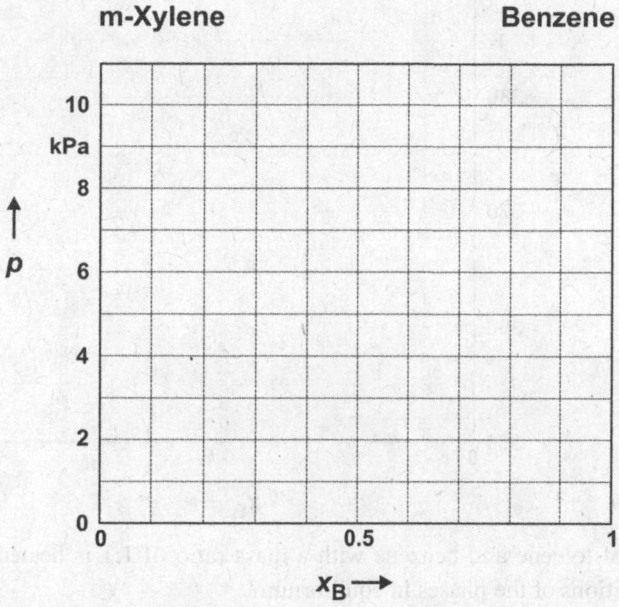

1.14.7 Boiling temperature diagram of toluene and benzene as well as distillation

Toluene (substance A) and benzene (substance B) are also (almost) indifferent to each other. The following table shows the vapor pressures of pure toluene and pure benzene at different temperatures. The standard boiling temperature of toluene is 384 K, that of benzene 353 K.

T/K	$p_A^\bullet$/kPa	$p_B^\bullet$/kPa
358	46	116
363	54	133
368	63	155
373	74	179
378	86	206

a) In order to construct a boiling temperature diagram for the toluene/benzene system, we have to calculate the mole fractions of benzene in the liquid phase (boiling point curve) and in the gaseous phase (dew point curve) for the temperatures given in the table above. The constant total pressure is supposed to be 100 kPa.

Plot the boiling temperature diagram into the coordinate grid where x_B represents the mole fraction of benzene. Label the diagram completely and shade possible two-phase regions.

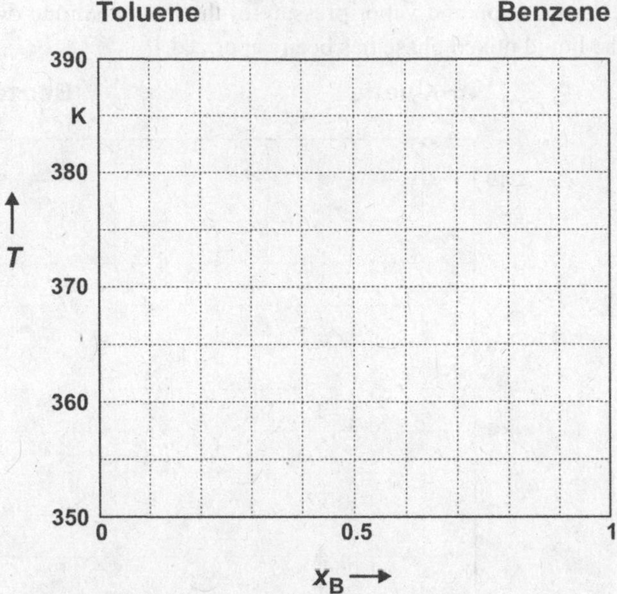

b) A mixture of toluene and benzene with a mass ratio of 1:1 is heated to 365 K. What are the compositions of the phases in equilibrium?

c) Solvent waste containing benzene as well as toluene and boiling at 375 K shall be recycled by distillation in the laboratory. For this purpose, a distillation column with five theoretical plates is used. What is the purity (referred to mole fraction) of the recovered benzene?

1.14.8 Boiling temperature diagram with azeotropic maximum

In the adjoining boiling temperature diagram of the substances A and B, x_B^l represents the mole fraction of B in the liquid (phase l) and x_B^g that in the vapor (phase g).

a) Trace the boiling point curve and the dew point curve in different colors. Label the diagram completely and shade the two-phase regions.

b) Determine the temperature at which a liquid mixture with a mole fraction of B of 0.7 boils. What is the composition of the resulting vapor?

c) Determine the temperature at which a vapor with a B-fraction of 0.7 condenses.

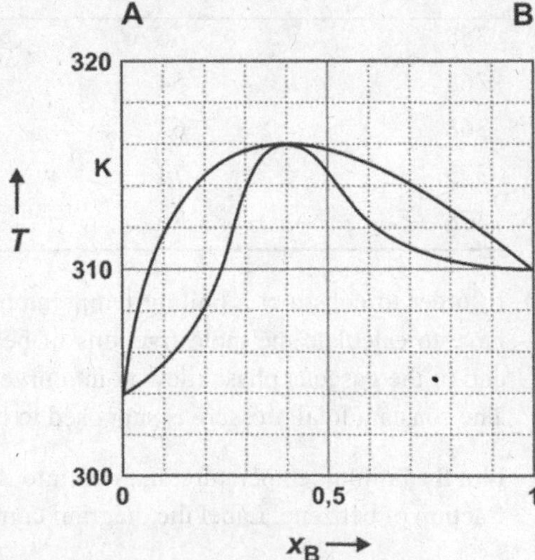

What is the composition of the resulting condensate?

d) What is the composition of the azeotropic mixture and at what temperature does it boil? What is the composition of the resulting vapor?

e) What can be said about the "compatibility" of the liquids A and B?

1.14.9 Distillation of mixtures of water and ethanol

The best-known system that shows a boiling temperature diagram with an azeotrope is the system water (substance A) / ethanol (substance B) because it plays an important role in the production of spirits. The boiling temperature diagram can be approximately described by the data in the table at the end.

a) Plot the boiling temperature diagram into the coordinate grid below. x_B represents the mole fraction of ethanol.

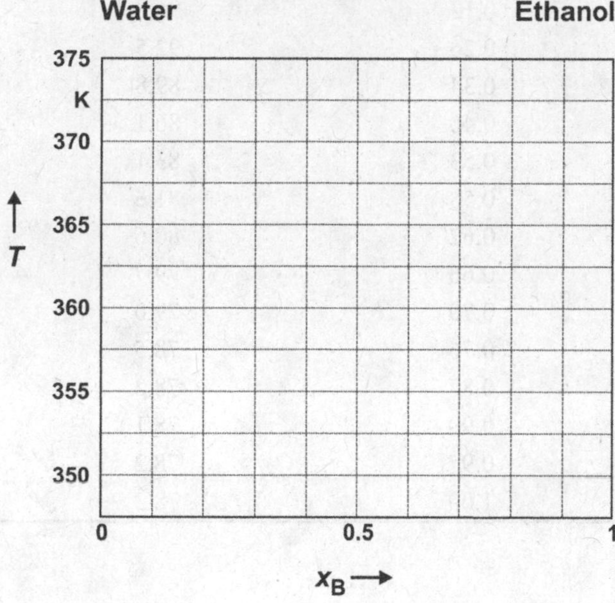

Label the diagram completely and shade possible two-phase regions.

b) In the spirits trade, the "alcoholic" content (ethanol content) of a beverage is not given as mole fraction or mass fraction, but in "vol%", i.e. in the so-called volume concentration $\sigma_B = V_B/V_{total}$. V_B is the volume of a constituent B *prior to* the mixing process and V_{total} is the actual total volume of the mixture *after* the mixing process. In the present case, the total volume is not equal to the sum of the volumes of the pure components due to the volume contraction when mixing water and ethanol, but the effect is so small that it can be neglected here.

Estimate the volume concentration σ_B obtained by a simple distillation of a water-ethanol mixture with a volume concentration of ethanol of 10 % (this corresponds approximately to the alcoholic content of wine). Which beverage, for example, shows such a concentration? [Densities (at 20 ° C): water: $\rho_A = 0.998\ \text{g cm}^{-3}$; ethanol: $\rho_B = 0.791\ \text{g cm}^{-3}$]

c) By distillation, a volume concentration of ethanol of 80% is to be achieved (this corresponds approximately to the "alcoholic" content of "Stroh Rum"). How many theoretical plates are required for this purpose?

Necessary data [from: Kadlec P, Henke S, Bubník Z (2010) Properties of ethanol and ethanol-water solutions–Tables and Equations. Sugar Industry 135:607–613; data for a pressure of 1 atm (101325 Pa)]:

w_B^l	w_B^g	ϑ / °C
0.00	0.00	100.0
0.01	0.12	96.8
0.03	0.26	92.5
0.05	0.34	89.8
0.10	0.46	86.1
0.20	0.53	83.1
0.30	0.58	81.6
0.40	0.62	80.6
0.50	0.66	79.7
0.60	0.70	79.0
0.70	0.76	78.5
0.80	0.82	78.2
0.90	0.90	78.1
0.97	0.97	78.2
1.00	1.00	78.3

1.15 Interfacial Phenomena

1.15.1 Surface energy

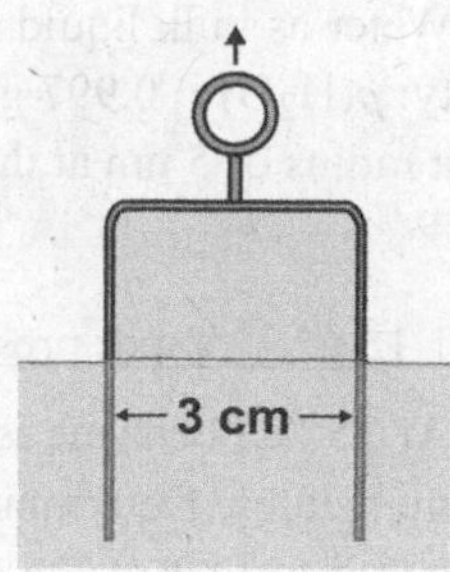

A U-shaped wire frame (with a distance of 3 cm between the two legs of the frame) is dipped into a diluted soap solution ($\sigma = 30\ \text{mN m}^{-1}$) and then pulled upwards in such a way that a plane soap lamella is formed. What is the surface energy required to produce a rectangular lamella with a size of $3 \times 4\ \text{cm}^2$?

1.15.2* Atomization

A volume of 1 L of water is to be atomized into fine spherical droplets with a diameter of 1 μm at 25 °C. Calculate the surface energy necessary for the process.

1.15.3 Capillary pressure

Determine the excess pressure p_σ in a spherical droplet of water with a radius of 250 nm at 283 K.

1.15.4* Floating drops and bubbles

The smaller a drop, the more its shape is influenced by the interfacial tensions, while weight and buoyancy forces recede. What shape would the drop on the right take (it is supposed to be very small) if (please complete the pictures accordingly!):

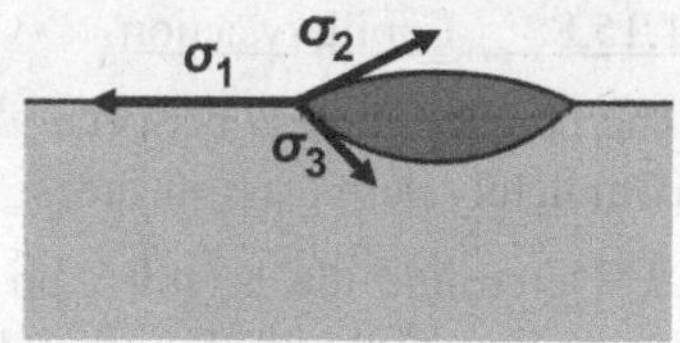

a) a_1) $\sigma_1 = \sigma_2 = \sigma_3$, a_2) $\sigma_1 > \sigma_2 + \sigma_3$, a_3) $\sigma_2 > \sigma_1 + \sigma_3$, a_4) $\sigma_3 > \sigma_1 + \sigma_2$?

b) What position and shape (please sketch) do air bubbles take at a water surface ($\sigma = 73\ \text{mN m}^{-1}$) and what is the excess pressure in them,

b_1) if they are very small, $r = 0.1$ mm, b_2) if they are large, $r = 10$ mm?

1.15.5 Vapor pressure of small droplets (I)

Water as bulk liquid shows at 298 K a saturation vapor pressure of $p_{\text{lg},r=\infty} = 3167$ Pa [Density: $\rho(H_2O) = 0.997$ g cm^{-3}]. Calculate the vapor pressure of a spherical droplet of water with a radius of 5 nm at the same temperature.

1.15.6* Vapor pressure of small droplets (II)

At 25 °C, benzene is atomized in tiny spherical droplets, which comprise an average of 200 molecules. Determine the factor by which the vapor pressure of the droplets increases compared to that of the bulk liquid. The surface tension of benzene at the given temperature is 28.2 mN m^{-1}, its density 0.876 g cm^{-3}.

1.15.7 Determination of the surface tension

Capillary action can be used to determine the surface tension of liquids that completely wet glass (capillary rise method). The surface tension of an ethanol-water mixture with a mass fraction of 30 % of ethanol is to be determined at 20 °C (density of the mixture: ρ = 0.955 g cm^{-3}). The liquid rises at this temperature in a glass capillary tube with an internal diameter of 0.2 mm up to a height of 3.54 cm.

1.15.8* Capillary action

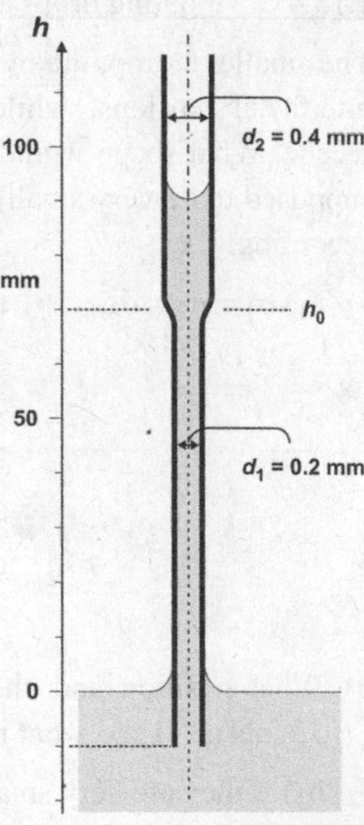

The contact angle θ between glass and water is almost zero for completely pure glass surfaces.

a) Determine the height h up to which water would rise at a temperature of 0 °C in the glass tube on the right. What would it look like at 100 °C? The surface tension σ of water at 0 °C is 76 mN m^{-1}; at 100 °C, however, it is 59 mN m^{-1}. Its density ρ at 0 °C is 1000 kg m^{-3} and at 100 °C 958 kg m^{-3}.

b) What is the radius r of curvature of the resulting water surface in the capillary in both cases?

c) Plot the course of the pressure along the tube axis from h = −20 ... +100 mm into the coordinate grid below, more precisely $\Delta p = p - p_{\text{air}}$ for the cases under a) (1 mm water column corresponds to approximately 10 Pa).

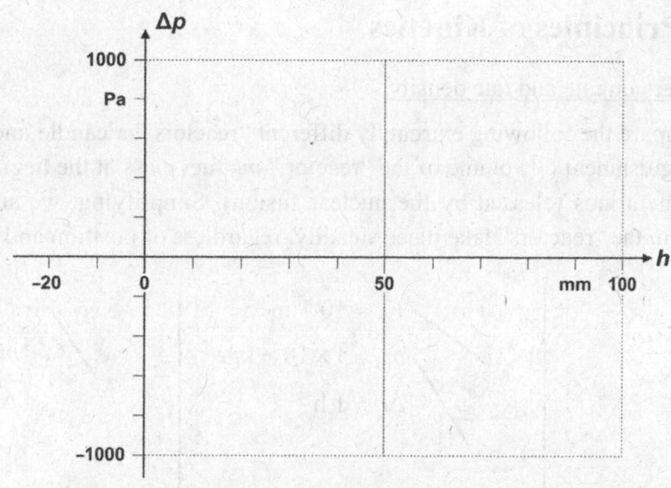

1.15.9 Fractional coverage

50.0 g of activated charcoal with a specific surface area of 900 m^2 g^{-1} adsorb 46 cm^3 of nitrogen at a pressure of 490 kPa and a temperature of 190 K. Determine the fractional coverage of the surface if the area occupied by a nitrogen molecule is 0.16 nm^2.

1.15.10 LANGMUIR isotherm (I)

The adsorption of a gas on a surface is described at 298 K by a LANGMUIR isotherm with a constant $\overset{\circ}{K} = 0.65$ kPa^{-1}. Calculate the pressure at which the gas occupies half of the surface.

1.15.11* LANGMUIR isotherm (II)

Carbon monoxide (CO) is to be adsorbed at 273 K on charcoal. The mass of gas adsorbed at a pressure of 400 mbar is 31.45 mg, at a pressure of 800 mbar, however, 51.34 mg. The adsorption process is supposed to obey the LANGMUIR isotherm.

a) Determine the constant $\overset{\circ}{K}$ as well as the mass m_{mono}, which corresponds to a monomolecular adsorption layer (meaning a complete coverage of the surface).

b) Calculate the fractional coverage of the surface at the two pressures.

1.16 Basic Principles of Kinetics

1.16.1 Conversion rate and rate density

We roughly compare the following extremely different "reactors," a candle and the sun, from ignition to extinguishment (V volume of the "reactor," m_0 fuel mass at the beginning, Δt burning duration, ν neutrinos released by the nuclear fusion). Simplifying, we suppose that the transformations in the "reactors" take place steadily, regardless of position and time.

	Candle	Sun
	$V \approx 10^{-6}\ \mathrm{m^3}$	$V \approx 10^{27}\ \mathrm{m^3}$
	$m_0 \approx 1\times10^{-2}\ \mathrm{kg}$	$m_0 \approx 2\times10^{30}\ \mathrm{kg}$
	$\Delta t \approx 1\ \mathrm{h}$	$\Delta t \approx 5\times10^{18}\ \mathrm{s}$
Conversion formula	$2\,(CH_2)+3\,O_2 \rightarrow 2\,CO_2+2\,H_2O$	$4\,{}^1H \rightarrow {}^4He + 2\,\nu$
Chemical drive $\mathcal{A}$	1.2×10^6 G	3×10^{12} G
Total conversion $\Delta\xi$	.... mol	.... mol
Conversion rate ω	.... $\mathrm{mol\,s^{-1}}$	.... $\mathrm{mol\,s^{-1}}$
Rate density r	.... $\mathrm{mol\,m^{-3}\,s^{-1}}$	.... $\mathrm{mol\,m^{-3}\,s^{-1}}$
Heating power $P = \mathcal{A}\times\omega$	.... W	.... W
Power density $\varphi = \mathcal{A}\times r$	.... $\mathrm{W\,m^{-3}}$	.... $\mathrm{W\,m^{-3}}$

1.16.2 Rate law

For a reaction

$$B + B' + B'' \rightarrow \text{products}\,,$$

the following values are determined for the rate density r at different concentrations of the reactants B, B′ and B″:

No.	$c_B/\mathrm{kmol\,m^{-3}}$	$c_{B'}/\mathrm{kmol\,m^{-3}}$	$c_{B''}/\mathrm{kmol\,m^{-3}}$	$r/\mathrm{mol\,m^{-3}}$
1	0.2	0.2	0.2	3.2
2	0.3	0.2	0.2	4.8
3	0.2	0.3	0.2	7.2
4	0.3	0.2	0.4	9.6

a) Determine the order of the reaction with respect to the individual reactants B, B′ and B″ and establish the corresponding rate law.

b) Find the value of the rate coefficient k.

1.16.3* Rate law for advanced students

The figure illustrates the dependence $c_{B'}(t)$ for batches with different initial concentrations c_B of the reaction

$$2\,B + B' \rightarrow D\,.$$

Since we have $c_B >> c_{B'}$, c_B remains almost constant during the reaction process.

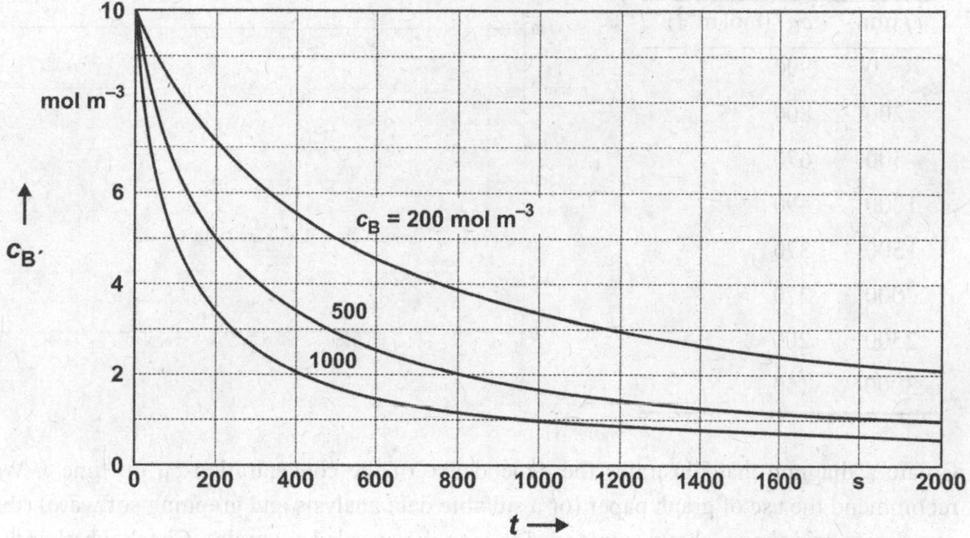

a) Estimate the (differential) change in concentration $\dot{c}_{B'} = dc_{B'}/dt$ and the rate density r at $t = 0$ for $c_B/\text{mol m}^{-3}$: 200 500 1000

	200	500	1000
$\dot{c}_{B',0}/\text{mol m}^{-3}\,\text{s}^{-1}$	....	....	....
$r_0/\text{mol m}^{-3}\,\text{s}^{-1}$	....	....	....

b) What is the rate density r in mol m^{-3} s^{-1} at the following concentrations:

$c_B/\text{mol m}^{-3}$	200	500	1000
$c_{B'}/\text{mol m}^{-3} = 2$	–	....	....
5	....	....	....
10	....	....	....

c) Deduce the rate law of the reaction from these results. Estimate the rate coefficient k.

1.16.4 Decomposition of dinitrogen pentoxide

Dinitrogen pentoxide $N_2O_5|s$ (substance B) exists as colorless crystals that sublime slightly above room temperature. The salt decomposes easily—often explosively—into brown nitrogen dioxide and oxygen.

a) What are the chemical drives $\mathcal{A}^\ominus$ for sublimation and decomposition under standard conditions?

$$N_2O_5|s \rightarrow N_2O_5|g, \qquad N_2O_5|s \rightarrow 2\,NO_2|g + \tfrac{1}{2}\,O_2|g.$$

b) An experimental study of the decomposition of N_2O_5 in a CCl_4 solution at 45 °C gives the following values:

t / min	c_B / (mol m^{-3})
0	900
200	800
500	670
1000	490
1500	370
2000	270
2500	200
3000	150

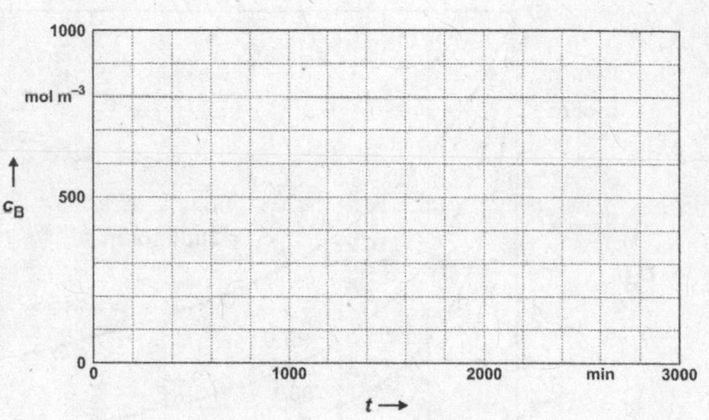

Create a diagram that illustrates the dependence of the concentration c_B on time t. We recommend the use of graph paper (or a suitable data analysis and graphing software) (the coordinate grid shown above is intended to be a downscaled example). Check whether the data in the table can be represented by an exponential function. (Use the criterion given in Appendix A1.1 of the textbook "Physical Chemistry from a Different Angle" for this purpose).

c) Verify the order of the reaction with respect to dinitrogen pentoxide determined in part b) graphically by means of a suitable plot.

d) Determine the rate coefficient k with the aid of the diagram in part c). What is the half-life $t_{1/2}$?

1.16.5 Dissociation of ethane

The dissociation of ethane (substance B) into methyl radicals [which is important in cracking (process in petroleum refining by which long-chain hydrocarbon molecules are broken up into lighter molecules with a shorter chain length)],

$$C_2H_6 \rightarrow \bullet CH_3 + \bullet CH_3\,,$$

represents a first-order reaction. The rate coefficient k at 700 °C is 0.033 min^{-1}.

a) By what percentage has the initial amount $n_{B,0}$ decreased after 2 hours?

b) What is the half-life $t_{1/2}$ of ethane?

1.16.6 Decomposition of dibenzoyl peroxide

The decomposition of dibenzoyl peroxide (substance B) in benzene as solvent also follows a first-order rate law. At 70 °C, the molar concentration decreases from an initial value of 200 mmol L^{-1} to a value of 143 mmol L^{-1} within 5 hours. What is the rate coefficient k of this reaction?

1.16.7 Radiocarbon dating

Radiocarbon dating was developed by Willard Frank LIBBY in the late 1940s who received the Noble Prize in chemistry for his work in 1960. This method for determining the age of carbon-containing archaeological samples is based on the fact that the amount of bound radioactive ^{14}C atoms in dead organisms decreases by the decay to ^{14}N atoms according to a first-order rate law. The half-life of this decay process is 5,730 years (abbreviation: a).

a) Calculate the rate coefficient k of the decay process.

b) A wooden beam found during an archaeological excavation contains only 77 % of the carbon-14 which is present in fresh wood. How long ago was the artifact made?

1.16.8 Decomposition of hydrogen iodide

Hydrogen iodide (substance B) decomposes in the gas phase into hydrogen and iodine,

$$2\,HI \rightarrow H_2 + I_2\,.$$

The reaction follows a second-order rate law. The rate coefficient k is 1.0×10^{-2} L mol^{-1} s^{-1} at 500 °C. How long does it take for the concentration of B to drop to

a) one-half,

b) one-eighth of its initial value of 20 mol m^{-3}?

1.16.9 Alkaline saponification of esters

The saponification of ethyl acetate in alkaline solution,

$$\begin{array}{lcccc} \text{Symbol} & B & B' & D & D' \\ & CH_3COOC_2H_5 + & OH^- \rightarrow & CH_3COO^- + & C_2H_5OH, \end{array}$$

obeys a second-order rate law as well. Calculate the rate coefficient k and the half-life $t_{1/2}$ if the initial concentrations of the reactants are both 100 mol m^{-3} and it takes 60 min for the concentrations to decrease to 62 mol m^{-3}.

1.17 Composite Reactions

1.17.1 A "sweet" equilibrium reaction

The transition of α-D-glucose (substance B) into the isomeric β-D-glucose (substance D) in aqueous solution is an equilibrium reaction,

$$\alpha\text{-D-Glucose} \underset{-1}{\overset{+1}{\rightleftarrows}} \beta\text{-D-Glucose}.$$

Both the forward and backward reactions obey a first-order rate law. In 1968, H. C. CURTIS, J. A. VOLLMIN and M. MÜLLER determined for this transition process at 37 °C (and pH 6) a rate coefficient k_{+1} for the forward reaction of 0.0525 min^{-1} and a conventional equilibrium constant $\mathring{K}_c$ of 1.64.

a) Determine the rate coefficient k_{-1} for the backward reaction.

b) Calculate the basic value $\mathring{\mathcal{A}}$ of the chemical drive.

c*) How long will it take until the concentration of β-D-glucose is just the same as that of α-D-glucose if we start at $t = 0$ with pure α-D-glucose?

1.17.2 Equilibrium reaction

An equilibrium reaction

$$\text{B} \underset{-1}{\overset{+1}{\rightleftarrows}} \text{D}$$

is supposed to be monomolecular in both directions. The initial concentration $c_{\text{B},0}$ of 1.00 mol m^{-3} has decreased to c_B = 0.70 kmol m^{-3} after 10 min. The equilibrium concentration $c_{\text{B,eq}}$ is 0.20 kmol m^{-3}.

a) Calculate the conventional equilibrium constant $\mathring{K}_c$.

b*) Determine the rate coefficients k_{-1} and k_{+1}.

1.17.3 Decomposition of acetic acid

Acetic acid (substance B) decomposes at high temperatures (550 until 950 °C) simultaneously into methane (substance D) and carbon dioxide or into ketene (substance D′) and water:

$$CH_3COOH \overset{1}{\rightarrow} CH_4 + CO_2 ,$$

$$CH_3COOH \overset{2}{\rightarrow} CH_2CO + H_2O .$$

The rate coefficients of the competitive first-order reactions at 1179 K are k_1 = 3.74 s^{-1} and k_2 = 4.65 s^{-1}, respectively.

a) How long will it take to consume 90 % of the acetic acid?

b) Determine the concentration ratio of the reaction products methane and ketene.

c) What is the maximum yield $\eta_{D',max}$ of ketene (in percentage of the initial concentration of acetic acid) obtainable under the given conditions?

1.17.4* Parallel reactions

A substance B reacts simultaneously to the products D and D′ by two parallel monomolecular elementary reactions 1 and 2:

$$B \begin{array}{l} \overset{1}{\nearrow} D \\ \underset{2}{\searrow} D' \end{array} .$$

The initial concentration of B is $c_{B,0} = 0.50\ \text{kmol m}^{-3}$. After 40 min, the concentration of B has decreased to $0.05\ \text{kmol m}^{-3}$; in the same time, the product D′ with a concentration $c_{D'} = 0.10\ \text{kmol m}^{-3}$ is formed. Calculate the rate coefficients k_1 and k_2.

1.17.5* Consecutive reactions

The simplest case of a consecutive reaction is a series of two monomolecular elementary reactions

$$B \overset{1}{\rightarrow} I \overset{2}{\rightarrow} D .$$

An example of such a process is a radioactive decay series. At the beginning, only the reactant B should be present in the concentration $c_{B,0}$. Then, the change in concentration of the intermediate substance I (which is formed by reaction 1 and decomposes by reaction 2) with time can be described as follows [Eq. (17.27)]:

$$c_I(t) = \frac{k_1}{k_2 - k_1} c_{B,0}(e^{-k_1 t} - e^{-k_2 t}) .$$

a) Deduce from the equation above an expression for the time t_{max} at which the concentration of the intermediate substance I reaches its maximum value.

b) The initial concentration $c_{B,0}$ of B is supposed to be $1.00\ \text{kmol m}^{-3}$, the rate coefficient k_1 $0.010\ \text{s}^{-1}$ and the rate coefficient k_2 $0.006\ \text{s}^{-1}$. Calculate t_{max} for this particular case.

c) Specify the maximum concentration $c_{I,max}$ of I.

1.17.6* Construction of $c(t)$ curves for a multistep reaction

The multistep reaction

$$B \overset{1}{\rightarrow} I \overset{2}{\rightarrow} I' \overset{3}{\rightarrow} D$$

is supposed to obey the following rate laws:

$r_1 = k_1 \times c_B \times c_I$ with $k_1 = 5 \times 10^{-6}\ \text{mol}^{-3}\ \text{m}^3\ \text{s}^{-1}$,

$r_2 = k_2 \times c_I \times c_{I'}$ with $k_2 = 100 \times 10^{-6}\ \text{mol}^{-3}\ \text{m}^3\ \text{s}^{-1}$ and

$r_3 = k_3 \times c_{I'}$ with $k_3 = 5 \times 10^{-3}\ \text{s}^{-1}$.

a) Complement the equations for the (differential) change in concentration $\dot{c} = \text{d}c/\text{d}t$ of the four substances B, I, I′, D:

$\dot{c}_B =$ .... $=$ .... ,

$\dot{c}_I = r_1 - r_2$ .... $= k_1 c_B c_I - k_2 c_I c_{I'}$ .... ,

$\dot{c}_{I'} =$ .... $=$ .... ,

$\dot{c}_D =$ .... $=$ ....

b) The concentration curves $c(t)$ of the four substances can be plotted step by step by calculating the slopes $\dot{c}$ of the curves using the equations above and considering them constant during a short period of time Δt. Repeating the step with the new values for the concentration at the endpoints of the short, straight segment of the curve gives approximately the curves in question as polylines. Extend the two polylines for the substances I and I′ in the figure below with a time interval Δt of 100 s. To simplify matters, it is assumed that the concentration $c_B = 1000\ \text{mol}\,\text{m}^{-3}$ is kept constant by continuously replacing the consumed substance B.

t / s	c_I / mol m^{-3}	$c_{I'}$ / mol m^{-3}	$k_1 c_B c_I$ / mol m^{-3} s^{-1}	$k_2 c_I c_{I'}$ / mol m^{-3} s^{-1}	$k_3 c_{I'}$ / mol m^{-3} s^{-1}	$\dot{c}_I \times \Delta t$ / mol m^{-3}	$\dot{c}_{I'} \times \Delta t$ / mol m^{-3}
0	50.0	35.0	0.250	0.175	0.175	7.5	0
100	57.5	35.0	0.288	0.201	0.175	8.7	2.6
200	66.2	37.6	....	....	....	....	....
300	....	....	....	....	....	....	....
400	....	....	....	....	....	....	....
500	....	....	....	....	....	....	....
600	....	....	....	....	....	....	....
700	....	....	....	....	....	....	....
800	....	....	....	....	....	....	....
900	....	....	....	....	....	....	....
1000	....	....	–	–	–	–	–

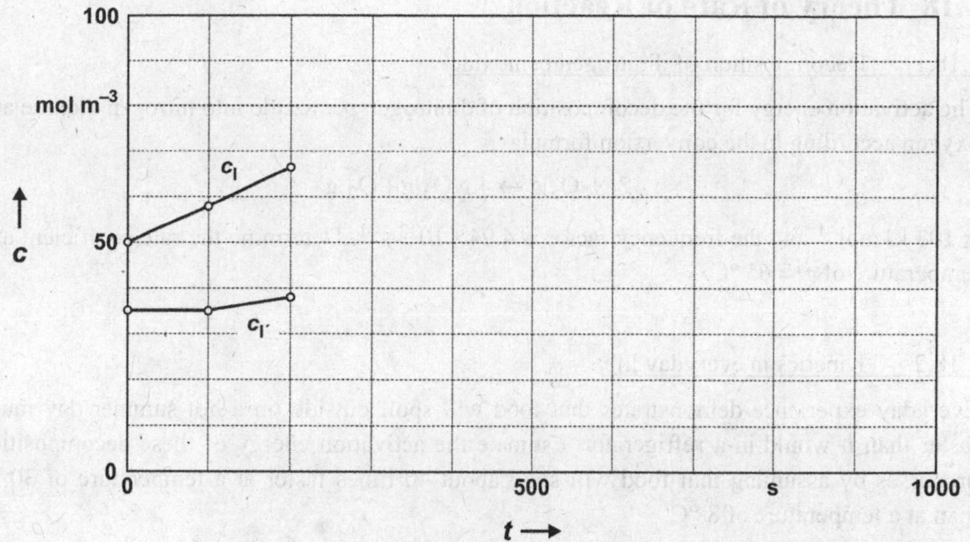
100
mol m⁻³
50
c
0
0
500
s
1000
t
c_I
c_I'

1.18 Theory of Rate of Reaction

1.18.1 Decomposition of dinitrogen pentoxide

The activation energy for the decomposition of dinitrogen pentoxide into nitrogen dioxide and oxygen according to the conversion formula

$$2\ N_2O_5|g \rightarrow 4\ NO_2|g + O_2|g$$

is 103 kJ mol^{-1} and the frequency factor is 4.94×10^{13} s^{-1}. Determine the rate coefficient at a temperature of $\vartheta = 65$ °C.

1.18.2 Kinetics in everyday life

Everyday experience demonstrates that food will spoil outside on a hot summer day much faster than it would in a refrigerator. Estimate the activation energy of these decomposition processes by assuming that food will spoil about 40 times faster at a temperature of 30 °C than at a temperature of 8 °C.

1.18.3 Acid-catalyzed hydrolysis of benzylpenicillin

The acid-catalyzed hydrolysis of benzylpenicillin (so-called penicillin G, discovered in 1928 by the Scottish physician and microbiologist Sir Alexander FLEMING) is a first-order reaction. At a temperature of 60 °C and a pH value of 4, a half-life of 18.3 min^{-1} was determined. The activation energy W_A is 87.14 kJ mol^{-1}.

a) Determine the frequency factor k_∞.

b) What is the half-life of this reaction at 30 °C and pH 4?

1.18.4 Rate of the decomposition of dintrogen tetroxide

A dinitrogen tetroxide molecule can decompose into two nitrogen dioxide molecule if it collides with an appropriate collision partner, such as a nitrogen molecule:

$$\begin{array}{lll} \text{Symbol} & \text{B} & \text{B}' \\ & N_2O_4|g & + N_2|g \rightarrow 2\ NO_2|g + N_2|g. \end{array}$$

The rate law for this process is

$$r = k \times c_B \times c_{B'} \qquad \text{with} \qquad k = k_\infty \mathrm{e}^{-\frac{W_A}{RT}},$$

where the two parameters are $k_\infty = 2 \times 10^{11}$ m^3 mol^{-1} s^{-1} and $W_A = 46$ kJ mol^{-1}, respectively. Since the nitrogen is not consumed, its concentration remains constant with the result that N_2O_4 decomposes in a pseudo first-order reaction. It vanishes exponentially, at least as long as only a little bit of NO_2 is present and thus the backward reaction plays no role.

a) What is the rate density r if N_2O_4 decomposes under room conditions, more precisely if we have $c_B = 1\ \text{mol m}^{-3}$, $c_{B'} = 40\ \text{mol m}^{-3}$ and $T = 298\ \text{K}$?

b) What would be the half-life $t_{1/2}$ if the backward reaction were ignored?

1.18.5 Hydrogen iodide equilibrium

The equilibrium involving hydrogen, iodine and hydrogen iodide is based on two opposing second-order reactions, the formation of hydrogen iodide from the elements and its decomposition back into the elements,

$$H_2|g + I_2|g \underset{-1}{\overset{+1}{\rightleftarrows}} 2\,HI|g\,.$$

Already in 1899, it was thoroughly investigated by the German physical chemist Max BODENSTEIN. Based on BODENSTEIN's data, the equilibrium constant at a temperature of 356 °C is 76.4, at a temperature of 427 °C, however, 52.0. The ARRHENIUS parameters for the forward reaction are $W_{A,+1} = 165.7\ \text{kJ mol}^{-1}$ and $k_{\infty,+1} = 3.90\times 10^{11}\ \text{dm}^3\ \text{mol}^{-1}\ \text{s}^{-1}$, respectively. Determine the ARRHENIUS parameters $W_{A,-1}$ and $k_{\infty,-1}$ for the backward reaction.

1.18.6 Collision theory

An important quantity in the collision theory is the fraction q of all particles having a minimum energy of W_{min}.

a) Calculate this fraction at a temperature of 300 K if the value of W_{min} is 60 kJ mol^{-1}.

b) How does the fraction of particles with sufficient kinetic energy change when the temperature is increased by 100 K?

1.18.7* Nitrogen oxides in air

Colorless nitrogen monoxide is easily oxidized by oxygen to brown nitrogen dioxide. The process can be described as a single-step trimolecular reaction with a "transition substance" ‡ consisting of six atoms:

	2 NO	+ O_2	→ ‡	→ 2 NO_2
$\mathring{\mu}$(298 K)/kG	2×87.6	……..	+240	……..
	…….. (sum)			……..
α/GK^{-1}	2×(−211)	……..	−356	……..
$\mathring{\mu}$(398 K)/kG	……..	……..	……..	……..
	…….. (sum)			……..

In this case, the basic value $\mathring{\mu}\ (=\overset{\bullet}{\mu})$ corresponds to the chemical potential of the pure substance.

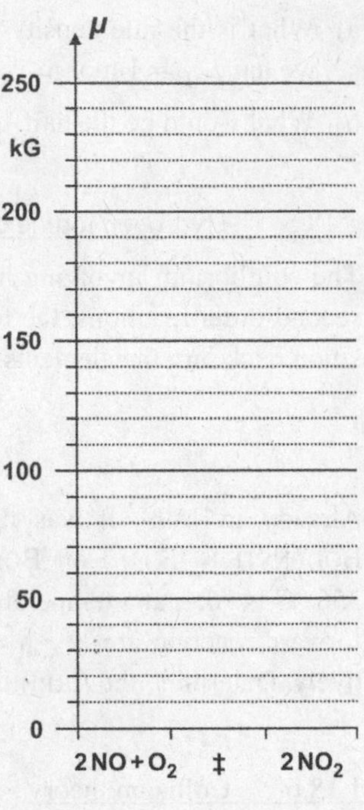

a) Add the data missing above (for example with the help of Table A2.1 in the Appendix of the textbook "Physical Chemistry from a Different Angle"). Subsequently, calculate the values for a temperature of 398 K. Plot the chemical potentials $\overset{\circ}{\mu}(2\ NO + O_2)$, $\overset{\circ}{\mu}(\ddagger)$, $\overset{\circ}{\mu}(2\ NO_2)$ for both temperatures into the diagram on the right side.

b) What is the basic value $\overset{\circ}{\mathcal{A}}_\ddagger$ of the chemical drive and the equilibrium number $\overset{\circ}{\mathcal{K}}_\ddagger$ for the activation step as well as the basic value $\overset{\circ}{\mathcal{A}}$ of the chemical drive and the equilibrium number $\overset{\circ}{\mathcal{K}}$ for the overall process at both temperatures?

c) The equilibrium number $\overset{\circ}{\mathcal{K}}_\ddagger = \overset{\circ}{\mathcal{K}}_{\ddagger p}$ can be converted into the equilibrium number $\overset{\circ}{\mathcal{K}}_{\ddagger c}$ by multiplying with the factor $(p^\ominus/c^\ominus RT)^\nu$ [with $\nu = \nu(\mathrm{NO}) + \nu(\mathrm{O_2}) + \nu(\ddagger)$]. First, justify this assertion. The rate coefficient k can then be calculated from the latter equilibrium number according to $k = \kappa_\ddagger \times (k_B T/h)\overset{\circ}{\mathcal{K}}_{\ddagger c}$ with the dimension factor $\kappa_\ddagger = (c^\ominus)^\nu$. Determine $\overset{\circ}{\mathcal{K}}_{\ddagger c}$ and k at both temperatures.

d) What is the value of the potential $\mu = \mu(2\ NO + O_2)$ in air with a NO content of $x(\mathrm{NO}) = 0.001$ at 298 K at the beginning ($\xi = 0$) and after oxidation of 90 % of the NO ($\xi = 0.9 \times \xi_{max}$) and what is the value of the potential $\mu = \mu(2\ NO_2)$ at the end (ξ_{final})? (Bear the composition of air in mind). Plot these values into the diagram as well.

e) What is the half-life of NO at a NO content of $x(\mathrm{NO}) = 0.001$ in air (at standard pressure)? Because of the large surplus of oxygen, the reaction can be described as pseudo "bimolecular" with a rate coefficient $k' = k \times c(\mathrm{O_2})$. The half-life can then be calculated accordingly.

f) How long does it take for the NO content to drop to 1/1000 of its initial value?

1.19 Catalysis

1.19.1 Catalytic decomposition of hydrogen peroxide by iodide

The decomposition of hydrogen peroxide in aqueous solution into oxygen and water requires a molar activation energy W_A of 76 kJ mol^{-1}, which is why it runs so slowly at room temperature. When iodide ions are added as catalyst, the threshold reduces to 59 kJ mol^{-1}. By what factor is the reaction accelerated at 298 K if we assume that the frequency factor does not change?

1.19.2 Catalytic decomposition of acetylcholine

Acetylcholinesterase, an enzyme in the nervous system, catalyzes the hydrolysis of the neurotransmitter acetylcholine into acetate and choline. The MICHAELIS constant K_M for acetylcholinesterase is 9×10^{-5} mol L^{-1} and its turnover number k_2 is 1.4×10^4 s^{-1}. Determine the initial rate density r_0 of the enzymatic splitting of acetylcholine when the initial concentration $c_{S,0}$ of substrate is 10^{-2} mol L^{-1} and the enzyme concentration is $c_{E,0}$ 5×10^{-9} mol L^{-1}. What is the maximum initial rate density $r_{0,max}$?

1.19.3 Application of MICHAELIS-MENTEN kinetics

For a simple enzymatic reaction obeying MICHAELIS-MENTEN kinetics, the following rate coefficients were experimentally determined:

$k_1 = 1.0\times10^5$ m^3 mol^{-1} s^{-1}, $k_{-1} = 4\times10^4$ s^{-1} and $k_2 = 8\times10^5$ s^{-1}.

a) Determine the dissociation constant $\overset{\circ}{K}_{diss}$ of the enzyme-substrate complex.

b) What are the MICHAELIS constant K_M and the catalytic efficiency (k_2/K_M) of the enzyme used?

c) Is K_M useful as a measure of substrate affinity of the enzyme under the given circumstances? Under what conditions would this be the case?

1.19.4 Hydration of carbon dioxide by carbonic anhydrase

Carbonic anhydrase, a zinc-containing enzyme, catalyzes the hydration of carbon dioxide to hydrogen carbonate ions:

$$CO_2|g + H_2O|l \rightleftarrows HCO_3^-|w + H^+|w\,.$$

The significantly more soluble hydrogen carbonate can be efficiently transported in the blood stream to the lungs. Howard DEVOE et al. investigated in 1961 the enzyme kinetics of carbonic anhydrase [De Voe H, Kistiakowsky GB (1961) The Enzymic Kinetics of Carbonic Anhydrase from Bovine and Human Erythrocytes. J Am Chem Soc 83:274-280]. They determined the following initial rate densities r_0 as a function of different substrate concentrations $c_{S,0}$ at a concentration of the enzyme (derived from bovine erythrocytes) of 2.8×10^{-9} mol L^{-1}, a pH value of 7.1, and a temperature of 0.5 °C.

$c_{S,0}$ / (mmol L^{-1})	r_0 / (mol L^{-1} s^{-1})
1.25	2.78×10^{-5}
2.50	4.98×10^{-5}
5.00	8.13×10^{-5}
20.00	1.54×10^{-4}
40.00	1.89×10^{-4}

a) Determine the MICHAELIS constant K_M of carbonic anhydrase as well as the maximum initial rate density $r_{0,max}$ under the selected conditions by means of a suitably chosen plot.
b) What is the catalytic efficiency?

1.20 Transport Phenomena

1.20.1 Flux of glucose

The concentration of glucose (substance B) in a chamber is $c_{B,1} = 1.0 \text{ mmol L}^{-1}$, that in a second one $c_{B,2} = 0.2 \text{ mmol L}^{-1}$. Both chambers are separated from each other by an aqueous diffusion path with a length l of 1 mm and a cross section of 5 cm². The diffusion coefficient D_B of glucose in water at 25 °C is $0.67 \times 10^{-9} \text{ m}^2 \text{ s}^{-1}$. Determine the matter flux J_B (assuming a linear concentration gradient).

1.20.2 Diffusion of carbon in iron

One side of a sheet of iron (body-centered cubic structure) with a thickness of 3 mm is exposed to a carbon-containing atmosphere at a temperature of 750 °C. Carbon (substance B) begins to migrate into the iron and after a while equilibrium is established. The flux density j_B of carbon under stationary conditions is $28.3 \times 10^{-6} \text{ mol m}^{-2} \text{ s}^{-1}$. After quenching the sheet to room temperature, a carbon concentration $c_{B,1}$ of 1850 mol m⁻³ is measured on one side and a carbon concentration $c_{B,2}$ of 185 mol m⁻³ on the other side.

a) Calculate the diffusion coefficient D_B of carbon in the iron sample at the given temperature.

b) The diffusion coefficient depends on temperature. This temperature dependence can be described for the thermally activated diffusion in solids by an ARRHENIUS approach:

$$D_B = D_{B,0} \times \exp\left(-\frac{W_A}{RT}\right),$$

where $D_{B,0}$ represents the "frequency factor" and W_A the activation energy of the diffusion process.

If the above experiment is repeated at 900 °C, a diffusion coefficient D_B of $1.7 \times 10^{-10} \text{ m}^2 \text{ s}^{-1}$ is found. What are the values for $D_{B,0}$ and W_A in this case?

1.20.3* Diffusion of sucrose

Imagine a glass of cold tea sweetened with a teaspoon of sugar (substance B) (about 10 g). The result should be a linear concentration gradient of sucrose from the bottom to 0 at the surface.

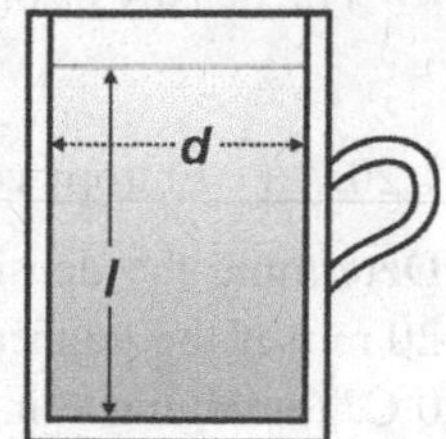

a) What would be the concentration $c_{B,0}$ of the sugar if it were uniformly distributed in the tea glass? What is consequently its concentration $c_{B,1}$ at the bottom? The dimensions of the glass should be $l = 7$ cm and $d = 5$ cm.

b) Determine the flux density j_B and the migration velocity $v_{B,1}$ near the bottom. The diffusion coefficient D_B of sucrose under the given conditions is approximately 0.5×10^{-9} m² s⁻¹.

c) How long does one have to wait until the sugar is approximately homogeneously distributed (without stirring)?

1.20.4 Oil as lubricant

A cuboid-shaped block was placed on a thin film of oil on a horizontal plate. Determine the force that must act on the block so that it moves with a velocity v_0 of 0.3 m s^{-1}. The oil has a dynamic viscosity η of 0.1 Pa s, the thickness d of the lubricant film is 0.1 mm and the rectangular contact area A of the block 0.6×0.3 m.

1.20.5 Molecular radius and volume

Myoglobin is the red pigment in muscle tissue, hemoglobin is the red blood pigment in the erythrocytes. Both proteins are members of the group of globins, (nearly) spherical proteins, and are responsible for the transport of oxygen. The diffusion coefficient of myoglobin in water was determined to be 0.113×10^{-9} m^2 s^{-1} at 20 °C, that of hemoglobin to be 0.069×10^{-9} m^2 s^{-1}. The (dynamic) viscosity of water at this temperature is 1.002 mPa s.

a) Estimate the radius and the volume of the globular ("globe-like") protein molecules.
b) Interpret the result by taking into account the structure of the two globins.

1.20.6 Sinking of pollutant particles in air

Nearly spherical pollutant particles with a diameter d_p of 16 µm and a density ρ_p of 2.5 g cm^{-3} sink downwards in the air in the Earth's gravitational field.

a) Estimate the (constant) sinking velocity v_p of the particles in force equilibrium. The (dynamic) viscosity η_a of air is 18.5 µPa s at 300 K.

b) How long does it take for such a particle to sink 50 m downwards when there is no wind?

c) What changes in the result if the buoyant force F_b $(= \rho_a \times V_p \times g)$ is taken into account? The density ρ_a of air at 300 K is 1.2 kg m^{-3}.

1.20.7 Entropy conduction through a copper plate

Determine the density j_S of the entropy flux through a copper plate with a thickness d of 20 mm if the temperature on one side of the plate is ϑ_1 = 50 °C and on the other side ϑ_2 = 0 C° (assuming a linear temperature gradient).

1.20.8* Entropy loss in buildings †

Polystyrene boards with a thickness of 5 cm are fixed on the outside of a solid brick wall with a thickness of 12 cm. What is the reduction of the entropy loss by this insulation measure? The temperature difference between the surface of the brick wall on the inside and the surface of the polystyrene cladding on the outside is supposed to be always 20 K. The entropy conductivity of solid bricks (B) is $\sigma_{S,B} = 0.003\ \mathrm{Ct\,K^{-1}\,s^{-1}\,m^{-1}}$, that of polystyrene (P) $\sigma_{S,P} = 0.00013\ \mathrm{Ct\,K^{-1}\,s^{-1}\,m^{-1}}$.

a) First calculate the entropy flux density $j_{S,B}$ for the non-insulated wall.

b) Estimate the entropy flux density $j_{S,BP,1}$ for the insulated wall; for simplicity's sake, assume that it consists only of the polystyrene cladding. How could this approach be justified?

c) For a more detailed calculation of the entropy flux density $j_{S,BP,2}$ for the insulated wall, the temperature profile through the wall has to be known. In order to determine this profile, assume that the entropy flux flowing through the bricks ($J_{S,B}$) is equal to the entropy flux flowing through the polystyrene boards ($J_{S,P}$). Subsequently, compare the entropy flux density $j_{S,BP,2}$ with the entropy flux density $j_{S,B}$ from a) and the entropy flux density $j_{S,BP,1}$ from b).

   Hint: Unfortunately, the minus sign is missing in the expression for the entropy flux in the textbook "Physical Chemistry from a Different Angle."

In the present exercise, only the entropy transfer by conduction (in matter at rest) is considered. An entropy transfer by convection (as it occurs, for example, in thin air boundary layers on both sides of the wall) or also by radiation remain out of consideration.

1.21 Electrolyte Solutions

1.21.1 Conductivity of tap water

Tap water is, roughly speaking, a thin $Ca(HCO_3)_2$ solution with additions of other salts (in the following an exemplary composition):

	Ca^{2+}	Mg^{2+}	Na^+	SO_4^{2-}	Cl^-	HCO_3^-
$c\,/\,\text{mol m}^{-3}$	1.0	0.2	0.5	0.3	0.5	....
$u\,/\,10^{-8}\,\text{m}^2\,\text{V}^{-1}\,\text{s}^{-1}$	....	....	....	....	....	....

a) What concentration of hydrocarbonate ions is required for an electrically neutral solution?

b) Add the (electric) mobilities u at 25 °C (e.g. from Table 21.2 in the textbook "Physical Chemistry from a Different Angle") and calculate the conductivity (specific conductance) σ of the tap water at 25 °C.

c)* What conductivity results from the rule of thumb "$\Delta\sigma/\sigma$ = 2 % per degree" for boiling water?

1.21.2 Migration of Zn^{2+} ions

In a homogeneous electric field $E = 5\ \text{V cm}^{-1}$, the migration velocity v of Zn^{2+} ions in aqueous solution at 25 °C is $2.74\times10^{-5}\ \text{m s}^{-1}$.

a) Estimate the radius r of the hydrated Zn^{2+} ion. The dynamic velocity η of water at the given temperature is 0.890 mPa s.

b) What is the diffusion coefficient D of the ion?

1.21.3 Dissociation of formic acid in aqueous solution

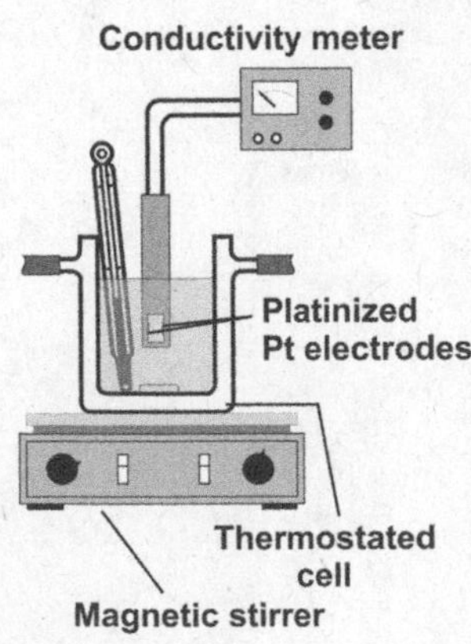

An electrolyte solution between two electrodes behaves like an ohmic resistor, where $R = \rho\times Z$. Z represents a quantity that depends on the geometric structure of the measuring cell and is referred to as cell constant. In the ideal case of a cuboidal electrolytic trough with a cross section A and a length l and with two electrodes attached to its end walls, one obtains $Z = l/A$ [see. Eq. (21.25)]. In practice [see adjacent figure; so-called "platinum double electrode" with two platinized platinum electrodes and corresponding conductivity meter], however, the cell constant Z is determined by using a calibrating solution of known σ value (usually a solution of potassium chloride).

[Since the conductivity is temperature-dependent, it is advisable to adjust the temperature carefully (as indicated in the figure by the double-walled measuring cell with connections)].

a) For calibrating a conductivity measuring cell a KCl calibration solution (C) with a concentration $c_{\mathrm{C}} = 0.100\ \mathrm{mol\,L^{-1}}$ was used. The measured resistance R_{C} at 25 °C was 40.0 Ω. The tabulated value for the conductivity (specific conductance) σ_{C} of such a solution at the given temperature is $12.88\ \mathrm{mS\,cm^{-1}}$. What is the value of the cell constant Z?

b) Subsequently, the measuring cell was filled with an aqueous solution of formic acid of the concentration $c = 0.010\ \mathrm{mol\,L^{-1}}$. The measurement of the resistance R at 25 °C resulted in 1026.3 Ω. Determine the conductivity σ and the molar conductivity Λ of the solution.

c) Calculate the degree of dissociation α as well as the corresponding equilibrium constant $K_c^{\ominus}$ and equilibrium number $\mathcal{K}_c^{\ominus}$. The limiting molar conductivity Λ^0 of formic acid is $404.3\ \mathrm{S\,cm^2\,mol^{-1}}$.

d) Determine the standard value $\mu_{\mathrm{p}}^{\ominus}$ of the proton potential of the acid-base pair formic acid/formate ($HCOOH/HCOO^-$) and the proton potential μ_{p} in the formic acid solution.

1.21.4* Solubility product of lead sulfate

A saturated solution of $PbSO_4$ in distilled water wasw filled into the conductivity measuring cell presented in Exercise 1.21.3. Subsequently, a resistance R of 10340 Ω was measured. The conductivity of distilled water at 25 °C is $\sigma_{\mathrm{W}} = 1.80\ \mathrm{\mu S\,cm^{-1}}$. The tabulated values of the limiting molar conductivities of the involved ions are $\Lambda^0(Pb^{2+}) = 142\ \mathrm{S\,cm^2\,mol^{-1}}$ and $\Lambda^0(SO_4^{2-}) = 160\ \mathrm{S\,cm^2\,mol^{-1}}$. Determine the solubility product $\mathcal{K}_{\mathrm{sd}}^{\ominus}$ of $PbSO_4$.

1.21.5 Limiting molar conductivity of silver bromide

Calculate the limiting molar conductivity of AgBr at 25 °C by means of the following values determined at the same temperature: $\Lambda^0(\mathrm{NaBr}) = 12.82\ \mathrm{mS\,m^2\,mol^{-1}}$, $\Lambda^0(\mathrm{NaNO_3}) = 12.15\ \mathrm{mS\,m^2\,mol^{-1}}$, and $\Lambda^0(\mathrm{AgNO_3}) = 13.33\ \mathrm{mS\,m^2\,mol^{-1}}$.

1.21.6 Experimental determination of the limiting molar conductivity

An aqueous electrolyte solution was filled in a conductivity measuring cell and then progressively diluted. At each step, the resistance of the solution was measured (see the table below). By means of a calibration solution, a cell constant Z of $0.3130\ \mathrm{cm^{-1}}$ was determined.

$c\,/\,\mathrm{mol\,L^{-1}}$	0.05	0.02	0.01	0.005	0.001	0.0005
$R\,/\,\Omega$	81.38	192.6	373.7	730.3	3537	7018

a) How can one decide whether the electrolyte is strong or weak?

b) Determine depending on the answer under a) the limiting molar conductivity Λ^0 of the electrolyte.

1.21.7 Migration of permanganate ions in the electric field

The migration velocity of permanganate ions in a thin aqueous $KMnO_4$ solution is measured in an experimental setup with two electrodes at a distance of 15 cm, to which an electric voltage of 20 V is applied. The value obtained at 25 °C is $v = -8.5 \times 10^{-4}$ cm s^{-1}. (For the sake of simplicity, assume that the value is approximately also valid for a vanishingly small concentration.) The limiting molar conductivity of the K^+ ion is 73.5 S cm^2 mol^{-1}. Calculate the transport number of the MnO_4^- ion.

1.21.8 HITTORF transport numbers

A hydrochloric acid solution with an initial concentration c_0 of 0.1000 kmol m^{-3} is electrolyzed in a so-called HITTORF cell with Pt electrodes. After a given duration of electrolysis, one measured in the cathode compartment a concentration of 0.0979 kmol m^{-3}, in the anode compartment, however, a concentration of 0.0904 kmol m^{-3}.

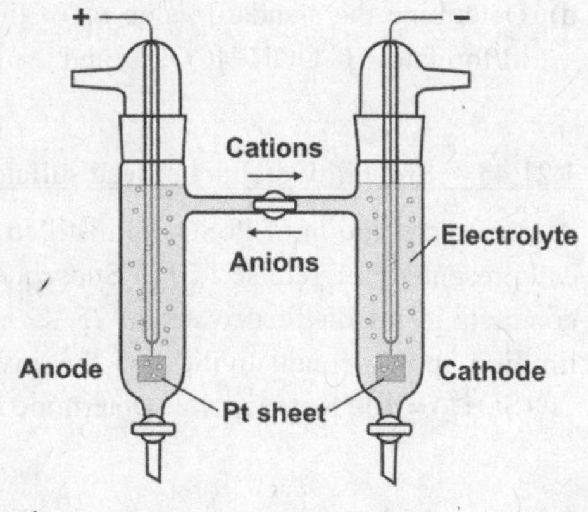

a) Determine the transport numbers of the H^+ and Cl^- ions.

b) What are the ionic conductivities of the H^+ and Cl^- ions at infinite dilution if the limiting molar conductivity Λ^0(HCl) is 426.0 S cm^{-2} mol^{-1}?

1.21.9* HITTORF transport numbers for advanced students

A potassium hydroxide solution with a mass fraction w_0 of 0.2 % of KOH is filled in a HITTORF cell with Pt electrodes. Subsequently, the solution is electrolyzed at a constant current of 80 mA for one hour. After the electrolysis, 25.00 g of the solution in the cathode compartment are drained off and titrated; they contain 0.0615 g of KOH. What are the transport numbers of the K^+ and the OH^- ions? Assume that the cathode compartment contains a total of 100.00 g of solution and take into account that the K^+ ions are not discharged at the cathode.

1.22 Electrode Reactions and Galvani Potential Differences

1.22.1 Electron potential

Formulate the half-reaction for the composite redox system in which the gas NO and NO_3^- ions are involved (in acidic solution). What is the standard value of the electron potential?

1.22.2 Galvani potential difference and difference of chemical potentials in electrochemical equilibrium

What is the difference between the chemical potentials of zinc ions in metal [$\mu(Zn^{2+}|m)$] and in aqueous solution [$\mu(Zn^{2+}|w)$] if there is a Galvani potential difference (Galvani "voltage") $U_{S \to Me}$ of +0.3 V between both phases?

1.22.3 Galvani potential difference of redox electrodes

Formulate the half-reaction for the composite redox system in which Cr^{3+} and $Cr_2O_7^{2-}$ ions are involved (in acidic solution) and set up NERNST's equation for the potential difference $\Delta\varphi$.

1.22.4 Concentration dependence of the Galvani potential difference of metal-metal ion electrodes

How does the Galvani potential difference $\Delta\varphi$ of a Cu/Cu^{2+} electrode change at 25 °C if the concentration of the Cu^{2+} ions is increased by a factor of 5?

1.22.5 Diffusion voltage

a) A NaCl solution with a concentration of 0.3 $kmol\,m^{-3}$ (solution I) is supposed to border another NaCl solution with a concentration of 0.5 $kmol\,m^{-3}$ (solution II). The phase boundary is stabilized by a diaphragm. Calculate the diffusion voltage U_{diff} (= $-\Delta\varphi_{diff}$) in the steady-state at 25 °C. The electric mobility of the Na^+ ion is $5.2 \times 10^{-8}\,m^2\,V^{-1}\,s^{-1}$, that of the Cl^- ion is $-7.9 \times 10^{-8}\,m^2\,V^{-1}\,s^{-1}$.

b) What would be the diffusion voltage U_{diff} if KCl solutions of the same concentrations as the NaCl solutions in part a) were used? The electric mobility of the K^+ ion is $7.6 \times 10^{-8}\,m^2\,V^{-1}\,s^{-1}$.

1.22.6 Membrane voltage

An ion-selective membrane that is only permeable to K^+ and Cl^- ions separates an external pure KCl solution (solution I) from an internal solution, which, in addition to K^+ and Cl^-, also contains a positively charged protein $Prot^{5+}$ (solution II).

a) Formulate the electroneutrality rule for both solutions.

b) What is the internal Cl^- concentration if the external Cl^- concentration is 150 mol m^{-3}? The ratio $c_{K^+}(I)/c_{K^+}(II)$ after establishing of the equilibrium is supposed to be 1.08.

c) Calculate the membrane voltage (DONNAN voltage) U_{mem} $(= -\Delta\varphi_{mem})$ at 25 °C.

1.22.7* Membrane voltage for advanced students

At a certain pH value, ribonuclease carries a negative net charge of 3. 100 mL of a solution of ribonuclease (as sodium salt) with a concentration of 3 mol m^{-3} (solution II) are enclosed by a membrane that is only permeable to small ions. Outside are 100 mL of a NaCl solution with a concentration of 50 mol m^{-3} (solution I).

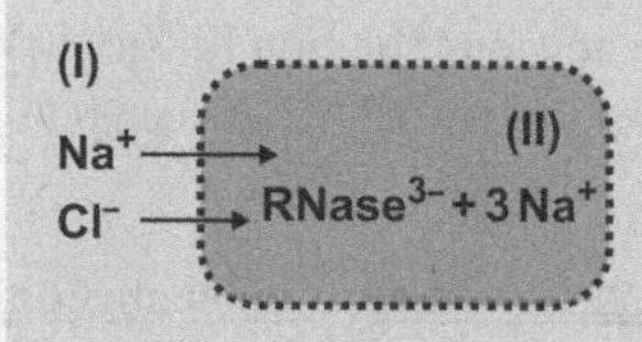

a) What are the final concentrations of ions on both sides of the membrane?

b) Calculate the membrane voltage U_{mem} at 25 ° C

1.23 Redox Potentials and Galvanic Cells

1.23.1 NERNST's equation of a half-cell

Calculate the standard values $E^{\ominus}$ of the redox potentials from the chemical potentials of the substances involved and formulate the corresponding NERNST equation in the following cases:

a) Half-cell $Cl^-|AgCl|Ag$

Conceptually, one should imagine that the half-cells are always supplemented by a standard hydrogen electrode (SHE), which is connected to the half-cell in question by means of a salt bridge (abbreviation: ¦¦) (in order to avoid diffusion voltage). Therefore, $Cl^-|AgCl|Ag$ is to be understood as the short name for the complete cell $Pt|H_2|H^+$¦¦$Cl^-|AgCl|Ag|Pt$.

b) Half-cell $MnO_4^-, Mn^{2+}|Pt$

c) Half-cell $HCOOH|CO_2|Pt$

The chemical potential of formic acid in aqueous solution is −372.4 kG under standard conditions. Carbon dioxide should be present as a gas.

d) What changes in the case of the half-cell c) if one considers an alkaline solution instead of an acidic solution [$\mu^{\ominus}(HCOO^-|w) = -351.0$ kG]?

1.23.2 Concentration dependence of redox potentials

Calculate the redox potentials E of the following half cells at 25 °C:

a) Half-cell $Fe^{2+}, Fe^{3+}|Pt$

The concentration of Fe^{2+} ions is $c(Fe^{2+}) = 0.005\ kmol\,m^{-3}$, that of Fe^{3+} ions $c(Fe^{3+}) = 0.010\ kmol\,m^{-3}$.

b) Half-cell $H^+, Cl^-, ClO_4^-\ |Pt$

The ions Cl^- and ClO_4^- should be present in the following concentrations: $c(Cl^-) = 0.005\ kmol\,m^{-3}$ and $c(ClO_4^-) = 0.020\ kmol\,m^{-3}$. The pH value is 3.0, the standard value $E^{\ominus}$ of the redox potential +1.389 V.

c) Half-cell $OH^-, ClO_3^-, ClO_4^-\ |Pt$

The ion concentrations should be $c(ClO_3^-) = 0.010\ kmol\,m^{-3}$, $c(ClO_4^-) = 0.020\ kmol\,m^{-3}$ and $c(OH^-) = 0.010\ kmol\,m^{-3}$. For the standard value $E^{\ominus}$ of the redox potential, +0.36 V can be found.

1.23.3 Redox potential of gas electrodes

A gas electrode (whereby the term "electrode" is to be understood here in a wider sense) is a half-cell in which a gas is in equilibrium with its ions in a solution in the presence of an inert metal. In the present case, a platinum sheet is to be immersed in a KOH solution with a pH value of 9.0 at a temperature of 298 K and bathed in oxygen gas at a pressure of 250 mbar. Determine the redox potential E of this oxygen electrode. The conversion formula for the electrode process is (insert the conversion numbers into the blank boxes in the formula):

$$\square\, OH^-|w \rightarrow \square\, O_2|g + \square\, H_2O|l + \square\, e^-.$$

The standard value $E^\ominus$ of the redox potential is 0.401 V, the so-called "ionic product" K_W of pure water is 1.0×10^{-14} kmol^2 m^{-6} [$K_W = c(H^+) \times c(OH^-)$].

1.23.4 Calculation of redox potentials from redox potentials

The standard values $E^\ominus$ of the redox potentials of the redox pairs Cr/Cr^{2+} and Cr/Cr^{3+} are 0.913 V and 0.744 V, respectively. Calculate $E^\ominus(Cr^{2+}/Cr^{3+})$.

1.23.5 Solubility product of silver iodide

Calculate the solubility product $K_{sd}^\ominus$ of AgI and its saturation concentration c_{sd} in aqueous solution at 25 °C from the standard values $E^\ominus$ of the redox potentials of the silver-silver ion electrode [$E^\ominus(Ag/Ag^+) = +0.7996$ V] and the silver-silver iodide electrode [$E^\ominus(Ag + I^-/AgI) = -0.1522$ V].

1.23.6 Concentration cell

One silver electrode is immersed in a silver nitrate solution of the concentration $c_1(AgNO_3) =$ 0.0005 kmol m^{-3} and a second silver electrode in one of the concentration $c_2(AgNO_3) =$ 0.0100 kmol m^{-3}. The electrolyte solutions are connected by a salt bridge. Calculate the difference ΔE of the redox potentials for the electrochemical cell at 25 °C (under equilibrium conditions).

1.23.7 Galvanic cell

The following electrochemical cell with aqueous electrolyte solutions is given:
$Pt|Cr^{2+},Cr^{3+}||Fe^{2+},Fe^{3+}|Pt$.

a) Formulate the two half-reactions as well as the total reaction of the cell.

b) Find the relationship for the concentration dependence of the potential difference ΔE of the cell at 25 ° C.

c) Specify the standard value $\Delta E^\ominus$.

d) Calculate the chemical drive $\mathcal{A}^\ominus$ of the cell reaction. In which direction does the reaction occur spontaneously? Which electrode will act as cathode under standard conditions?

1.23.8* Potentiometric redox titration

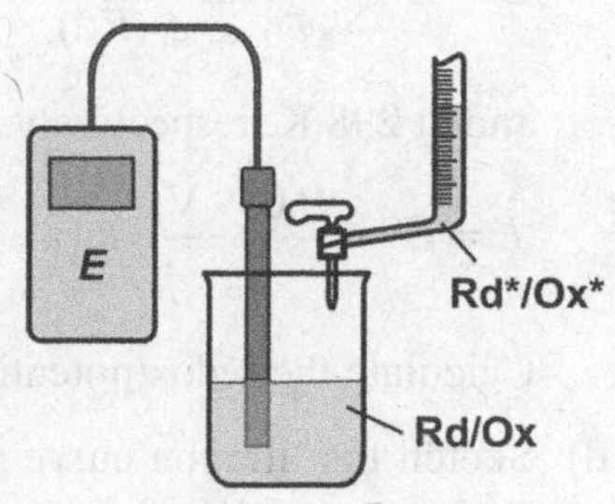

The redox potential E in a solution can be easily measured with a combination electrode (which combines a measuring and reference electrode in a single unit) and a suitable measuring device (high impedance voltmeter), provided that the electron exchange between the measuring electrode and the redox pairs in the solution is not too strongly inhibited. The redox potential changes in characteristic steps during a titration of one or more redox pairs Rd/Ox, Rd′/Ox′ ... in an analyte solution with another redox pair Rd*/Ox* in the used standard solution (titrator).

100 mL of an acidic Fe^{2+} solution with a concentration of 0.010 kmol m^{-3} are titrated with an acidic Ce^{4+} solution with a concentration of 1.000 kmol m^{-3}. The redox potential of the pair Ce^{3+}/Ce^{4+} under standard conditions (in a matrix of perchloric acid with a concentration of 1 kmol m^{-3}) is +1.70 V. What will the corresponding titration curve look like?

a) For the redox pair Fe^{2+}/Fe^{3+} relevant for the redox potential before the equivalence point, NERNST's equation can be used in the following form, in which $\alpha_\xi = \xi/\xi_{max}$ represents the "degree of conversion" (in this special case called "degree of titration" τ) [see also Equation (6.12) in the textbook "Physical Chemistry from a Different Angle"]:

$$E = \overset{\circ}{E} + \frac{RT}{\nu_e \mathcal{F}} \times \ln \frac{c_r(\mathrm{Ox})}{c_r(\mathrm{Rd})} = \overset{\circ}{E} + \frac{E_N}{\nu_e} \times \lg \frac{\alpha_\xi}{1-\alpha_\xi} .$$

The factor E_N at the standard temperature $T^{\ominus} = 298$ K is 0.059 V, i.e. we have

$$E = E^{\ominus} + \frac{0.059\ \mathrm{V}}{\nu_e} \times \lg \frac{\tau}{1-\tau} .$$

Calculate the redox potentials E (at 298 K) for $\tau = {}^1\!/_{101}$, ${}^1\!/_{11}$, ${}^1\!/_{2}$, ${}^{10}\!/_{11}$ and ${}^{100}\!/_{101}$. These values are especially suitable for a fast construction of the titration curve because they result in integer multiples of the factor E_N.

b) What volume V_{ep} of Ce^{4+} solution is required to reach the equivalence point? Calculate the redox potential E_{ep} at this point.

c) For an exact determination of the equivalence point from the titration diagram, one should continue to add the titrator until τ reaches a value of 2. However, one should keep in mind that a reaction only takes place up to the equivalence point ($\tau = 1$), i.e. the titrator added afterwards remains unchanged in the solution.

After the equivalence point, the redox pair Ce^{3+}/Ce^{4+} determines the redox potential. Since the Ce^{3+} concentration remains constant during further addition of Ce^{4+} solution, but the amount of Ce^{4+} ions increases steadily, NERNST's equation takes the following form:

$$E = \overset{\circ}{E} + \frac{RT}{\nu_e \mathcal{F}} \times \ln \frac{c_r(\text{Ox})}{c_r(\text{Rd})} = \overset{\circ}{E} + \frac{E_N}{\nu_e} \times \lg(\alpha_\xi - 1)$$

and at 298 K, respectively

$$E = E^\ominus + \frac{0.059\ \text{V}}{\nu_e} \times \ln(\tau - 1)\,.$$

Calculate the redox potential E for $\tau = {}^{101}\!/_{100}$, ${}^{11}\!/_{10}$ and 2.

d) Sketch the titration curve meaning plot the redox potential E against the added volume V (in mL) of Ce^{4+} solution.

2 Solutions

2.1 Introduction and First Basic Concepts

2.1.1 Molar concentration

Molar concentration c_B of the glucose solution:

The molar concentration c_B of a dissolved substance B results from the quotient of the amount of solute n_B and the volume of solution V_S:

$$c_B = \frac{n_B}{V_S} . \qquad \text{Eq. (1.9)}$$

Amount n_B:

The amount n_B of glucose can be calculated from its mass m_B by means of the molar mass $M_B = M(C_6H_{12}O_6) = 180.0 \times 10^{-3}\ \text{kg mol}^{-1}$:

$$n_B = \frac{m_B}{M_B} = \frac{45 \times 10^{-3}\ \text{kg}}{180.0 \times 10^{-3}\ \text{kg mol}^{-1}} = 0.25\ \text{mol} . \qquad \text{Eq. (1.6)}$$

$$c_B = \frac{0.25\ \text{mol}}{500 \times 10^{-6}\ \text{m}^3} = 500\ \text{mol m}^{-3} = \mathbf{0.50\ kmol\ m^{-3}} .$$

2.1.2 Mass fraction and mole fraction

Mass fraction w_B of sodium chloride:

The mass fraction w_B of a component B in a mixture corresponds to the quotient of its mass m_B and the total mass m_{total} of all the substances present in the mixture:

$$w_B = \frac{m_B}{m_{total}} . \qquad \text{Eq. (1.8)}$$

Mass m_S of the solution:

In order to calculate the mass m_S ($= m_{total}$) of the solution, its (mass) density $\rho = m/V$ is used:

$$m_S = \rho_S \times V_S = 1099\ \text{kg m}^{-3} \times (100 \times 10^{-6}\ \text{m}^3) = 109.9 \times 10^{-3}\ \text{kg} .$$

$$w_B = \frac{15.4 \times 10^{-3}\ \text{kg}}{109.9 \times 10^{-3}\ \text{kg}} = \mathbf{0.140} .$$

G. Job and R. Rüffler, *Physical Chemistry from a Different Angle Workbook*,
https://doi.org/10.1007/978-3-030-28491-6_2

Mole fraction x_B of sodium chloride:

The mole fraction x_B of a particular substance B is calculated in the same way as the mass fraction w_B; only the masses are replaced by the amounts:

$$x_B = \frac{n_B}{n_{total}} = \frac{n_B}{n_B + n_A} . \qquad \text{Eq. (1.7)}$$

Amount n_B of sodium chloride:

$$n_B = \frac{m_B}{M_B} = \frac{15.4 \times 10^{-3}\ \text{kg}}{58.5 \times 10^{-3}\ \text{kg mol}^{-1}} = 0.263\ \text{mol} .$$

Mass m_A of water:

The mass m_A of water corresponds to the difference of the total mass of the solution and the mass of sodium chloride:

$$m_A = m_S - m_B = (109.9 \times 10^{-3}\ \text{kg}) - (15.4 \times 10^{-3}\ \text{kg}) = 94.5 \times 10^{-3}\ \text{kg} .$$

Amount n_A of water:

$$n_A = \frac{m_A}{M_A} = \frac{94.5 \times 10^{-3}\ \text{kg}}{18.0 \times 10^{-3}\ \text{kg mol}^{-1}} = 5.25\ \text{mol} .$$

$$x_B = \frac{0.263\ \text{mol}}{0.263\ \text{mol} + 5.25\ \text{mol}} = \mathbf{0.048} .$$

2.1.3 Mass fraction

Mass m_A of water in "hard liquor:"

We start from Equation (1.8) for the mass fraction w_B of a substance B (here ethanol) in a mixture:

$$w_B = \frac{m_B}{m_{total}} = \frac{m_B}{m_B + m_A} .$$

We solve the equation for m_A,

$$m_A = \frac{m_B - w_B \times m_B}{w_B} ,$$

and obtain:

$$m_A = \frac{(100 \times 10^{-3}\ \text{kg}) - 0.335 \times (100 \times 10^{-3}\ \text{kg})}{0.335} = \mathbf{0.199\ kg} .$$

2.1.4 Composition of a gas mixture

One mole of any gas and therefore also one mole of air has a volume V of about 24.8 L at standard conditions. 21 % of the molecules and consequently 0.21 moles are oxygen.

Molar concentration c_B of oxygen:

$$c_B = \frac{n_B}{V} = \frac{0.21\ \text{mol}}{0.0248\ \text{m}^3} = \mathbf{8.5\ mol\ m^{-3}}\,.$$

Mass concentration β_B of oxygen:

The mass concentration β_B of a dissolved substance B corresponds to the quotient of the mass m_B of this substance and the volume of solution V. The same applies to gas mixtures:

$$\beta_B = \frac{m_B}{V}\,. \qquad \text{Eq. (1.10)}$$

Mass m_B of oxygen:

We obtain with the molar mass $M_B = 32.0\times10^{-3}\ \text{kg mol}^{-1}$ of oxygen:

$$m_B = n_B \times M_B = 0.21\ \text{mol}\times(32.0\times10^{-3}\ \text{kg mol}^{-1}) = 6.72\times10^{-3}\ \text{kg}\,.$$

$$\beta_B = \frac{6.72\times10^{-3}\ \text{kg}}{0.0248\ \text{m}^3} = \mathbf{0.271\ kg\ m^{-3}}\,.$$

Mole fraction x_B of oxygen:

$$x_B = \frac{n_B}{n_{total}} = \frac{0.21\ \text{mol}}{1\ \text{mol}} = \mathbf{0.21}\,.$$

Mass fraction w_B of oxygen:

$$w_B = \frac{m_B}{m_{total}} = \frac{m_B}{m_B + m_A}\,.$$

Mass m_A of nitrogen:

The mass m_A of nitrogen in the air can be calculated from its amount $n_A = n_{total} - n_B = 1\ \text{mol} - 0.21\ \text{mol} = 0.79\ \text{mol}$ by means of the molar mass $M_A = 28.0\times10^{-3}\ \text{kg mol}^{-1}$:

$$m_A = n_A \times M_A = 0.79\ \text{mol}\times(28.0\times10^{-3}\ \text{kg mol}^{-1}) = 22.12\times10^{-3}\ \text{kg}\,.$$

$$w_B = \frac{6.72\times10^{-3}\ \text{kg}}{(6.72\times10^{-3}\ \text{kg}) + (22.12\times10^{-3}\ \text{kg})} = \mathbf{0.233}\,.$$

2.1.5* Converting measures of composition

Calculation process:

$$\frac{M_A c_B}{\rho - c_B(M_B - M_A)} \xrightarrow{1)} \frac{M_A c_B}{(c_A M_A + c_B M_B) - (c_B M_B - c_B M_A)} \xrightarrow{2)}$$

$$\frac{M_A c_B}{c_A M_A + c_B M_A} \xrightarrow{3)} \frac{c_B}{c_A + c_B} \xrightarrow{4)} \frac{n_B/V}{n_A/V + n_B/V} \xrightarrow{5)} \frac{n_B}{n_A + n_B} = x_B\,.$$

1) First, we have to find an appropriate expression for the mass density ρ of the solution. Starting point is the defining equation of ρ:

$$\rho = \frac{m_A + m_B}{V}\,.$$

The masses can be substituted by amounts [Eq. (1.5)],

$$\rho = \frac{n_A M_A + n_B M_B}{V} = \frac{n_A M_A}{V} + \frac{n_B M_B}{V}\,,$$

and the quotients n/V by the molar concentrations c [Eq. (1.9)],

$$\rho = c_A M_A + c_B M_B\,.$$

2) The terms $c_B M_B$ in the denominator cancel each other.
3) M_A cancels out.
4) The concentrations c are substituted by the quotients n/V.
5) V cancels out.

2.1.6 Description of a reaction process

a) Conversion numbers ν_i:

$$\nu_B = \mathbf{-1}\,, \qquad \nu_{B'} = \mathbf{-3}\,, \qquad \nu_D = \mathbf{+2}\,.$$

The conversion numbers ν_i are negative for reactants and positive for products.

b) Conversion $\Delta\xi$:

The conversion $\Delta\xi$ of any reaction $\mathcal{R}$ results from the change of the (time-dependent) extent of reaction ξ:

$$\Delta\xi = \xi(10^h 40^m) - \xi(10^h 10^m) = 19\text{ mol} - 13\text{ mol} = \mathbf{6\ mol}\,.$$

c) Changes Δn_i in amount:

The change in amount of one of the substances involved in the reaction can be calculated from conversion $\Delta\xi$ by inserting its value into Equation (1.17), a variant of the basic stoichiometric equation:

$\Delta n_i \quad = \nu_i \times \Delta\xi \,.$

$\Rightarrow \quad \Delta n_B \quad = (-1) \times 6 \text{ mol} = \mathbf{-6 \text{ mol}},$
$\Delta n_{B'} = (-3) \times 6 \text{ mol} = \mathbf{-18 \text{ mol}},$
$\Delta n_D \quad = (+2) \times 6 \text{ mol} = \mathbf{+12 \text{ mol}}.$

Changes Δm_i in mass:

$\Delta m_i \quad = \Delta n_i \times M_i \,.$ cf. Eq. (1.5)

Molar masses M_i:

$M_B = 28.0 \times 10^{-3} \text{ kg mol}^{-1}$, $M_{B'} = 2.0 \times 10^{-3} \text{ kg mol}^{-1}$ and $M_D = 17.0 \times 10^{-3} \text{ kg mol}^{-1}$

$\Rightarrow \quad \Delta m_B \quad = -6 \text{ mol} \times (28.0 \times 10^{-3} \text{ kg mol}^{-1}) \quad = \mathbf{-0.168 \text{ kg}}$
$\Delta m_{B'} \quad = -18 \text{ mol} \times (2.0 \times 10^{-3} \text{ kg mol}^{-1}) \quad = \mathbf{-0.036 \text{ kg}}$
$\Delta m_D \quad = +12 \text{ mol} \times (17.0 \times 10^{-3} \text{ kg mol}^{-1}) \quad = \mathbf{+0.204 \text{ kg}}.$

2.1.7 Application of the basic stoichiometric equation in titration

a) Conversion formula and basic stoichiometric equation:

Conversion formula: $H_2SO_4 + 2\, NaOH \rightarrow Na_2SO_4 + 2\, H_2O$

Basic equation: $\Delta\xi = \dfrac{\Delta n_B}{\nu_B} = \dfrac{\Delta n_{B'}}{\nu_{B'}}$ Eq. (1.15)

b) Change $\Delta n_{B'}$ in amount of NaOH until the equivalence point is reached:

The amount $n_{B',0}$ of NaOH in 24.4 mL of the sodium hydroxide solution can be calculated by means of Equation (1.9) (V_S is the volume of the solution in question).

$$c_{B',0} \quad = \frac{n_{B',0}}{V_{S,B'}} \quad \Rightarrow$$

$$n_{B',0} \quad = c_{B',0} \times V_{S,B'} = (0.1 \times 10^3 \text{ kmol m}^{-3}) \times (24.4 \times 10^{-6} \text{ m}^3) = 2.44 \times 10^{-3} \text{ mol}\,.$$

This amount of NaOH is completely consumed until the equivalence point is reached meaning the change $\Delta n_{B'}$ in amount is

$$\Delta n_{B'} \quad = 0 - n_{B',0} = 0 - (2.44 \times 10^{-3} \text{ mol}) = -2.44 \times 10^{-3} \text{ mol}\,.$$

Amount $n_{B,0}$ in the sulfuric acid solution:

The change Δn_B in amount of sulfuric acid until the equivalence point is reached results from the basic stoichiometric equation:

$$\frac{\Delta n_B}{\nu_B} = \frac{\Delta n_{B'}}{\nu_{B'}} \quad \Rightarrow$$

$$\Delta n_{\mathrm{B}} = \frac{\nu_{\mathrm{B}} \times \Delta n_{\mathrm{B'}}}{\nu_{\mathrm{B'}}} = \frac{(-1)\times(-2.44\times10^{-3}\ \mathrm{mol})}{(-2)} = -1.22\times10^{-3}\ \mathrm{mol}.$$

Consequently, we obtain for the amount $n_{\mathrm{B},0}$ of sulfuric acid in the initial solution:

$$\Delta n_{\mathrm{B}} = 0 - n_{\mathrm{B},0} \quad \Rightarrow$$

$$n_{\mathrm{B},0} = -\Delta n_{\mathrm{B},0} = -(-1.22\times10^{-3}\ \mathrm{mol}) = +1.22\times10^{-3}\ \mathrm{mol}.$$

Concentration $c_{\mathrm{B},0}$ of the sulfuric acid solution:

$$c_{\mathrm{B},0} = \frac{n_{\mathrm{B},0}}{V_{\mathrm{S,B}}} = \frac{1.22\times10^{-3}\ \mathrm{mol}}{250\times10^{-6}\ \mathrm{m}^3} = \mathbf{4.88\ mol\ m^{-3}}.$$

c) Mass $m_{\mathrm{B},0}$ of sulfuric acid:

$$m_{\mathrm{B},0} = n_{\mathrm{B},0} \times M_{\mathrm{B}}.$$

With the molar mass $M_{\mathrm{B}} = 98.1\times10^{-3}\ \mathrm{kg\,mol^{-1}}$ of sulfuric acid we obtain:

$$m_{\mathrm{B},0} = (1.22\times10^{-3}\ \mathrm{mol})\times(98.1\times10^{-3}\ \mathrm{kg\,mol^{-1}}) = 120\times10^{-6}\ \mathrm{kg} = \mathbf{120\ mg}.$$

2.1.8 Application of the basic stoichiometric equation in precipitation analysis

a) Conversion $\Delta\xi$:

The conversion $\Delta\xi$ results from the basic stoichiometric equation:

$$\Delta\xi = \frac{\Delta n_{\mathrm{D}}}{\nu_{\mathrm{D}}}.$$

Change Δn_{D} in amount of barium sulfate:

The amount n_{D} of barium sulfate after complete precipitation can be calculated from its mass m_{D} by means of its molar mass $M_{\mathrm{D}} = M(\mathrm{BaSO_4}) = 233.4\times10^{-3}\ \mathrm{kg\,mol^{-1}}$:

$$n_{\mathrm{D}} = \frac{m_{\mathrm{D}}}{M_{\mathrm{D}}} = \frac{467\times10^{-6}\ \mathrm{kg}}{233.4\times10^{-3}\ \mathrm{kg\,mol^{-1}}} = 2.00\times10^{-3}\ \mathrm{mol}.$$

Consequently, the change Δn_{D} in amount is $\Delta n_{\mathrm{D}} = n_{\mathrm{D}} - 0 = 2.00\times10^{-3}\ \mathrm{mol}$.

$$\Delta\xi = \frac{2.00\times10^{-3}\ \mathrm{mol}}{(+1)} = \mathbf{2.00\times10^{-3}\ mol}.$$

b) Change Δn_{B} in amount of Ba^{2+}:

$$\Delta\xi = \frac{\Delta n_{\mathrm{B}}}{\nu_{\mathrm{B}}} \quad \Rightarrow \quad \Delta n_{\mathrm{B}} = \nu_{\mathrm{B}} \times \Delta\xi = (-1)\times(2.00\times10^{-3}\ \mathrm{mol}) = \mathbf{-2.00\times10^{-3}\ mol}.$$

c) The concentration of Ba^{2+} in the volumetric flask (*1*) is identically equal to the concentration of Ba^{2+} in the 200 mL volumetric pipette (*2*).

Amount $n_{B,2}$ of Ba^{2+} in the pipette:

$$\Delta n_B = 0 - n_{B,2} \quad \Rightarrow$$

$$n_{B,2} = -\Delta n_B = -(-2.00\times10^{-3}\text{ mol}) = +2.00\times10^{-3}\text{ mol}.$$

Concentration $c_{B,1}$ of the Ba^{2+} solution in the volumetric flask:

$$c_{B,1} = c_{B,2} = \frac{n_{B,2}}{V_2} = \frac{2.00\times10^{-3}\text{ mol}}{200\times10^{-6}\text{ m}^3} = \mathbf{10.00\ mol\ m^{-3}}.$$

d) Amount $n_{B,1}$ of Ba^{2+} in the full volumetric flask:

$$c_{B,1} = \frac{n_{B,1}}{V_1} \quad \Rightarrow$$

$$n_{B,1} = c_{B,1}\times V_1 = 10.00\text{ mol m}^{-3}\times(500\times10^{-6}\text{ m}^3) = 5.00\times10^{-3}\text{ mol} = \mathbf{500\ mmol}.$$

2.1.9 Basic stoichiometric equation with participation of gases

a) Conversion formula and basic stoichiometric equation:

	Symbol	B	B′	D	D′	D″
Conversion formula:		$BaCO_3\vert s$	$+\,2\,H^+\vert w$	$\rightarrow Ba^{2+}\vert w$	$+\,CO_2\vert g$	$+\,H_2O\vert l$

$$\text{Basic equation:} \quad \Delta\xi = \frac{\Delta n_B}{\nu_B} = \frac{\Delta n_{B'}}{\nu_{B'}} = \frac{\Delta n_D}{\nu_D} = \frac{\Delta n_{D'}}{\nu_{D'}} = \frac{\Delta n_{D''}}{\nu_{D''}}$$

b) Amount $n_{B'}$ of HNO_3 required to dissolve the carbonate:

The change $\Delta n_{B'}$ in amount of H^+ and therefore of HNO_3 due to the dissolution reaction results from the basic stoichiometric reaction:

$$\frac{\Delta n_B}{\nu_B} = \frac{\Delta n_{B'}}{\nu_{B'}} \quad \Rightarrow$$

$$\Delta n_{B'} = \frac{\nu_{B'}\times\Delta n_B}{\nu_B}.$$

Change Δn_B in amount of $BaCO_3$:

The amount n_B, which corresponds to a mass m_B of 1×10^{-3} kg of solid $BaCO_3$, can be calculated by means of its molar mass $M_B = M(BaCO_3) = 197.3\times10^{-3}\text{ kg mol}^{-1}$:

$$n_B = \frac{m_B}{M_B} = \frac{1\times10^{-3}\ \text{kg}}{197.3\times10^{-3}\ \text{kg mol}^{-1}} = 5.07\times10^{-3}\ \text{mol}.$$

Consequently, the change Δn_B in amount is $\Delta n_B = 0 - n_B = -5.07\times10^{-3}$ mol.

$$\Delta n_{B'} = \frac{(-2)\times(-5.07\times10^{-3}\ \text{mol})}{(-1)} = -10.14\times10^{-3}\ \text{mol}.$$

Therefore, one obtains for the amount $n_{B'}$ of HNO_3 required to dissolve the carbonate $n_{B'} = -\Delta n_{B'} = +10.14\times10^{-3}$ mol.

Volume $V_{S,B'}$ of nitric acid:

$$c_{B'} = \frac{n_{B'}}{V_{S,B'}} \quad \Rightarrow$$

$$V_{S,B'} = \frac{n_{B'}}{c_{B'}} = \frac{10.14\times10^{-3}\ \text{mol}}{2\times10^{3}\ \text{mol m}^{-3}} = -5.07\times10^{-6}\ \text{m}^{-3} = \mathbf{-5.07\ mL}.$$

Thus, at least 5.07 mL of acid are necessary to completely dissolve the barium carbonate.

c) Volume $V_{D'}$ of CO_2 released in the course of the reaction:

Again, we start from the basic stoichiometric equation:

$$\frac{\Delta n_B}{\nu_B} = \frac{\Delta n_{D'}}{\nu_{D'}} \quad \Rightarrow$$

$$\Delta n_{D'} = \frac{\nu_{D'}\times\Delta n_B}{\nu_B} = \frac{(+1)\times(-5.07\times10^{-3}\ \text{mol})}{(-1)} = 5.07\times10^{-3}\ \text{mol}.$$

According to the additional information in the hint, the amount $n_{D'}$ ($= \Delta n_{D'}$) of carbon dioxide corresponds to a volume $V_{D'}$ of $(5.07\times10^{-3}\ \text{mol})\times(24.8\times10^{-3}\ \text{m}^3\ \text{mol}^{-1}) = 126\times10^{-6}\ \text{m}^3 = \mathbf{126\ mL}$.

2.2 Energy

2.2.1 Energy expenditure for stretching a spring

a) Energy W_1 for stretching the unstretched spring:

The energy W_1 for stretching the unstretched spring can be calculated by means of Equation (2.4) (For simplicity, the energy W_0 of the unstretched spring is set equal to zero.):

$$W_1 = \tfrac{1}{2} D \times (l - l_0)^2 .$$

$$W_1 = \tfrac{1}{2} \times (1 \times 10^5 \text{ N m}^{-1}) \times (0.1 \text{ m})^2 = 500 \text{ N m} = \mathbf{500\ J} .$$

b) Energy W_2 for stretching the already stretched spring:

The energy W_2 for stretching the already stretched spring results from the difference between the energy $W(l_2)$ necessary for stretching the spring by 60 cm and the energy $W(l_1)$ necessary for stretching the spring by 50 cm:

$$W_2 = W(l_2) - W(l_1) = \tfrac{1}{2} D \times (l_2 - l_0)^2 - \tfrac{1}{2} D \times (l_1 - l_0)^2 = \tfrac{1}{2} D \times [(l_2 - l_0)^2 - (l_1 - l_0)^2] .$$

$$W_2 = \tfrac{1}{2} \times (1 \times 10^5 \text{ N m}^{-1}) \times [(0.6 \text{ m})^2 - (0.5 \text{ m})^2] = 5500 \text{ N m} = \mathbf{5500\ J} .$$

It will become more and more strenuous to extend the spring by a small distance Δl the further the spring is already pre-stretched.

c) Extension Δl of the spring:

Starting point is HOOKE's law [Eq. (2.3)]:

$$F = D \times (l - l_0) = D \times \Delta l .$$

Solving for Δl results in:

$$\Delta l = \frac{F}{D} .$$

Gravitational force (weight) F_g of the person:

The gravitational force F_g of the person is given by

$$F_g = m \times g = 50 \text{ kg} \times 9.81 \text{ m s}^{-2} = 490.5 \text{ kg m s}^{-2} = 490.5 \text{ N} ,$$

where g is the gravitational acceleration (acceleration of free fall).

$$\Delta l = \frac{490.5 \text{ N}}{1 \times 10^5 \text{ N m}^{-1}} = 4.9 \times 10^{-3} \text{ m} \approx \mathbf{5\ mm} .$$

2.2.2 Energy and change of volume (I)

To simplify matters, the substances are again characterized by symbols in the following.

Symbol B B′ D D′ D″

Conversion formula: $BaCO_3|s + 2\,H^+|w \rightarrow Ba^{2+}|w + CO_2|g + H_2O|l$

Basic equation: $\Delta\xi = \frac{\Delta n_B}{\nu_B} = \frac{\Delta n_{B'}}{\nu_{B'}} = \frac{\Delta n_D}{\nu_D} = \frac{\Delta n_{D'}}{\nu_{D'}} = \frac{\Delta n_{D''}}{\nu_{D''}}$

a) Increase $\Delta V_{D'}$ in volume because of the produced carbon dioxide gas:

Again, we start from the basic stoichiometric equation (cf. Solution 2.1.9):

$$\frac{\Delta n_B}{\nu_B} = \frac{\Delta n_{D'}}{\nu_{D'}} \Rightarrow$$

$$\Delta n_{D'} = \frac{\nu_{D'} \times \Delta n_B}{\nu_B} = \frac{\nu_{D'} \times (0 - n_B)}{\nu_B} = -\frac{\nu_{D'} \times m_B}{M_B \times \nu_B}.$$

$$\Delta n_{D'} = -\frac{(+1) \times (20 \times 10^{-3}\ \text{kg})}{(197.3 \times 10^{-3}\ \text{kg mol}^{-1}) \times (-1)} = 0.1014\ \text{mol}.$$

According to the additional information given in the hint, the amount $n_{D'}$ (= $\Delta n_{D'}$) of carbon dioxide produced during the reaction corresponds to a volume $V_{D'}$ of $(0.1014\ \text{mol}) \times (24.8 \times 10^{-3}\ \text{m}^3\ \text{mol}^{-1}) = 0.00251\ \text{m}^3$. Therefore, the increase in volume is $\Delta V_{D'}\ (= V_{D'} - 0) = \mathbf{0.00251\ m^3}$.

b) Change ΔW in energy because of the increase in volume:

The change ΔW in energy results from the defining equation for pressure [Eq. (2.6)],

$$p = -\frac{dW}{dV},$$

by rearranging:

$$dW = -p dV$$

or for small finite changes

$$\Delta W = -p \times \Delta V \qquad \text{for } p = \text{const.}$$

$$\Delta W = -(100 \times 10^3\ \text{Pa}) \times 0.00251\ \text{m}^3 = -(100 \times 10^3\ \text{N m}^{-2}) \times 0.00251\ \text{m}^3$$

$$= -251\ \text{N m} = \mathbf{-251\ J}.$$

c) In order to make the change ΔV in volume visible, one could carry out the experiment in an Erlenmeyer flask instead of the beaker and close its opening with a rubber balloon (see the figure to Exercise 1.1.9).

2.2.3 Energy and change of volume (II)

Symbol B B′ D D′

Conversion formula: $2\ C_8H_{18}|l + 25\ O_2|g \rightarrow 16\ CO_2|g + 18\ H_2O|g$

Basic equation: $\Delta\xi = \dfrac{\Delta n_B}{\nu_B} = \dfrac{\Delta n_{B'}}{\nu_{B'}} = \dfrac{\Delta n_D}{\nu_D} = \dfrac{\Delta n_{D'}}{\nu_{D'}}$

Conversion $\Delta\xi$:

$$\Delta\xi = \frac{\Delta n_B}{\nu_B} = \frac{0 - n_B}{\nu_B} = -\frac{m_B}{M_B \times \nu_B} = -\frac{V_B \times \rho_B}{M_B \times \nu_B}.$$

$$\Delta\xi = -\frac{(1\times10^{-3}\ m^3)\times 700\ kg\,m^3}{(114.0\times10^{-3}\ kg\,mol^{-1})\cdot(-2)} = 3.07\ mol.$$

Changes in amount of substance Δn(gases) and in volume ΔV(gases) of the gases involved in the reaction:

The change in volume due to the reaction of the liquid phase octane ($\Delta V_B = -1\ L = -1\times10^{-3}\ m^3$) can be neglected compared to the change in volume due to the consumed and produced gases, since the latter changes are many times larger.

Changes in amount of substance of the gases involved:

$$\Delta n_i = \nu_i \times \Delta\xi$$

$$\Rightarrow \quad \Delta n_{B'} = -25\times3.07\ mol = -76.75\ mol,$$

$$\Delta n_D = +16\times3.07\ mol = +49.12\ mol, \quad \Delta n_{D'} = +18\times3.07\ mol = +55.26\ mol.$$

During the reaction, 49.12 moles of carbon dioxide and 55.26 moles of water vapor were produced, but also 76.75 mol of oxygen disappeared.

Change in amount of substance Δn(gases):

$$\Delta n(\text{gases}) = \Delta n_{B'} + \Delta n_D + \Delta n_{D'} = -76.75\ mol + 49.12\ mol + 55.26\ mol = 27.63\ mol.$$

This corresponds to a change in volume ΔV(gases) of $27.63\ mol \times (24.8\times10^{-3}\ m^3\ mol^{-1}) = 0{,}685\ m^3$.

Change in energy ΔW because of the increase in volume:

$$\Delta W = -p\times\Delta V = -(100\times10^3\ Pa)\times0.685\ m^3 = -68500\ J = \mathbf{-68.5\ kJ}.$$

2.2.4 Energy of a body in motion

a) Kinetic energy W_{kin} of the car:

The kinetic energy W_{kin} of the car results according to Equation (2.9) in

$W_{kin} = \frac{1}{2} m \times v^2$.

(Since the car is accelerated from rest, we have $W_0 = 0$.)

$$W_{kin} = \frac{1}{2} \times 1500\ \text{kg} \times (50\ \text{km h}^{-1})^2 = \frac{1}{2} 1500\ \text{kg} \times (13.9\ \text{m s}^{-1})^2$$
$$= 145 \times 10^3\ \text{kg m}^2\,\text{s}^{-2} = 145 \times 10^3\ \text{N m} = 145 \times 10^3\ \text{J} = \mathbf{145\ kJ}.$$

b) Height h of the car on the ramp:

The kinetic energy W_{kin} in the moving car is used to raise the vehicle in the gravitational field of the Earth. The energy stored in this way is called potential energy, W_{pot}. Friction is supposed to be negligible, that means that no energy is diverted for other purposes and we have:

$$W_{kin} + W_{pot} = \text{const.} \qquad \text{Eq. (2.14)}$$

or more detailed

$$\underbrace{W_{pot}(\text{final})}_{m \times g \times h} + \underbrace{W_{kin}(\text{final})}_{0} = \underbrace{W_{pot}(\text{initial})}_{0} + \underbrace{W_{kin}(\text{initial})}_{\frac{1}{2} m \times v^2} \quad \Rightarrow \quad h = \frac{W_{kin}(\text{initial})}{m \times g}.$$

$$h = \frac{145 \times 10^3\ \text{kg m}^2\,\text{s}^{-2}}{1500\ \text{kg} \times 9.81\ \text{kg m s}^{-2}} = \mathbf{9.85\ m}.$$

2.2.5 Falling without friction

a) Speed v_E of the elephant:

The distance traveled downward counts negatively if we use height h as spatial coordinate in order to describe the jumping process. The speed of a falling body, in this case the elephant, then results according to Equation (2.13) in

$$v_E = -g \times \Delta t_E = -9.81\ \text{m s}^{-2} \times 2\ \text{s} = 19.62\ \text{m s}^{-2} \approx \mathbf{-20\ m\,s^{-1}}.$$

Momentum p_E of the elephant:

The momentum of a moving body can be calculated by means of Equation (2.11):

$$p_E = m_E \times v_E = 2000\ \text{kg} \times (-19.62\ \text{m s}^{-1}) \approx -40000\ \text{kg m s}^{-1} \approx \mathbf{-40000\ N\,s}.$$

b) Duration Δt_M of the fall and speed v_M of the mouse:

Without friction all bodies fall with the same speed i.e. also the elephant and the mouse. Duration of the fall and speed are equal in both cases.

$$\Delta t_M = \mathbf{2\ s}; \qquad v_M \approx \mathbf{-20\ m\,s^{-1}}.$$

Momentum p_M of the mouse:

$$p_M = m_M \times v_M = 0.02\ \text{kg} \times (-19.62\ \text{m s}^{-1}) \approx -0.40\ \text{kg m s}^{-1} \approx \mathbf{-0.40\ N\,s}\,.$$

c) Height h_0 of the bridge:

The energy principle (law of conservation of energy) [cf. Eq. (2.13)] can be used in order to calculate the height h_0 of the bridge:

$$\tfrac{1}{2} m \times v^2 = m \times g \times (h_0 - h)\,.$$

The water surface is chosen as zero level, $h = 0$. Solving for h_0 results in

$$h_0 = \frac{\tfrac{1}{2} v^2}{g} = \frac{\tfrac{1}{2}(-19.62\ \text{m s}^{-1})^2}{9.81\ \text{m s}^{-2}} \approx \mathbf{20\ m}\,.$$

2.2.6 Energy of a raised body

a) Potential energy W_{pot} of the usable amount of water:

The potential energy W_{pot} results according to Equation (2.13) in:

$$W_{pot} = m \times g \times (h_u - h_l)$$

Mass m of the water:

$$m = \rho \times V = 1000\ \text{kg m}^{-3} \times (8 \times 10^6\ \text{m}^3) = 8 \times 10^9\ \text{kg}\,.$$

$$W_{pot} = (8 \times 10^9\ \text{kg}) \times 9.81\ \text{m s}^{-2} \times (800\ \text{m} - 400\ \text{m})$$

$$= 31.4 \times 10^{12}\ \text{kg m}^2\,\text{s}^{-2} = 31.4 \times 10^{12}\ \text{N m} = 31.4 \times 10^{12}\ \text{J} = \mathbf{31.4\ TJ}\,.$$

b) Time period Δt required to consume the stored energy:

Power P is the amount of energy consumed in a certain time period divided by this time period:

$$P = \frac{W_{pot}}{\Delta t}\,.$$

Solving for Δt results in the time period in question:

$$\Delta t = \frac{W_{pot}}{P} = \frac{31.4 \times 10^{12}\ \text{J}}{800 \times 10^6\ \text{J s}^{-1}} = 39250\ \text{s} \approx \mathbf{11\ h}\,.$$

2.2.7 Energy expenditure for climbing

a) Energy expenditure W_T of the tourist (T) for climbing:

The sea level represents the zero level.

$$W_T \quad = m_T \times g \times h_T = 80\ \text{kg} \times 9.81\ \text{m s}^{-2} \times 100\ \text{m} = 78500\ \text{J} = \mathbf{78{,}5\ kJ}.$$

The climbing results in an increase of the potential energy for the tourist.

b) Climbing height Δh_A of the astronaut (A):

$$W_A \quad = W_T = m_A \times g_{Moon} \times \Delta h_A \quad \Rightarrow$$

$$\Delta h_A \quad = \frac{W_T}{m_A \times g_{Moon}} = \frac{78500\ \text{kg m}^2\,\text{s}^{-2}}{200\ \text{kg} \times 1.60\ \text{m s}^{-2}} \approx \mathbf{245\ m}.$$

2.3 Entropy and Temperature

2.3.1 Ice calorimeter

a) Amount of entropy S released during the reaction of 20 g of the Fe-S mixture:

We have:

$$\frac{V}{V_{\mathrm{I}}} = \frac{S}{S_{\mathrm{I}}}.$$

Solving for S results in:

$$S = S_{\mathrm{I}} \times \frac{V}{V_{\mathrm{I}}} = 1\,\mathrm{Ct} \times \frac{63 \times 10^{-6}\ \mathrm{m}^3}{0.82 \times 10^{-6}\ \mathrm{m}^3} = 76.8\ \mathrm{Ct}.$$

Amount $n(\mathrm{FeS})$ of iron sulfide produced during the reaction of the mixture:

$$n(\mathrm{FeS}) = \frac{m(\mathrm{FeS})}{M(\mathrm{FeS})} = \frac{20 \times 10^{-3}\ \mathrm{kg}}{87.9 \times 10^{-3}\ \mathrm{kg\,mol^{-1}}} = 0.228\ \mathrm{mol}.$$

Amount of entropy S' released during the reaction of 1 mole of Fe with 1 mole of S (to 1 mole of FeS):

$$\frac{S'}{S} = \frac{n'(\mathrm{FeS})}{n(\mathrm{FeS})} \quad \Rightarrow \quad S' = S \times \frac{n'(\mathrm{FeS})}{n(\mathrm{FeS})} = 76.8\ \mathrm{Ct} \cdot \frac{1\ \mathrm{mol}}{0.228\ \mathrm{mol}} = \mathbf{336.8\ Ct}.$$

b) Amount of entropy S_{g} generated by the immersion heater:

The amount of entropy S_{g} generated in the immersion heater results from the equation $W_{\mathrm{b}} = T \times S_{\mathrm{g}}$ [Eq. (3.14)] by solving for S_{g}. The "burnt" energy W_{b} can be calculated from the power P of the immersion heater and its switch-on duration Δt:

$$S_{\mathrm{g}} = \frac{W_{\mathrm{b}}}{T} = \frac{P \times \Delta t}{T} = \frac{1000\ \mathrm{J\,s^{-1}} \times 27\ \mathrm{s}}{273\ \mathrm{K}} = 98.9\ \mathrm{J\,K^{-1}} = 98.9\ \mathrm{Ct}.$$

The entropy unit Carnot (Ct) corresponds to Joule/Kelvin ($\mathrm{J\,K^{-1}}$).

Volume V of melt water:

$$\frac{V}{V_{\mathrm{I}}} = \frac{S_{\mathrm{g}}}{S_{\mathrm{I}}} \quad \Rightarrow$$

$$V = V_{\mathrm{I}} \times \frac{S_{\mathrm{g}}}{S_{\mathrm{I}}} = 0.82 \times 10^{-6}\ \mathrm{m}^3 \times \frac{98.9\ \mathrm{Ct}}{1\ \mathrm{Ct}} = 81.1 \times 10^{-6}\ \mathrm{m}^3 = \mathbf{81.1\ mL}.$$

2.3.2 Measuring the entropy content

a) From the heating curve for the 1 kg block of copper, we first have to determine the temperatures that are obtained after every additional increase in energy of 5 kJ (gray values in the upper row, determined always in the middle of the interval). The procedure was illustrated in the given diagram by means of four examples (black values). The particular increase in entropy $\Delta S_{g,i}$ can be calculated according to Equation (3.9):

$$\Delta S_{g,i} = \frac{\Delta W_i}{T}.$$

For example, we obtain for the first value:

$$\Delta S_{g,1} = \frac{\Delta W_1}{T} = \frac{5000\ \text{J}}{68\ \text{K}} = 74\ \text{J}\,\text{K}^{-1} = 74\ \text{Ct}.$$

Further values are inserted into the diagram (lower row).

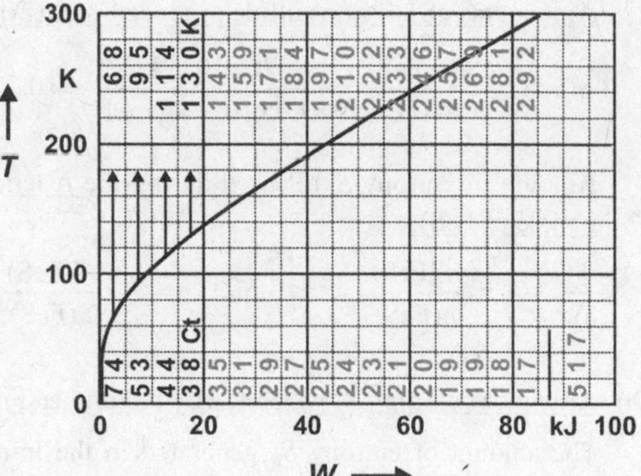

b) Entropy S of the block in total:

To obtain the total entropy contained in the block at 300 K, we have to add up the 17 values:

$$S \approx \mathbf{517\ Ct}.$$

Molar entropy S_m of copper:

The above entropy value refers to 1 kg of copper. The molar entropy of copper then results in [cf. Eq. (3.7)]:

$$S_m = \frac{S}{n}.$$

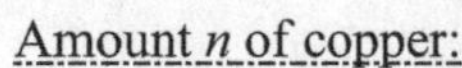
Amount n of copper:

$$n = \frac{m}{M} = \frac{1\ \text{kg}}{63.5\times10^{-3}\ \text{kg}\,\text{mol}^{-1}} = 15.75\ \text{mol}.$$

$$S_m \approx \frac{517\ \text{Ct}}{15.75\ \text{mol}} = \mathbf{33\ Ct\,mol^{-1}}.$$

This value corresponds quite well with the literature value of 33.150 Ct mol^{-1} [from: Haynes W M et al (ed) (2015) CRC Handbook of Chemistry and Physics, 96th edn. CRC Press, Boca Raton].

2.3.3 Entropy of water

The different entropy values are read off the diagram.

a) Entropy ΔS needed to heat the water:

The increase in entropy needed to heat 1 g of lukewarm water (20 °C) to a boil corresponds to the "length" of the vertical black double arrow plotted at 293 K:

$\Delta S \quad \approx \mathbf{1\ Ct}$.

b) Molar zero-point entropy $S_{0,\text{m}}$:

First, the zero-point entropy, i. e. the entropy value at 0 K, has to be determined from the diagram:

$S_0 \quad \approx 0.2\ \text{Ct}$.

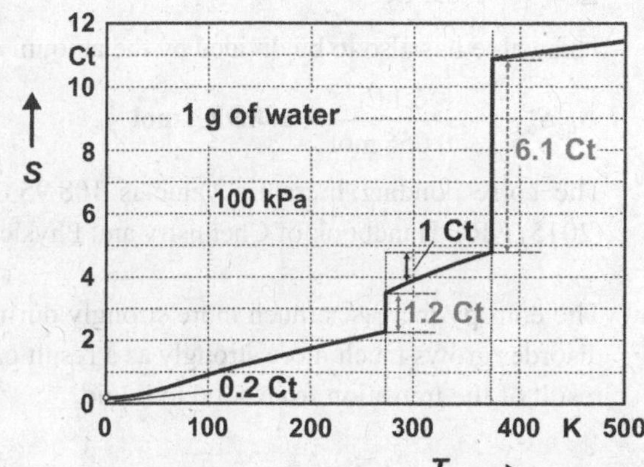

In order to calculate the molar zero-point entropy we have to divide the value per gram of water by the amount of substance:

$$S_{0,\text{m}} = \frac{S_0}{n} .$$

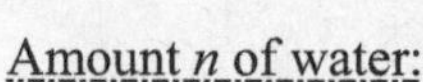

Amount n of water:

$$n = \frac{m}{M} = \frac{1\times10^{-3}\ \text{kg}}{18\times10^{-3}\ \text{kg mol}^{-1}} = 0.056\ \text{mol} .$$

$$S_{0,\text{m}} \approx \frac{0.2\ \text{Ct}}{0.056\ \text{mol}} = \mathbf{3.6\ Ct\ mol^{-1}} .$$

b) Molar entropy of fusion $\Delta_{\text{sl}}S^{\ominus}_{\text{eq}}$:

The molar entropy of fusion $\Delta_{\text{sl}}S^{\ominus}_{\text{eq}}$ [$\equiv \Delta S_{\text{sl}}(T^{\ominus}_{\text{sl}})$] meaning the change $\Delta_{\text{sl}}S$ of entropy per mole of substance due to the melting process at the standard melting point $T^{\ominus}_{\text{sl}}$ [see also Section 3.9 (p. 75) in the textbook "Physical Chemistry from a Different Angle"] can be determined by means of the "length" of the vertical part in the diagram at 273 K (melting point of water at the standard pressure $p^{\ominus}$ of 100 kPa). This is illustrated by the gray double arrow:

$\Delta S \quad \approx 1.2\ \text{Ct}$.

Subsequently, this value has to be divided by the amount of water:

$$\Delta_{\text{sl}}S^{\ominus}_{\text{eq}} \approx \frac{1.2\ \text{Ct}}{0.056\ \text{mol}} = \mathbf{21.4\ Ct\ mol^{-1}} .$$

The literature value is 22.00 Ct mol^{-1} [from: Haynes W M et al (ed) (2015) CRC Handbook of Chemistry and Physics, 96th edn. CRC Press, Boca Raton].

c) Molar entropy of vaporization $\Delta_{lg}S^{\ominus}_{eq}$:

The molar entropy of vaporization $\Delta_{lg}S^{\ominus}_{eq}$ [$\equiv \Delta S_{lg}(T^{\ominus}_{lg})$] can be determined in an analogous manner ["length" of the vertical part in the diagram at 373 K (boiling point of water at standard pressure $p^{\ominus}$ of 100 kPa)]. This is illustrated by the dashed gray double arrow:

$$\Delta S \approx 6.1\,\text{Ct}\,.$$

This value has also to be divided by the amount of water:

$$\Delta_{lg}S^{\ominus}_{eq} \approx \frac{6.1\,\text{Ct}}{0.056\,\text{mol}} = \mathbf{108.9\,Ct\,mol^{-1}}\,.$$

The corresponding literature value is 108.95 Ct mol^{-1} [from: Haynes W M et al (ed) (2015) CRC Handbook of Chemistry and Physics, 96th edn. CRC Press, Boca Raton].

e) The entropy increases much more strongly during vaporization than during melting, since disorder grows much more strongly as a result of the transition from liquid to gas than as a result of the transition from solid to liquid.

2.3.4* Entropy and entropy capacity

a) Power P_{in} delivered by the sun:

In order to calculate the power P_{in} delivered by the sun, we have to multiply the power density P''_{in} of the solar radiation by the area A of the pond:

$$P''_{in} = \frac{P_{in}}{A} \quad \Rightarrow \quad P_{in} = P''_{in} \times A = 500\,\text{J}\,\text{s}^{-1}\,\text{m}^{-2} \times 6000\,\text{m}^2 = 3\times10^6\,\text{J}\,\text{s}^{-1}\,.$$

Energy W transferred into the pond water:

The efficiency η_{abs} corresponds to the ratio between the energy W transferred into the pond water and the energy W_{in} delivered by the sun:

$$\eta_{abs} = \frac{W}{W_{in}} = \frac{W}{P_{in} \times \Delta t} \quad \Rightarrow$$

$$W = P_{in} \times \Delta t \times \eta_{abs}\,.$$

$$W = (3\times10^6\,\text{J}\,\text{s}^{-1}) \times 4\,\text{h} \times 0.8 = (3\times10^6\,\text{J}\,\text{s}^{-1}) \times 14400\,\text{s} \times 0.8$$

$$= 3.46\times10^{10}\,\text{J} = \mathbf{34.6\,GJ}\,.$$

b) Increase in entropy ΔS of the pond water:

The energy W transferred into the pond water is "burnt" thereby generating entropy:

$$\Delta S = \frac{W}{T}. \qquad \text{cf. Eq. (3.4)}$$

$$\Delta S = \frac{34.6 \times 10^9 \text{ J}}{298 \text{ K}} = 116 \times 10^6 \text{ JK}^{-1} = \mathbf{116\ MCt}.$$

c) Increase in temperature ΔT of the water:

Starting point of the calculation is the defining equation for the molar entropy capacity $\mathcal{C}_m$ [Eq. (3.13)]. By means of this equation, we can determine the increase in temperature ΔT of the water caused by the increase in entropy ΔS:

$$\mathcal{C}_m = \frac{1}{n} \times \frac{\Delta S}{\Delta T} \quad \Rightarrow$$

$$\Delta T = \frac{1}{n} \times \frac{\Delta S}{\mathcal{C}_m}.$$

Volume V of water in the pond:

$$V = A \times l = 6000 \text{ m}^2 \times 1 \text{ m} = 6000 \text{ m}^3.$$

Mass m of water:

$$m = \rho \times V = 997 \text{ kg m}^{-3} \times 6000 \text{ m}^3 = 6 \times 10^6 \text{ kg}.$$

Amount n of water:

$$n = \frac{m}{M} = \frac{6 \times 10^6 \text{ kg}}{18.0 \times 10^3 \text{ kg mol}^{-1}} = 330 \times 10^6 \text{ mol}.$$

$$\Delta T = \frac{1}{330 \times 10^6 \text{ mol}} \times \frac{116 \times 10^6 \text{ Ct}}{0.253 \text{ Ct mol}^{-1} \text{ K}^{-1}} = \mathbf{1.4\ K}.$$

2.3.5 Heat pump

Energy W_t applied to transfer the entropy S_t:

The energy W_t applied to transfer the entropy S_t results from the power P of the heat pump and the switch-on duration Δt of one hour:

$$W_t = P \times \Delta t = (3 \times 10^3 \text{ J s}^{-1}) \times 3600 \text{ s} = 10.8 \times 10^6 \text{ J} = 10.8 \text{ MJ}.$$

Amount of entropie S_t transferred from the air into the water:

The transferred amount of entropy S_t can be calculated by means of Equation (3.27):

$$W_t = (T_2 - T_1) \times S_t \quad \Rightarrow$$

$$S_t = \frac{W_t}{T_2 - T_1} = \frac{10.8\times10^6\ \text{J}}{300\ \text{K} - 290\ \text{K}} = 1.08\times10^6\ \text{J}\,\text{K}^{-1} = \mathbf{1080\ kCt}.$$

2.3.6. Heat pump versus electric heating

a) Amount of entropy S_t transferred in one day from outside into the house:

The heat pump transfers 30 Ct in 1 s from outside into the house. Therefore, it transfers in one day (= 86400 s) 2.6×10^6 Ct:

$$S_t = 2.6\times10^6\ \text{Ct} = \mathbf{2.6\ MCt}.$$

Energy consumption W_t of the heat pump:

$$W_t = (T_2 - T_1)\times S_t = (298\ \text{K} - 273\ \text{K})\times(2.6\times10^6\ \text{J}\,\text{K}^{-1}) = 65\times10^6\ \text{J} = \mathbf{65\ MJ}.$$

b) Amount of entropy S_g generated in the house in one day:

The same amount of entropy is to be generated in the house as was transferred into it in part a) of the exercise.

$$S_g = 2.6\times10^6\ \text{Ct} = \mathbf{2.6\ MCt}.$$

Energy consumption W_b by electric heating:

The energy W_b has to be "burnt" in order to generate the amount of entropy S_g:

$$W_b = T_2\times S_g. \qquad \text{cf. Eq. (3.14)}$$

$$W_b = 298\ \text{K}\times(2.6\times10^6\ \text{J}\,\text{K}^{-1}) = 775\times10^6\ \text{J} = \mathbf{775\ MJ}.$$

Conclusion: Electric heating consumes far more energy than a heat pump (approximately the twelvefold).

2.3.7* Thermal power plant (I)

a) Expenditure of energy W_1 for generating the entropy $S_g = 1$ Ct:

In order to generate the entropy S_g in the steam boiler, the energy W_1 has to be expended, i.e. "burnt."

$$W_1 = T_1\times S_g. \qquad \text{cf. Eq. (3.14)}$$

$$= 800\ \text{K}\times1\ \text{J}\,\text{K}^{-1} = \mathbf{800\ J}.$$

Useful energy W_{use}:

In the steam power plant, the energy W_{use} (= $-W_t$) is used, which can be gained by transferring the entropy S_t (= S_g) from the steam boiler into the cooling tower:

$$W_{use} = (T_1 - T_2)\times S_g. \qquad \text{cf. Eq. (3.27)}$$

$$W_{use} = (800\,\text{K} - 300\,\text{K}) \times 1\,\text{J K}^{-1} = \mathbf{500\,J}.$$

b) Efficiency η in the ideal case:

$$\eta_{ideal} = \frac{W_{use}}{W_1} = \frac{500\,\text{J}}{800\,\text{J}} = 0.625 = \mathbf{62.5\,\%}.$$

c) Expenditure of energy W_1 for generation of entropy in the firing:

A useful energy W_{use} of 1 kJ should be gained:

$$\eta_{real} = \frac{W_{use}}{W_1} \Rightarrow W_1 = \frac{W_{use}}{\eta_{real}} = \frac{1000\,\text{J}}{0.40} = 2500\,\text{J}.$$

Entropy S'_g generated in the firing:

$$S'_g = \frac{W_1}{T_1} = \frac{2500\,\text{J}}{800\,\text{K}} = 3.1\,\text{J K}^{-1} = \mathbf{3.1\,Ct}.$$

Expenditure of energy W_2 for the transfer of entropy to the "repository," meaning the environment

$$W_2 = W_1 - W_{use} = 2500\,\text{J} - 1000\,\text{J} = 1500\,\text{J}.$$

Amount of entropy S''_g with which the environment is eventually burdened:

$$S''_g = \frac{W_2}{T_2} = \frac{1500\,\text{J}}{300\,\text{K}} = 5.0\,\text{J K}^{-1} = \mathbf{5.0\,Ct}.$$

d) Energy loss W'_2 during the firing meaning the generation of entropy S'_g:

$$W'_2 = T_2 \times S'_g = 300\,\text{K} \times 3.1\,\text{J K}^{-1} = 930\,\text{J}.$$

Referred to the expended energy W_1 of 2500 J, the percentage of the firing at the energy loss is therefore **37 %** (930 J/2500 J = 0.37). **23 %** (= 60 % – 37 %) are allotted to the rest of the plant.

2.3.8* Thermal power plant (II)

a) Mass m_1 of hard coal, which has to be burnt in the ideal case:

The mass ratio m_1/m_0 of hard coal corresponds to the energy ratio $W_{use,1}/W_{use,0}$:

$$\frac{m_1}{m_0} = \frac{W_{use,1}}{W_{use,0}} \Rightarrow m_1 = \frac{m_0 \times W_{use,1}}{W_{use,0}} = \frac{1\,\text{kg} \times (1100 \times 10^6\,\text{J})}{35 \times 10^6\,\text{J}} = \mathbf{31.4\,kg}.$$

b) Mass m_2 of hard coal, which has to be burnt in the real case:

The real efficiency η_{real} corresponds to the mass ratio m_1/m_2.

$$\eta_{real} = \frac{m_1}{m_2} \quad \Rightarrow \quad m_2 = \frac{m_1}{\eta_{real}} = \frac{31.4\ \text{kg}}{0.40} = \mathbf{78.5\ kg}.$$

c) Hence, 60 % of the usable energy gets already lost in the power plant under generation of entropy. Please mark the parts with a cross where the energy is mainly lost.

boiler ☒, turbine ☐, generator ☐, condenser + cooling tower ☐.

At what point does the generated entropy mainly leave the power plant?

boiler ☐, turbine ☐, generator ☐, condenser + cooling tower ☒.

What are the parts through which the entropy is transported without noticeable increase in its amount?

boiler ☐, turbine ☒, generator ☐, condenser + cooling tower ☒.

d) Expenditure of energy W_1 for generation of entropy in the firing:

The index 1 characterizes in the case of W_1 (as in the figure in Exercise 1.3.7) the entry, the boiler.

$$\eta_{real} = \frac{W_{use}}{W_1} \quad \Rightarrow \quad W_1 = \frac{W_{use}}{\eta_{real}} = \frac{1100\times 10^6\ \text{J}}{0.40} = 2750\times 10^6\ \text{J} = 2750\ \text{MJ}.$$

Expenditure of energy W_2 for the transfer of entropy to the "repository," meaning the environment

The index 2 signifies the exit, the cooling tower.

$$W_2 = W_1 - W_{use} = (2750\times 10^6\ \text{J}) - (1100\times 10^6\ \text{J}) = 1650\times 10^6\ \text{J} = 1650\ \text{MJ}.$$

Entropy S_g, which is ultimately generated (in all places together):

$$S_g = \frac{W_2}{T_a} = \frac{1650\times 10^6\ \text{J}}{300\ \text{K}} = 5.5\times 10^6\ \text{Ct} = \mathbf{5.5\ MCt}.$$

2.3.9 Types of power plants

a) Transferred amount of entropy S_t:

$$W_{use} = (T_1 - T_2)\times S_t \quad \Rightarrow$$

$$S_t = \frac{W_{use}}{T_1 - T_2}.$$

Coal-fired power plant (A):

$$S_t(A) = \frac{1200\times10^6\ \mathrm{J}}{800\ \mathrm{K}-320\ \mathrm{K}} = 2.5\times10^6\ \mathrm{JK^{-1}} = \mathbf{2.5\ MCt}.$$

Nuclear power plant (B):

$$S_t(B) = \frac{1200\times10^6\ \mathrm{J}}{550\ \mathrm{K}-320\ \mathrm{K}} = 5.2\times10^6\ \mathrm{JK^{-1}} = \mathbf{5.2\ MCt}.$$

b) Energy W_1 for generating the entropy S_g (= S_t):

$$W_1 = T_1 \times S_g.$$

Coal-fired power plant (A):

$$W_1(A) = 800\ \mathrm{K}\times(2.5\times10^6\ \mathrm{JK^{-1}}) = 2000\times10^6\ \mathrm{J} = \mathbf{2000\ MJ}.$$

Nuclear power plant (B):

$$W_1(B) = 550\ \mathrm{K}\times(5.2\times10^6\ \mathrm{JK^{-1}}) = 2860\times10^6\ \mathrm{J} = \mathbf{2860\ MJ}.$$

c) Ideal efficiency η_{ideal}:

$$\eta_{ideal} = \frac{W_{use}}{W_1}$$

Coal-fired power plant (A):

$$\eta_{ideal}(A) = \frac{1200\ \mathrm{MJ}}{2000\ \mathrm{MJ}} = 0.60 = \mathbf{60\ \%}.$$

Nuclear power plant (B):

$$\eta_{ideal}(B) = \frac{1200\ \mathrm{MJ}}{2860\ \mathrm{MJ}} = 0.42 = \mathbf{42\ \%}.$$

The live steam temperatures at the entry to the steam turbine are in nuclear power plants lower than in coal-fired power plants due to inherent design limitations in the nuclear reactors. Therefore, nuclear power plants have a lower efficiency compared to coal-fired power plants.

d) In a hydroelectric power plant (C), only a small amount of entropy is generated by friction. The efficiency is much higher than in coal-fired and nuclear power plants (up to 95 %).

2.3.10 Entropy generation during entropy conduction

a) Amount of entropy S'_g generated in the heating wire (h) in one second ($\Delta t = 1$ s):

$$S'_g = \frac{W_{b,h}}{T_1} = \frac{P \times \Delta t}{T_1} = \frac{1000\ \text{J s}^{-1} \times 1\ \text{s}}{1000\ \text{K}} = 1\ \text{J K}^{-1} = \mathbf{1\ Ct}.$$

b) Amount of entropy S_g generated along the path from the heating wire to the water:

When the amount of entropy S'_g flows through a conducting connection (c) the additional amount of entropy S_g is generated. This results from "burning" of energy $W_{b,c}$ released by the transfer of the amount of entropy S'_g from a higher temperature T_1 to a lower temperature T_2:

$$S_g = \frac{W_{b,c}}{T_2} \quad \text{with} \quad W_{b,c} = (T_1 - T_2) \times S'_g \quad \Rightarrow \qquad \text{cf. Eq. (3.30)}$$

$$S_g = \frac{(T_1 - T_2) \times S'_g}{T_2} = \frac{(1000\ \text{K} - 373\ \text{K}) \times 1\ \text{Ct}}{373\ \text{K}} = \mathbf{1.7\ Ct}.$$

c) Entropy S_{total} that in total flows into the water per second:

$$S_{total} = S_g + S'_g = 1\ \text{Ct} + 1.7\ \text{Ct} = \mathbf{2.7\ Ct}.$$

2.4 Chemical Potential

2.4.1 Stability of states

In general, the state with the lowest chemical potential is stable for given conditions (in this case standard conditions). The corresponding μ value is printed in bold.

a) Graphite — Diamond

	Graphite	Diamond
$\mu^\ominus$/kG	**0**	2.9

Graphite is the stable modification of carbon at standard conditions.

b) Rhombic sulfur — Monoclinic sulfur

	Rhombic sulfur	Monoclinic sulfur
$\mu^\ominus$/kG	**0**	0.07

At standard conditions, rhombic sulfur is stable.

c) Solid iodine — Liquid iodine — Iodine vapor

	Solid iodine	Liquid iodine	Iodine vapor
$\mu^\ominus$/kG	**0**	3.3	19.3

At standard conditions, the solid state of iodine is stable.

d) Water ice — Water — Water vapor

	Water ice	Water	Water vapor
$\mu^\ominus$/kG	–236.6	**–237.1**	–228.6

At standard conditions, the liquid state of water is stable.

e) Ethanol — Ethanol vapor

	Ethanol	Ethanol vapor
$\mu^\ominus$/kG	**–174.6**	–167.9

The liquid state of ethanol is stable at standard conditions.

2.4.2 Predicting Reactions

A reaction can run spontaneously if its chemical drive $\mathcal{A}$ is positive. The drive $\mathcal{A}$ of a reaction

$$|\nu_B|B + |\nu_{B'}|B' + ... \rightarrow \nu_D D + \nu_{D'} D' + ...$$

is defined in the following way:

$$\mathcal{A} = |\nu_B|\mu(B) + |\nu_{B'}|\mu(B') + ... - \nu_D\mu(D) - \nu_{D'}\mu(D') - ... ,$$

abbreviated: $\mathcal{A} = \sum_{\text{initial}} |\nu_i|\mu_i - \sum_{\text{final}} \nu_j\mu_j$.

a) Binding of carbon dioxide by quicklime (CaO)

Conversion formula: $CO_2|g + CaO|s \rightarrow CaCO_3|s$

$\mu^\ominus$/kG: –394.4 –603.3 –1128.8

$$\mathcal{A}^\ominus = \{[(-394.4) + (-603.3)] - [-1128.8]\}\ \text{kG} = \mathbf{+131.1\ kG}\ .$$

The drive is positive, i.e. the reaction runs spontaneously.

b) Combustion of ethanol vapor (while water vapor is produced)
Conversion formula: $C_2H_6O|g + 3\ O_2|g \rightarrow 2\ CO_2|g \quad + 3\ H_2O|g$
$\mu^{\ominus}/\text{kG}$: $-167.9 \quad 3\times 0 \quad 2\times(-394.4) \quad 3\times(-228.6)$
$$\mathcal{A}^{\ominus} = \{[(-167.9)+3\times 0]-[2\times(-394.4)+3\times(-228.6)]\}\ \text{kG} = \mathbf{+1306.7\ kG}\,.$$
The drive is positive, i.e. the reaction runs spontaneously.

c) Decomposition of silver oxide into its elements
Conversion formula: $2\ Ag_2O|s \rightarrow 4\ Ag|s + O_2|g$
$\mu^{\ominus}/\text{kG}$: $2\times(-11.3) \quad 4\times 0 \quad 0$
$$\mathcal{A}^{\ominus} = \{[2\times(-11.3)]-[4\times 0+0]\}\ \text{kG} = \mathbf{-22.6\ kG}\,.$$
The drive is negative, therefore the reaction does not run spontaneously.

d) Reduction of hematite (Fe_2O_3) with carbon (graphite) to iron (thereby releasing carbon monoxide gas)
Conversion formula: $Fe_2O_3|s + 3\ C|\text{Graphit} \rightarrow 2\ Fe|s + 3\ CO|g$
$\mu^{\ominus}/\text{kG}$: $-741.0 \quad 3\times 0 \quad 2\times 0 \quad 3\cdot(-137.2)$
$$\mathcal{A}^{\ominus} = \{[(-741.0)+3\times 0]-[2\times 0+3\times(-137.2)]\}\ \text{kG} = \mathbf{-329.4\ kG}\,.$$
The drive is negative, therefore the reaction does not run spontaneously.

2.4.3 Dissolving Behavior

Whether a substance dissolves easily or not in water, alcohol, etc. is a result of the difference of its chemical potential in the pure and in the dissolved state. If this difference is greater than zero (meaning there is a potential drop between the initial and the final state), the substance in question is easily soluble.

a) Cane sugar (sucrose)
Conversion formula: $C_{12}H_{22}O_{11}|s \rightarrow C_{12}H_{22}O_{11}|w$
$\mu^{\ominus}/\text{kG}$: $-1557.6 \quad > -1564.7$
Cane sugar dissolves by itself even in a solution which already contains 1 kmol m^{-3} of sugar (the tabulated standard value is valid for this concentration). Consequently, it is easily soluble as we know from using it every day (see also Experiment 4.11 in the textbook).

b) Table salt (sodium chloride)
Conversion formula: $NaCl|s \rightarrow Na^+|w + Cl^-|w$
$\mu^{\ominus}/\text{kG}$: $-384.0 > \underbrace{-261.9 \quad -131.2}_{-393.1}$
Table salt also dissolves easily in water.

c) Limestone (calcite)

Conversion formula: $CaCO_3|s \rightarrow Ca^{2+}|w + CO_3^{2-}|w$

$\mu^{\ominus}$/kG: $-1128.8 < \underbrace{-553.6 \quad -527.8}_{-1081.4}$

Limestone is poorly soluble in water.

d) Oxygen

Conversion formula: $O_2|g \rightarrow O_2|w$

$\mu^{\ominus}$/kG: $0 < 16.4$

Oxygen is not easily soluble in water.

e) Carbon dioxide

Conversion formula: $CO_2|g \rightarrow CO_2|w$

$\mu^{\ominus}$/kG: $-394.4 < -386.0$

Carbon dioxide is not easily soluble in water. Therefore, it tends to bubble out of carbonated liquids such as mineral water (into which carbon dioxide gas has been pressed) (see also Experiment 4.13 in the textbook).

f) Ammonia

Conversion formula: $NH_3|g \rightarrow NH_3|w$

$\mu^{\ominus}$/kG: $-16.5 > -26.6$

Ammonia is highly soluble in water. An impressive way of showing this excellent solubility is with a so-called fountain experiment (Experiment 4.12 in the textbook).

2.5 Influence of Temperature and Pressure on Transformations

2.5.1 Temperature dependence of chemical potential and transition temperature

a) Change $\Delta\mu$ in chemical potential of ethanol with temperature:

In order to describe the temperature dependence of chemical potential, we choose the simplest option, a linear approach:

$$\mu = \mu^{\ominus} + \alpha(T - T^{\ominus}) . \qquad \text{cf. Eq. (5.2)}$$

Here, the standard value $\mu^{\ominus}$ of the chemical potential is the initial value. However, what is required is the change in chemical potential with temperature:

$$\mu - \mu^{\ominus} = \Delta\mu = \alpha(T - T^{\ominus}) .$$

In this case, we only need the temperature coefficient $\alpha = -160.7\ \text{G K}^{-1}$ of liquid ethanol:

$$\Delta\mu = -160.7\ \text{G K}^{-1} \times (323\ \text{K} - 298\ \text{K}) = -160.7\ \text{G K}^{-1} \times 25\ \text{K}$$

$$= -4020\ \text{G} = \mathbf{-4.02\ kG} .$$

b) Standard boiling temperature $T_{\text{lg}}^{\ominus}$ of ethanol:

The standard boiling temperature $T_{\text{lg}}^{\ominus}$ of a substance can be calculated in exactly the same way as its melting temperature T_{sl} [Eq. (5.7)],

$$T_{\text{lg}}^{\ominus} = T^{\ominus} - \frac{\mathcal{A}^{\ominus}}{\alpha} ,$$

except that in this case the boiling process is taken as basis, more precisely, the boiling process of ethanol:

Conversion formula:	$C_2H_6O\vert l \rightarrow$	$C_2H_6O\vert g$
$\mu^{\ominus}/\text{kG}$	−174.6	−167.9
$\alpha/\text{G K}^{-1}$	−160.7	−281.6

Drive $\mathcal{A}^{\ominus}$ of the boiling process:

$$\mathcal{A}^{\ominus} = \sum_{\text{initial}} \mu_i^{\ominus} - \sum_{\text{final}} \mu_j^{\ominus} . \qquad \text{Eq. (4.3)}$$

$$= [(-174.6) - (-167.9)]\ \text{kG} = -6.7\ \text{kG} .$$

Temperature coefficient α:

The temperature coefficient α of the drive can be calculated by the same easy to remember procedure as the drive itself:

$$\alpha = \sum_{\text{initial}} \alpha_i - \sum_{\text{final}} \alpha_j . \qquad \text{Eq. (5.4)}$$

$$\alpha = [(-160.7\ \text{G K}^{-1}) - (-281.6\ \text{G K}^{-1})] = 120.9\ \text{G K}^{-1} .$$

$$T_{\text{lg}}^{\ominus} = 298\,\text{K} - \frac{(-6.7\times 10^3\,\text{G})}{120.9\,\text{G}\,\text{K}^{-1}} = 298\,\text{K} + 55\,\text{K} = \mathbf{353\,K}.$$

Literature value: 351.06 K [from: Haynes W M et al (ed) (2015) CRC Handbook of Chemistry and Physics, 96th edn. CRC Press, Boca Raton] (The tabulated values for boiling temperatures usually refer to "atmospheric pressure," i.e. a pressure of 101.325 kPa. Therefore, they have to be corrected to the standard pressure of 100 kPa, as has been done here.)

2.5.2 Decomposition and reaction temperatures

The decomposition and reaction temperatures can be calculated by means of formulas entirely equivalent to those for phase transition temperatures [see Eq. (5.7)].

a) Decomposition of limestone:

Conversion formula:	$CaCO_3\vert s \rightarrow$	$CaO\vert s$	$+ CO_2\vert g$.
$\mu^{\ominus}/\text{kG}$	−1128.8	−603.3	−394.4
$\alpha/\text{G}\,\text{K}^{-1}$	−92.7	−38.1	−213.8

$$T_{\text{D}} = T^{\ominus} - \frac{\mathcal{A}^{\ominus}}{\alpha}.$$

$$\mathcal{A}^{\ominus} = \{[-1128.8] - [(-603.3) + (-394.4)]\}\,\text{kG} = -131.1\,\text{kG}.$$

$$\alpha = \{[-92.7] - [(-38.1) + (-213.8)]\}\,\text{G}\,\text{K}^{-1} = 159.2\,\text{G}\,\text{K}^{-1}.$$

$$T_{\text{D}} = 298\,\text{K} - \frac{(-131.1\times 10^3\,\text{G})}{159.2\,\text{G}\,\text{K}^{-1}} = 298\,\text{K} + 824\,\text{K} = \mathbf{1122\,K}.$$

b) Reduction of magnetite with carbon:

Conversion formula:	$Fe_3O_4\vert s$	$+ 2\,C\vert s$	$\rightarrow 3\,Fe\vert s$	$+ 2\,CO_2\vert g$,
$\mu^{\ominus}/\text{kG}$	−1017.5	2×0	3×0	2×(−394.4)
$\alpha/\text{G}\,\text{K}^{-1}$	−145.3	2×(−5.7)	3×(−27.3)	2×(−213.8)

$$T_{\text{R}} = T^{\ominus} - \frac{\mathcal{A}^{\ominus}}{\alpha}.$$

$$\mathcal{A}^{\ominus} = \{[(-1017.5) + 2\times 0] - [3\times 0 + 2\times(-394.4)]\}\,\text{kG} = -228.7\,\text{kG}.$$

$$\alpha = \{[(-145.3) + 2\times(-5.7)] - [3\times(-27.3) + 2\times(-213.8)]\}\,\text{G}\,\text{K}^{-1} = 352.8\,\text{G}\,\text{K}^{-1}.$$

$$T_{\text{R}} = 298\,\text{K} - \frac{(-228.7\times 10^3\,\text{G})}{352.8\,\text{G}\,\text{K}^{-1}} = 298\,\text{K} + 648\,\text{K} = \mathbf{946\,K}.$$

2.5.3 Pressure dependence of chemical potential

a) Change $\Delta\mu_l$ in chemical potential of liquid water with pressure:

In order to describe the pressure dependence of the chemical potential of liquids (and solids), a linear approach is sufficient within the desired accuracy (as in the case of the temperature dependence) [cf. Eq. (5.8)]:

$$\mu_l = \mu_l^\ominus + \beta_l(p - p^\ominus) \quad \Rightarrow$$

$$\mu_l - \mu_l^\ominus = \Delta\mu_l = \beta_l(p - p^\ominus).$$

Since the pressure coefficient β_l of liquid water is 18.1 μG Pa^{-1}, we obtain:

$$\Delta\mu_l = (18.1\times10^{-6}\ \text{G Pa}^{-1})\times[(200\times10^3\ \text{Pa}) - (100\times10^3\ \text{Pa})] = \mathbf{1.8\ G}.$$

b) Change $\Delta\mu_g$ in chemical potential of water vapor with pressure:

In the case of an (ideal) gas such as water vapor, a logarithmic approach for the relationship between chemical potential and pressure has to be chosen [cf. Eq. (5.18)]:

$$\mu_g = \mu_g^\ominus + RT^\ominus \ln\frac{p}{p^\ominus}.$$

$$\mu_g - \mu_g^\ominus = \Delta\mu_g = RT^\ominus \ln\frac{p}{p^\ominus}.$$

$$\Delta\mu_g = 8.314\ \text{G K}^{-1}\times298\ \text{K}\times\ln\frac{200\times10^3\ \text{Pa}}{100\times10^3\ \text{Pa}} = 8.314\ \text{G K}^{-1}\times298\ \text{K}\times\ln 2$$

$$= 1720\ \text{G} = \mathbf{1.72\ kG}.$$

Conclusion: In the case of the gas, the increase in the chemical potential is about a thousand times greater than in the case of the liquid (with the same increase in pressure).

2.5.4 Behavior of gases at pressure change

Chemical potential μ_B of carbon dioxide:

The logarithmic approach has also to be used in order to describe the pressure dependence of the chemical potential of carbon dioxide gas:

$$\mu_B = \mu_B^\ominus + RT^\ominus \ln\frac{p}{p^\ominus}.$$

The chemical potential $\mu_B^\ominus$ of carbon dioxide gas at standard conditions is –394.4 kG.

$$\mu_B = (-394.4\times10^3\ \text{G}) + 8.314\ \text{G K}^{-1}\times298\ \text{K}\times\ln\frac{39\ \text{Pa}}{100\times10^3\ \text{Pa}}$$

$$= (-394.4\times10^3\ \text{G}) + 8.314\ \text{G K}^{-1}\times298\ \text{K}\times\ln 0.00039$$

$\mu_B = (-394.4\times10^3\ \text{G}) - (19.4\times10^3\ \text{G}) = \mathbf{-413.8\ kG}$.

2.5.5 "Boiling pressure"

"Boiling pressure" $p_{\text{lg}}^{\ominus}$ of water at room temperature:

The calculation is based on the boiling process of water:

Conversion formula: $H_2O|l \rightarrow H_2O|g$

$\mu^{\ominus}/\text{kG}$: −237.14 −228.58

The "boiling pressure" of water can be calculated analogously to the decomposition pressure of calcium carbonate at the end of Section 5.5 in the textbook "Physical Chemistry from a Different Angle." In the calculation, we ignore the pressure dependence of liquid water because, in comparison to water vapor, it is lower by three orders of magnitude (see Solution 2.5.3). Finally, we arrive at the following formula [cf. Eq. (5.19)]:

$$p_{\text{lg}}^{\ominus} = p^{\ominus} \exp\frac{\mathcal{A}^{\ominus}}{RT^{\ominus}}.$$

Drive $\mathcal{A}^{\ominus}$ of the boiling process:

$$\mathcal{A}^{\ominus} = [(-237.14) - (-228.58)]\ \text{kG} = -8.56\ \text{kG}.$$

$$p_{\text{lg}}^{\ominus} = (100\times10^3\ \text{Pa})\times\exp\frac{(-8.56\times10^3\ \text{G})}{8.314\ \text{G K}^{-1}\times298.15\ \text{K}} = (100\times10^3\ \text{Pa})\times\exp(-3.453) = \mathbf{3165\ Pa}.$$

Literature value: 3166 Pa (from: Cerbe G, Hoffmann H J (1990) Einführung in die Wärmelehre (Introduction to Thermodynamics), 9. edn. Carl Hanser Verlag, München)

2.5.6 Temperature and pressure dependence of a fermentation process

a) Temperature dependence of the fermentation process:

	B		B′	D	D′
Conversion formula:	$C_{12}H_{22}O_{11}|w$ +		$H_2O|l \rightarrow$	$4\ C_2H_6O|w$ +	$4\ CO_2|g$.
$\alpha/\text{G K}^{-1}$:	−435		−70	4×(−148)	4×(−214)

Temperature coefficient α of the process:

$$\alpha = \{[(-435)+(-70)] - [4\times(-148) + 4\times(-214)]\}\ \text{G K}^{-1} = \mathbf{943\ G K^{-1}}.$$

Change $\Delta\mathcal{A}$ in chemical drive with increase of temperature:

$$\mathcal{A} = \mathcal{A}^{\ominus} + \alpha(T - T^{\ominus}) \quad \Rightarrow \quad \text{Eq. (5.3)}$$

$$\mathcal{A} - \mathcal{A}^{\ominus} = \Delta\mathcal{A} = \alpha(T - T^{\ominus}).$$

$$\Delta\mathcal{A} = 943\ \text{G K}^{-1}\times(323\ \text{K} - 298\ \text{K}) = 943\ \text{G K}^{-1}\times25\ \text{K} = 23600\ \text{G} = \mathbf{23.6\ kG}.$$

b) A gas forming reaction such as the fermentation process benefits from the strongly negative temperature coefficients α of gases when the temperature rises; therefore, the chemical drive increases with increasing temperature.

c) Pressure dependence of the fermentation process:

The chemical potentials of the substances present in the liquid phase (cane sugar, water, alcohol) change so little with a change in pressure that their contributions can be neglected; only the increase in chemical potential $\Delta\mu_{D'}$ of the gas carbon dioxide has to be considered.

Increase $\Delta\mu_{D'}$ in chemical potential of the gas carbon dioxide:

$$\Delta\mu_{D'} = RT^{\ominus} \ln \frac{p}{p^{\ominus}} .$$

$$\Delta\mu_{D'} = 8.314\,\text{G K}^{-1} \times 298\,\text{K} \times \ln \frac{1.00\times10^{6}\,\text{Pa}}{100\times10^{3}\,\text{Pa}} = 8.314\,\text{G K}^{-1} \times 298\,\text{K} \times \ln 10$$

$$= 5700\,\text{G} = \mathbf{5.7\,kG} .$$

Change $\Delta\mathcal{A}$ in chemical drive with increase of pressure:

$$\Delta\mathcal{A} = -\nu_{D'} \times \Delta\mu_{D'} = -4 \times 5.7\,\text{kG} = \mathbf{-22.8\,kG} .$$

d) The chemical drive of a gas forming reaction is weakened by an increase in pressure.

2.5.7 Transition temperature and transition pressure

The transition from solid rhombic sulfur (rhom) to solid monoclinic sulfur (mono) should be considered:

	Symbol	α		β
Conversion formula:		S|rhom	→	S|mono

The chemical drive for this process is –75.3 G at 298 K and 100 kPa.

a) Since the chemical drive for the transition from rhombic to monoclinic sulfur is negative, the rhombic phase is stable under standard conditions.

b) Since the temperature coefficient of the chemical potential of rhombic sulfur is $-32.07\,\text{G K}^{-1}$, that of the chemical potential of monoclinic sulfur, however, $-33.03\,\text{G K}^{-1}$, the chemical potential of monoclinic sulfur decreases faster during heating than that of rhombic sulfur, which is stable at standard conditions. For this reason, the $\mu(T)$ curves, which are more or less linear according to the approximation used, must intersect at some point. Above this transition temperature, the monoclinic phase is stable.

Standard transition temperature $T^{\ominus}_{\alpha\beta}$ of sulfur:

$$T^{\ominus}_{\alpha\beta} = T^{\ominus} - \frac{\mathcal{A}^{\ominus}}{\alpha}. \qquad \text{cf. Eq. (5.7)}$$

$$\alpha = [(-32.07) - (-33.03)]\,\text{G K}^{-1} = 0.96\,\text{G K}^{-1}.$$

$$T^{\ominus}_{\alpha\beta} = 298\,\text{K} - \frac{(-75.3\,\text{G})}{0.96\,\text{G K}^{-1}} = 298\,\text{K} + 78\,\text{K} = \mathbf{376\,K}.$$

Literature value: 368.5 K [from: Haynes W M et al (ed) (2015) CRC Handbook of Chemistry and Physics, 96th edn. CRC Press, Boca Raton]

c) The pressure coefficient of the chemical potential of rhombic sulfur is 15.49 μG Pa^{-1}, that of the chemical potential of monoclinic sulfur, however, 16.38 μG Pa^{-1}. Therefore, the chemical potential of monoclinic sulfur increases more quickly with an increase in pressure than that of the rhombic sulfur, which is stable under standard conditions; the corresponding $\mu(p)$ curves cannot intersect.

2.5.8* Pressure dependence of chemical potential and shift of freezing point

a) Calculation of the pressure coefficient β of a substance from its density ρ:

Based on the memory aid "$\beta = V_m = V/n$" mentioned in Section 5.3, we can establish a relationship between the density of a substance and its pressure coefficient:

$$\rho = \frac{m}{V} = \frac{n \times M}{V} \quad \Rightarrow \quad \frac{V}{n} = V_m = \beta = \frac{M}{\rho}.$$

Pressure coefficient β_l of (liquid) water:

$$\beta_l = \frac{M}{\rho_l} = \frac{18.0 \times 10^{-3}\,\text{kg mol}^{-1}}{1000\,\text{kg m}^{-3}} = 18.0 \times 10^{-6}\,\text{m}^3\,\text{mol}^{-1}$$

$$= 18.0 \times 10^{-6}\,\text{G Pa}^{-1} = 18.0\,\mu\text{G Pa}^{-1}.$$

Since the conversion of the units is a little bit tricky, let us take a closer look at it:

$$\text{G Pa}^{-1} = \text{J mol}^{-1}\,\text{Pa}^{-1} = \text{N m mol}^{-1}\,\text{N}^{-1}\,\text{m}^2 = \text{m}^3\,\text{mol}^{-1}.$$

Pressure coefficient β_s of ice:

$$\beta_s = \frac{M}{\rho_s} = \frac{18.0 \times 10^{-3}\,\text{kg mol}^{-1}}{917\,\text{kg m}^{-3}} = 19.6 \times 10^{-6}\,\text{m}^3\,\text{mol}^{-1} = 19.6\,\mu\text{G Pa}^{-1}.$$

Calculation of the change in chemical potential with pressure:

$$\mu - \mu^{\ominus} = \Delta\mu = \beta(p - p^{\ominus}) \qquad \text{(see Solution 2.5.3)}$$

Change $\Delta\mu_l$ in potential of (liquid) water:

$$\Delta\mu_l = \beta_l(p-p^\ominus) = (18.0\times10^{-6}\ \mathrm{G\,Pa^{-1}})\times[(5.0\times10^6\ \mathrm{Pa})-(0.1\times10^6\ \mathrm{Pa})]$$

$$= (18.0\times10^{-6}\ \mathrm{G\,Pa^{-1}})\times(4.9\times10^6\ \mathrm{Pa}) = \mathbf{88.2\ G}.$$

Change $\Delta\mu_s$ in potential of ice:

$$\Delta\mu_s = \beta_s(p-p^\ominus) = (19.6\times10^{-6}\ \mathrm{G\,Pa^{-1}})\times[(5.0\times10^6\ \mathrm{Pa})-(0.1\times10^6\ \mathrm{Pa})]$$

$$= (19.6\times10^{-6}\ \mathrm{G\,Pa^{-1}})\times(4.9\times10^6\ \mathrm{Pa}) = \mathbf{96.0\ G}.$$

b) The chemical potential of ice is the same as that of ice-water [$\mu(H_2O|s) = \mu(H_2O|l)$] at 273 K (0 °C) and standard pressure (100 kPa). However, because of $\beta(H_2O|s) > \beta(H_2O|l)$, the value of $\mu(H_2O|s)$ increases above that of $\mu(H_2O|l)$ as the pressure increases, water becomes the stable phase, and the ice begins to melt (see also Experiment 5.5).

c) Shift ΔT_{sl} of the freezing point of water with increasing pressure:

The shift of the freezing point with increasing pressure results from Equation (5.15):

$$\Delta T_{sl} = -\frac{\beta_s-\beta_l}{\alpha_s-\alpha_l}\Delta p = -\frac{\beta_s-\beta_l}{\alpha}\Delta p.$$

$$\Delta T_{sl} = -\frac{(19.6\times10^{-6}\ \mathrm{G\,Pa^{-1}})-(18.0\times10^{-6}\ \mathrm{G\,Pa^{-1}})}{22.0\ \mathrm{G\,K^{-1}}}\times[(5.0\times10^6\ \mathrm{Pa})-(0.1\times10^6\ \mathrm{Pa})]$$

$$= -\frac{1.6\times10^{-6}\ \mathrm{G\,Pa^{-1}}}{22.0\ \mathrm{G\,K^{-1}}}\times(4.9\times10^6\ \mathrm{Pa}) = \mathbf{-0.36\ K}.$$

2.5.9 Decomposition pressure at elevated temperature

	Symbol B	D	D′
Conversion formula:	$2\,Ag_2O\|s$	$\rightarrow 4\,Ag\|s$	$+\,O_2\|g$
$\mu^\ominus/\mathrm{kG}$:	2×(−11.3)	4×0	0
$\alpha/\mathrm{G\,K^{-1}}$:	2×(−121.0)	4×(−42.6)	−205.2

The procedure corresponds to that used to calculate the decomposition pressure of calcium carbonate (see end of Section 5.5 in the textbook "Physical Chemistry from a Different Angle"). This implies that the pressure dependence of the chemical potentials of the solid substances can be neglected; only the pressure dependence of the chemical potential of the gas oxygen has to be considered. In order to calculate the decomposition pressure for a temperature different from the initial temperature $T^\ominus$, the drive $\mathcal{A}^\ominus$ in the exponent need to be converted to the new temperature. The linear formula for its temperature dependence used so far is generally good enough for this.

$$p_{D'} = p^{\ominus} \exp\frac{\mathcal{A}^{\ominus} + \alpha(T - T^{\ominus})}{RT}.$$

$$\mathcal{A}^{\ominus} = \{[2\times(-11.3)] - [4\times 0 + 0]\}\ \text{kG} = -22.6\ \text{kG}.$$

$$\alpha = \{[2\times(-121.0)] - [4\times(-42.6) + (-205.2)]\}\ \text{G K}^{-1} = 133.6\ \text{G K}^{-1}.$$

$$p_{D'} = (100\times 10^3\ \text{Pa})\times\exp\frac{(-22.6\times 10^3\ \text{G}) + 133.6\ \text{G K}^{-1}\times(400\ \text{K} - 298\ \text{K})}{8.314\ \text{G K}^{-1}\times 400\ \text{K}}$$

$$= (100\times 10^3\ \text{Pa})\times\exp(-2.70) = 6.7\times 10^3\ \text{Pa} = \mathbf{6.7\ kPa}.$$

2.6 Mass Action and Concentration Dependence of Chemical Potential

2.6.1 Concentration dependence of chemical potential

Molar concentration c_B of the glucose solution:

$$c_B = \frac{n_B}{V_S} .$$

Amount of substance n_B:

$$n_B = \frac{m_B}{M_B} = \frac{0.010 \text{ kg}}{180.0 \times 10^{-3} \text{ kg mol}^{-1}} = 0.056 \text{ mol} .$$

$$c_B = \frac{0.056 \text{ mol}}{500 \times 10^{-6} \text{ m}^3} = 112 \text{ mol m}^{-3} = \mathbf{0.112 \text{ kmol m}^{-3}} .$$

Chemical potential μ_B of glucose in the aqueous solution:

The chemical potential of glucose in the aqueous solution results from the mass action equation 1′ [Eq. (6.5)]. In this case, the standard value $\mu^\ominus \equiv \mathring{\mu}(p^\ominus, T^\ominus)$ is used as special basic value:

$$\mu_B = \mu_B^\ominus + RT^\ominus \ln \frac{c_B}{c^\ominus} .$$

$$\mu_B = (-917.0 \times 10^3 \text{ G}) + 8.314 \text{ G K}^{-1} \times 298 \text{ K} \times \ln \frac{0.112 \text{ kmol m}^{-3}}{1 \text{ kmol m}^{-3}}$$

$$= (-917.0 \times 10^3 \text{ G}) + 8.314 \text{ G K}^{-1} \times 298 \text{ K} \times \ln 0.112 = (-917.0 \times 10^3 \text{ G}) - (5.4 \times 10^3 \text{ G})$$

$$= -922.4 \times 10^3 \text{ G} = \mathbf{-922.4 \text{ kG}} .$$

2.6.2* Concentration dependence of a fermentation process

a) Drive $\mathcal{A}^\ominus$ for the fermentation process under standard conditions:

	Symbol	B	B′	D	D′
Conversion formula:		$C_{12}H_{22}O_{11}\vert w$ +	$H_2O\vert l$ →	$4\ C_2H_6O\vert w$ +	$4\ CO_2\vert g$.
$\mu^\ominus$/kG:		−1565	−237	4×(−181)	4×(−394)

$$\mathcal{A}^\ominus = \sum_{\text{initial}} |v_i| \mu_i^\ominus - \sum_{\text{final}} v_j \mu_j^\ominus . \qquad \text{cf. Eq. (4.3)}$$

$$= \{[(-1565) + (-237)] - [4 \times (-181) + 4 \times (-394)]\} \text{ kG} = \mathbf{498 \text{ kG}} .$$

b) Concentration $c_{B,1}$ of sugar in the initial solution:

According to the conversion formula, four alcohol molecules are formed from one sugar molecule, i.e. the initial concentration of sucrose is only one quarter of the final concentration of alcohol:

$$c_{B,1} = \tfrac{1}{4} c_{D,1} = \tfrac{1}{4} \times 1\ \text{kmol m}^{-3} = \mathbf{0.25\ kmol\ m^{-3}}.$$

c) Concentration $c_{B,2}$ of sugar when half of the initial sugar is fermented:

$$c_{B,2} = \tfrac{1}{2} c_{B,1} = \tfrac{1}{2} \times 0.25\ \text{kmol m}^{-3} = \mathbf{0.125\ kmol\ m^{-3}}.$$

Concentration $c_{D,2}$ of alcohol when half of the initial sugar is fermented:

$$c_{D,2} = 4c_{B,2} = 4 \times 0.125\ \text{kmol m}^{-3} = \mathbf{0.5\ kmol\ m^{-3}}.$$

Drive $\mathcal{A}$ at standard temperature:

In order to calculate the drive based on Equation (6.7), only the substances dissolved in water need to be taken into account. As a consequence, we obtain the following formula:

$$\mathcal{A} = \mathcal{A}^{\ominus} + RT^{\ominus} \ln \frac{(c_{B,2}/c^{\ominus})^{|\nu_B|}}{(c_{D,2}/c^{\ominus})^{\nu_D}}.$$

$$\mathcal{A} = (498 \times 10^3\ \text{G}) + 8.314\ \text{G K}^{-1} \times 298\ \text{K} \times \ln \frac{(0.125\ \text{kmol m}^{-3}/1\ \text{kmol m}^{-3})^{|-1|}}{(0.5\ \text{kmol m}^{-3}/1\ \text{kmol m}^{-3})^4}$$

$$= (498 \times 10^3\ \text{G}) + 8.314\ \text{G K}^{-1} \times 298\ \text{K} \times \ln \frac{(0.125)^1}{(0.5)^4}.$$

$$= (498 \times 10^3\ \text{G}) + 8.314\ \text{G K}^{-1} \times 298\ \text{K} \times \ln 2 \approx 500 \times 10^3\ \text{G} \approx \mathbf{500\ kG}.$$

2.6.3 Dependence of the chemical drive of ammonia synthesis on the gas composition

The concentration ratio in the equation for the concentration dependence of the drive [Eq. (6.7)] can be replaced by the (partial) pressure ratio [this approach is analogous to that used in the case of the mass action equations (see Section 6.5)].

Drive $\mathcal{A}$ of ammonia synthesis:

$$\mathcal{A} = \mathcal{A}^{\ominus} + RT^{\ominus} \ln \frac{[p_B/p^{\ominus}]^{|\nu_B|} \times [p_{B'}/p^{\ominus}]^{|\nu_{B'}|}}{[p_D/p^{\ominus}]^{\nu_D}}.$$

$$\mathcal{A} = (32.9 \times 10^3\ \text{G}) + 8.314\ \text{G K}^{-1} \times 298\ \text{K} \times \ln \frac{[25\ \text{kPa}/100\ \text{kPa}]^{|-1)|} \times [52\ \text{kPa}/100\ \text{kPa}]^{|-3)|}}{[75\ \text{kPa}/100\ \text{kPa}]^2}$$

$$\mathcal{A} = (32.9\times10^3\ \text{G})+8.314\ \text{G}\,\text{K}^{-1}\times298\ \text{K}\times\ln\frac{0.25\times0.52^3}{0.75^2}$$

$$= (32.9\times10^3\ \text{G})+8.314\ \text{G}\,\text{K}^{-1}\times298\ \text{K}\times\ln 0.065 = \mathbf{26.0\ kG}.$$

Although the drive has decreased from 32.9 kG to 26.0 kG, it is still positive, i.e. the forward reaction runs spontaneously.

2.6.4 Mass action law (I)

Conversion formula:	$Sn^{2+}\vert w$ +	$I_2\vert w$	⇄ $Sn^{4+}\vert w$ +	$2\ I^-\vert w$
$\mu^\ominus$/kG:	−27.2	+16.4	+2.5	2×(−51.6)

Drive $\mathcal{A}^\ominus$ of the reaction:

$$\mathcal{A}^\ominus = \{[(-27.2)+16.4]-[2.5+2\times(-51.6)]\}\ \text{kG} = \mathbf{+89.9\ kG}.$$

Equilibrium number $\mathcal{K}_c^\ominus$:

The equilibrium number $\mathcal{K}_c^\ominus$ can be calculated from the drive $\mathcal{A}^\ominus$ using Equation (6.18):

$$\mathcal{K}_c^\ominus = \exp\frac{\mathcal{A}^\ominus}{RT}.$$

$$\mathcal{K}_c^\ominus = \exp\frac{89.9\times10^3\ \text{G}}{8.314\ \text{G}\,\text{K}^{-1}\times298\ \text{K}} = \mathbf{5.74\times10^{15}}.$$

Conventional equilibrium constant $K_c^\ominus$:

The conventional equilibrium constant $K_c^\ominus$ is obtained by multiplying the equilibrium number $\mathcal{K}_c^\ominus$ by the dimension factor κ_c [Eq. (6.20)]:

$$K_c^\ominus = \mathcal{K}_c^\ominus\times\kappa_c.$$

Dimension factor κ_c:

$$\kappa_c = (c^\ominus)^{\nu_c} \quad \text{with} \quad \nu_c = \sum\nu_i = -1+(-1)+1+2 = 1.$$

$$\kappa_c = (1\ \text{kmol}\,\text{m}^{-3})^1 = 1\ \text{kmol}\,\text{m}^{-3}.$$

$$K_c^\ominus = (5.74\times10^{15})\times1\ \text{kmol}\,\text{m}^{-3} = \mathbf{5.74\times10^{15}\ kmol\ m^{-3}}.$$

Due to the strongly positive drive and thus the very high value for the equilibrium number (and the conventional equilibrium constant), (nearly) only final products are present in the equilibrium mixture.

2.6.5 Mass action law (II)

a) Chemical drive $\mathring{\mathcal{A}}$ of the formation of BrCl from the elements:

According to Equation (6.22), the drive $\mathring{\mathcal{A}}$ of the reaction at 500 K results in

$$\mathring{\mathcal{A}} = RT \ln \mathring{\mathcal{K}}_c = 8.314\ \mathrm{G\,K^{-1}} \times 500\ \mathrm{K} \times \ln 0.2 = \mathbf{-6.7\ kG}.$$

b) Conventional equilibrium constant $\mathring{K}_c$:

$$\mathring{K}_c = \mathring{\mathcal{K}}_c \times \kappa_c \quad \text{with} \quad \kappa_c = (c^{\ominus})^{\nu_c} = (c^{\ominus})^0 = 1 \quad \text{because of} \quad \nu_c = 1 + 1 - 2 = 0.$$

$$\mathring{K}_c = 0.2 \times 1 = 0.2.$$

Equilibrium concentration c_D of BrCl:

In this example, the mass action law is:

$$\mathring{K}_c = \frac{c_D^{\nu_D}}{c_B^{|\nu_B|} \times c_{B'}^{|\nu_{B'}|}} = \frac{c_D^2}{c_B \times c_{B'}}.$$

Solving for the concentration c_D of BrCl in the equilibrium mixture results in:

$$c_D = \sqrt{\mathring{K}_c \times c_B \times c_{B'}} = \sqrt{0.2 \times 1.45\ \mathrm{mol\,m^{-3}} \times 2.41\ \mathrm{mol\,m^{-3}}} = \mathbf{0.84\ mol\ m^{-3}}.$$

2.6.6 Composition of an equilibrium mixture (I)

Symbol α β

Conversion formula: α-D-Man|w ⇄ β-D-Man|w.

Equilibrium number $\mathcal{K}_c^{\ominus}$:

$$\mathcal{K}_c^{\ominus} = \exp\frac{\mathcal{A}^{\ominus}}{RT^{\circ}} = \exp\frac{(-1.7 \times 10^3\ \mathrm{G})}{8.314\ \mathrm{G\,K^{-1}} \times 298\ \mathrm{K}} = 0.50.$$

Conventional equilibrium constant $K_c^{\ominus}$:

$$K_c^{\ominus} = \mathcal{K}_c^{\ominus} \times \kappa_c \quad \text{with} \quad \kappa = (c^{\ominus})^{\nu_c} = (c^{\ominus})^0 = 1 \quad \text{because of} \quad \nu_c = -1 + 1 = 0.$$

$$K_c^{\ominus} = 0.5 \times 1 = 0.5.$$

"Density of conversion" c_ξ at equilibrium:

The equilibrium concentrations of α-D-mannose ($c_\alpha \equiv c_0 - c_\xi$) and β-D-mannose ($c_\beta \equiv c_\xi$) can be determined in the same manner as the equilibrium concentrations of α-D-glucose and β-D-glucose (see Section 6.4 in the textbook "Physical Chemistry from a Different Angle"). The basis for the calculation is the mass action law [Eq. (6.21)]:

$$K_c^{\ominus} = \frac{c_\beta{}^{\nu_\beta}}{c_\alpha{}^{|\nu_\alpha|}} = \frac{c_\beta}{c_\alpha} = \frac{c_\xi}{c_0 - c_\xi}.$$

Solving for c_ξ results in:

$$c_\xi = \frac{K_c^{\ominus} \times c_0}{K_c^{\ominus} + 1} = \frac{0.5 \times 0.1\ \text{kmol}\,\text{m}^{-3}}{0.5+1} = 0.033\ \text{kmol}\,\text{m}^{-3}.$$

According to this, the concentration c_β ($\equiv c_\xi$) of β-D-mannose is 0,033 kmol m^{-3}, the concentration c_α of α-D-mannose, however, is $c_0 - c_\xi$ = **0.067 kmol m^{-3}** Therefore, at equilibrium 67 % of all dissolved molecules are α-D-mannose molecules and only 33 % are β-D-mannose molecules.

2.6.7* Composition of an equilibrium mixture (II)

The best bet is to create first a kind of table as presented in Section 6.3 (Table 6.1).

Symbol	B		D	D′
	$2\ HI\|g$	⇄	$H_2\|g$ +	$I_2\|g$
$c_{i,0}/\text{mol m}^{-3}$	10		0	0
c_i	$c_0 - 2c_\xi$		c_ξ	c_ξ

"Density of conversion" c_ξ at equilibrium:

The basis is again the mass action law [Eq. (6.21)]:

$$\mathring{K}_c = \frac{c_D{}^{\nu_D} \times c_{D'}{}^{\nu_{D'}}}{c_B{}^{|\nu_B|}} = \frac{c_\xi \times c_\xi}{(c_0 - 2c_\xi)^2} = \left(\frac{c_\xi}{c_0 - 2c_\xi}\right)^2.$$

The equation has to be solved for c_ξ:

$$\frac{c_\xi}{c_0 - 2c_\xi} = \sqrt{\mathring{K}_c}$$

$$c_\xi = \sqrt{\mathring{K}_c} \times (c_0 - 2c_\xi)$$

$$c_\xi = \sqrt{\mathring{K}_c} \times c_0 - 2\sqrt{\mathring{K}_c} \times c_\xi$$

$$c_\xi + 2\sqrt{\mathring{K}_c} \times c_\xi = \sqrt{\mathring{K}_c} \times c_0$$

$$c_\xi \times \left(1 + 2\sqrt{\mathring{K}_c}\right) = \sqrt{\mathring{K}_c} \times c_0$$

$$c_\xi = \frac{\sqrt{\overset{\circ}{K}_c} \times c_0}{1+2\sqrt{\overset{\circ}{K}_c}} = \frac{\sqrt{0.0185} \times 10\ \text{mol m}^{-3}}{1+2\sqrt{0.0185}} = \mathbf{1.07\ mol\ m^{-3}}.$$

$c_\text{D}\ [= c(\text{H}_2)] = c_\xi = \mathbf{1.07\ mol\ m^{-3}}$.

$c_{\text{D}'}\ [= c(\text{I}_2)] = c_\xi = \mathbf{1.07\ mol\ m^{-3}}$.

$c_\text{B}\ [= c(\text{HI})] = c_0 - 2c_\xi = 10\ \text{mol m}^{-3} - (2 \times 1.07\ \text{mol m}^{-3}) = \mathbf{7.86\ mol\ m^{-3}}$.

2.6.8 Decomposition of silver oxide

	Symbol B		D	D′
Conversion formula:	$2\ Ag_2O|s$	$\rightleftarrows$	$4\ Ag|s$ +	$O_2|g$
$\mu^\ominus$/kG:	$2 \times (-11.3)$		4×0	0

In contrast to the *homogeneous* equilibria discussed so far we deal in this case with a *heterogeneous* equilibrium. Therefore, it should be noted that the mass action term $RT \ln c_\text{r}(\text{B})$ is omitted for pure solid substances, i.e., $\mu(\text{B}) = \overset{\circ}{\mu}(\text{B})$; these substances do not appear in the mass action law (and thus also not in the sum ν of the conversion numbers) (see also Section 6.6).

Drive $\mathcal{A}^\ominus$ of the reaction:

$$\mathcal{A}^\ominus = \{[2 \times (-11.3)] - [4 \times 0 + 0]\}\ \text{kG} = \mathbf{-22.6\ kG}.$$

Equilibrium number $\mathcal{K}_p^\ominus$:

$$\mathcal{K}_p^\ominus = \exp\frac{\mathcal{A}^\ominus}{RT^\ominus} = \exp\frac{(-22.6 \times 10^3\ \text{G})}{8.314\ \text{G K}^{-1} \times 298\ \text{K}} = \mathbf{1.1 \times 10^{-4}}. \qquad \text{cf. Eq. (6.18)}$$

Conventional equilibrium constant $K_p^\ominus$:

$$K_p^\ominus = \mathcal{K}_p^\ominus \times \kappa_p \quad \text{with} \quad \kappa_p = (p^\ominus)^{\nu_p} = (100\ \text{kPa})^1 \quad \text{because of} \quad \nu_p = 1.$$

(solid substances do not have to be considered)

$$K_p^\ominus = (1.1 \times 10^{-4}) \times (100 \times 10^3\ \text{Pa}) = \mathbf{11\ Pa}.$$

Decomposition pressure $p_{\text{D}'}$ of silver oxide:

As mentioned, we have to take into account that pure solid substances do not appear in the mass action law. Therefore, we have:

$$K_p^\ominus = p_{\text{D}'} = \mathbf{11\ Pa}.$$

The decomposition pressure is identical to the equilibrium constant.

This calculation is equivalent to the description presented at the end of Section 5.5 [Eq. (5.19)], because it is:

$$p \quad = p^{\ominus} \times \exp\frac{\mathcal{A}^{\ominus}}{RT^{\ominus}} = p^{\ominus} \times \mathcal{K}_p^{\ominus} = K_p^{\ominus}\ .$$

2.6.9 Solubility of silver chloride

Conversion formula: $AgCl|s \rightarrow Ag^+|w + Cl^-|w$

$\mu^{\ominus}/\text{kG}$: −109.8 +77.1 −131.2

This is also a *heterogeneous* equilibrium, more precisely, a *solubility equilibrium.*

Drive $\mathcal{A}^{\ominus}$ of the reaction:

$$\mathcal{A}^{\ominus} \quad = \{[-109.8]-[77.1+(-131.2)]\}\ \text{kG} = -55.7\ \text{kG}\ .$$

Equilibrium number $\mathcal{K}_{\text{sd}}^{\ominus}$:

$$\mathcal{K}_{\text{sd}}^{\ominus} \quad = \exp\frac{\mathcal{A}^{\ominus}}{RT^{\ominus}} = \exp\frac{(-55.7\times10^3\ \text{G})}{8.314\ \text{G}\,\text{K}^{-1}\times298\ \text{K}} = \mathbf{1.7\times10^{-10}}\ .$$

Conventional equilibrium constant $K_{\text{sd}}^{\ominus}$:

$$K_{\text{sd}}^{\ominus} \quad = \mathcal{K}_{\text{sd}}^{\ominus}\times\kappa_c \qquad \text{with} \quad \kappa_c = (c^{\ominus})^{\nu_c} = (1\ \text{kmol}\,\text{m}^{-3})^2 \quad \text{because of} \quad \nu_c = 1+1 = 2\ .$$

$$K_{\text{sd}}^{\ominus} \quad = (1.7\times10^{-10})\times(1\times10^3\ \text{mol}\,\text{m}^{-3})^2 = \mathbf{1.7\times10^{-4}\ mol^2\ m^{-6}}\ .$$

Saturation concentration c_{sd}:

$$K_{\text{sd}}^{\ominus} = c(Ag^+)\times c(Cl^-)$$

This special type of the mass action law is often called the "solubility product."
According to the conversion formula, one AgCl decomposes into one Ag^+ ion and one Cl^- ion. Therefore, we have

$$c_{\text{sd}} \quad = c(Ag^+) = c(Cl^-)\ .$$

Insertion into the mass action law results in

$$K_{\text{sd}}^{\ominus} \quad = {c_{\text{sd}}}^2$$

and consequently, the saturation concentration at 298 K is

$$c_{\text{sd}} \quad = \sqrt{K_{\text{sd}}^{\ominus}} = \sqrt{1.7\times10^{-4}\ \text{mol}^2\ \text{m}^{-6}} = \mathbf{0.013\ mol\ m^{-3}}\ .$$

As expected, the saturation concentration of silver chloride in water is very low.

2.6.10* Solubility product (I)

Conversion formula: $CaF_2|s \rightleftarrows Ca^{2+}|w + 2\,F^-|w$

a) Conventional equilibrium constant $\overset{\circ}{K}_{sd}$:

$$\overset{\circ}{K}_{sd} = \mathcal{K}_{sd}^{\ominus} \times \kappa_c \quad \text{with} \quad \kappa_c = (c^{\ominus})^{\nu_c} = (1\,\text{kmol}\,\text{m}^{-3})^3 \quad \text{because of} \quad \nu_c = 1 + 2 = 3\,.$$

$$\overset{\circ}{K}_{sd} = (3.45 \times 10^{-11}) \times (1 \times 10^3\,\text{mol}\,\text{m}^{-3})^3 = 0.0345\,\text{mol}^3\,\text{m}^{-9}\,.$$

Saturation concentration c_{sd}:

The equation for the conventional equilibrium constant is:

$$\overset{\circ}{K}_{sd} = c(Ca^{2+}) \times c(F^-)^2\,.$$

According to the conversion formula, one CaF_2 decomposes into one Ca^{2+} ion and two F^- ions. Therefore, we have

$$c(Ca^{2+}) = c_{sd} \quad \text{and} \quad c(F^-) = 2c_{sd}\,.$$

Inserting into the mass action law results in:

$$\overset{\circ}{K}_{sd} = c_{sd} \times (2c_{sd})^2 = c_{sd} \times 4c_{sd}^2 = 4c_{sd}^3\,.$$

Subsequently, we solve for c_{sd}:

$$c_{sd} = \sqrt[3]{\overset{\circ}{K}_{sd}/4} = \sqrt[3]{0.0345\,\text{mol}^3\,\text{m}^{-9}/4} = \mathbf{0.21\,mol\,m^{-3}}\,.$$

b) Concentration of Ca^{2+} ions:

We start from the mass action law and solve for $c(Ca^{2+})$:

$$c(Ca^{2+}) = \frac{\overset{\circ}{K}_{sd}}{c(F^-)^2} = \frac{0.0345\,\text{mol}^3\,\text{m}^{-9}}{(0.01 \times 10^{-3}\,\text{mol}\,\text{m}^{-3})^2} = \mathbf{3.45 \times 10^{-4}\,mol\,m^{-3}}\,.$$

As expected, the Ca^{2+} concentration was drastically reduced by the addition of F^- ions.

2.6.11* Solubility product (II)

a) Conversion formula: $Ca_3(PO_4)_2|s \rightleftarrows 3\,Ca^{2+}|w + 2\,PO_4^{3-}|w$.

Conventional equilibrium constant $\overset{\circ}{K}_{sd}$:

$$\overset{\circ}{K}_{sd} = \mathcal{K}_{sd}^{\ominus} \times \kappa_c \quad \text{with} \quad \kappa_c = (c^{\ominus})^{\nu_c} = (1\,\text{kmol}\,\text{m}^{-3})^5 \quad \text{because of} \quad \nu_c = 3 + 2 = 5\,.$$

$$\overset{\circ}{K}_{sd} = (2.07 \times 10^{-33}) \times (1 \times 10^3\,\text{mol}\,\text{m}^{-3})^5 = \mathbf{2.07 \times 10^{-18}\,mol^5\,m^{-15}}\,.$$

Saturation concentration c_{sd}:

In this case, the equation for the conventional equilibrium constant is:

$$K_{\text{sd}}^{\ominus} = c(\text{Ca}^{2+})^3 \times c(\text{PO}_4^{3-})^2 .$$

According to the conversion formula, one $Ca_3(PO_4)_2$ decomposes into three Ca^{2+} ions and two PO_4^{3-} ions. Therefore, we have

$$c(\text{Ca}^{2+}) = 3c_{\text{sd}} \quad \text{and} \quad c(\text{PO}_4^{3-}) = 2c_{\text{sd}} .$$

Inserting into the mass action law,

$$K_{\text{sd}}^{\ominus} = (3c_{\text{sd}})^3 \times (2c_{\text{sd}})^2 = 27{c_{\text{sd}}}^3 \times 4{c_{\text{sd}}}^2 = 108{c_{\text{sd}}}^5 ,$$

and solving for c_{sd} results in:

$$c_{\text{sd}} = \sqrt[5]{K_{\text{sd}}^{\ominus}/108} = \sqrt[5]{(2.07\times10^{-18}\ \text{mol}^5\ \text{m}^{-15})/108}$$

$$= 1.14\times10^{-4}\ \text{mol}\,\text{m}^{-3} = 0.114\ \text{mmol}\,\text{m}^{-3} .$$

<u>Mass m(TCP) of tricalcium phosphate in the solution:</u>

However, the searched value is the mass of tricalcium phosphate (TCP) in 500 mL of water. The molar mass of the phosphate is 310.2×10^{-3} kg mol^{-1}. (Due to the very low concentration of tricalcium phosphate, the volume of water can be set equal to the volume of the solution.)

$$c_{\text{sd}} = \frac{n(\text{TCP})}{V_{\text{S}}} = \frac{m(\text{TCP})}{M(\text{TCP})\times V_{\text{S}}} .$$

We solve for m(TCP):

$$m(\text{TCP}) = c_{\text{sd}} \times M(\text{TCP}) \times V_{\text{S}} .$$

$$m(\text{TCP}) = (0.113\times10^{-3}\ \text{mol}\,\text{m}^{-3})\times(310.2\times10^{-3}\ \text{kg}\,\text{mol}^{-1})\times(500\times10^{-6}\ \text{m}^3)$$

$$= 1.8\times10^{-8}\ \text{kg} = \mathbf{18\ \mu g} .$$

Theoretically, only the extremely low mass of 18 µg of calcium phosphate dissolves in 500 mL of water.

b) In fact, the experimental solubility of the salt is considerably higher as a result of "hydrolysis", meaning the reaction of the anions with water according to

$$\text{PO}_4^{3-}|\text{w} + \text{H}_2\text{O}|\text{l} \rightleftarrows \text{HPO}_4^{2-}|\text{w} + \text{OH}^-|\text{w} \quad \text{and}$$

$$\text{HPO}_4^{2-}|\text{w} + \text{H}_2\text{O}|\text{l} \rightleftarrows \text{PO}_4^{3-}|\text{w} + \text{OH}^-|\text{w} .$$

This so-called acid-base reaction reduces the anion concentration available to satisfy $K_{\text{sd}}^{\ominus}$. Therefore, the equilibrium shifts toward the products, i.e. more $Ca_3(PO_4)_2$ has to dissolve.

2.6.12 Oxygen content in water

Conversion formula: $\text{O}_2|\text{g} \rightleftarrows \text{O}_2|\text{w}$

Molar concentration $c(\mathrm{B|w})$ of oxygen in air-saturated water:

We start from HENRY's law [Eq. (6.37)] in order to calculate the oxygen content in the garden pond:

$$\overset{\circ}{K}_{\mathrm{gd}} = K_{\mathrm{H}}^{\ominus} = \frac{c(\mathrm{B|w})}{p(\mathrm{B|g})}.$$

Solving for $c(\mathrm{B|w})$ results in:

$$c(\mathrm{B|w}) = \overset{\circ}{K}_{\mathrm{gd}} \times p(\mathrm{B|g}).$$

The partial pressure of oxygen in the ambient air is about 21 kPa (since the percentage of oxygen in the air is about 21 %).

$$c(\mathrm{B|w}) = (1.3\times10^{-5}\ \mathrm{mol\,m^{-3}\,Pa^{-1}})\times(21\times10^{3}\ \mathrm{Pa}) = \mathbf{0.27\ mol\,m^{-3}}.$$

Mass concentration $\beta(\mathrm{B|w})$ of oxygen in air-saturated water:

The molar mass of oxygen is $32.0\times10^{-3}\ \mathrm{kg\,mol^{-1}}$.

$$c(\mathrm{B|w}) = \frac{n(\mathrm{B|w})}{V_{\mathrm{S}}} = \frac{m(\mathrm{B|w})}{M(\mathrm{B})\times V_{\mathrm{S}}} = \frac{\beta(\mathrm{B|w})}{M(\mathrm{B})}.$$

Solving for $\beta(\mathrm{B|w})$ results in:

$$\beta(\mathrm{B|w}) = c(\mathrm{B|w})\times M(\mathrm{B}) = 0.27\ \mathrm{mol\,m^{-3}}\times(32.0\times10^{-3}\ \mathrm{kg\,mol^{-1}})$$

$$= 8.6\times10^{-3}\ \mathrm{kg\,m^{-3}} = \mathbf{8.6\ mg\,L^{-1}}.$$

2.6.13 Solubility of CO_2

a) Conversion formula: $CO_2|g \rightleftarrows CO_2|w$

$\mu^{\ominus}/\mathrm{kG}$: −394.4 −386.0

Drive $\mathcal{A}^{\ominus}$ of the reaction:

$$\mathcal{A}^{\ominus} = [(-394.4)-(-386.0)]\ \mathrm{kG} = -8.4\ \mathrm{kG}.$$

Equilibrium number $\mathcal{K}_{\mathrm{gd}}^{\ominus}$:

$$\mathcal{K}_{\mathrm{gd}}^{\ominus} = \exp\frac{\mathcal{A}^{\ominus}}{RT^{\ominus}} = \exp\frac{(-8.4\times10^{3}\ \mathrm{G})}{8.314\ \mathrm{G\,K^{-1}}\times298\ \mathrm{K}} = 0.034.$$

HENRY's law constant $\overset{\circ}{K}_{\mathrm{gd}}$:

$$\overset{\circ}{K}_{\mathrm{gd}} = \mathcal{K}_{\mathrm{gd}}^{\ominus}\frac{c^{\ominus}}{p^{\ominus}} = 0.034\times\frac{1\times10^{3}\ \mathrm{mol\,m^{-3}}}{100\times10^{3}\ \mathrm{Pa}} = \mathbf{3.4\times10^{-4}\ mol\,m^{-3}\,Pa^{-1}}.$$

Concentration $c_1(B|w)$ of carbon dioxide in water:

$$K^{\ominus}_{gd} = \frac{c_1(B|w)}{p_1(B|g)}.$$

Solving for $c_1(B|w)$ results in:

$$c_1(B|w) = K^{\ominus}_{gd} \times p_1(B|g) = (3.4\times10^{-4}\ \text{mol m}^{-3}\ \text{Pa}^{-1})\times(100\times10^{3}\ \text{Pa})$$

$$= \mathbf{34\ mol\ m^{-3}}.$$

b) Amount $n_1(B)$ of carbon dioxide in water:

$$c_1(B|w) = \frac{n_1(B)}{V_S} \quad \Rightarrow$$

$$n_1(B) = c_1(B|w)\times V_S = 34\ \text{mol m}^{-3}\times(1\times10^{-3}\ \text{m}^3) = 34\times10^{-3}\ \text{mol} = 34\ \text{mmol}.$$

Volume $V_1(B|g)$ of the gas:

Based on the hint from Exercise 1.1.4, we know that one mole of any gas, be it pure or mixed, has a volume V of about 24.8 L at standard conditions. Therefore, the amount $n_1(B)$ of 34×10^{-3} moles of CO_2 corresponds to a volume $V_1(B|g)$ of the gas of $(34\times10^{-3}\ \text{moles})\times(24.8\times10^{-3}\ \text{m}^3\ \text{mol}^{-1}) = 0.84\times10^{-3}\ \text{m}^3 = \mathbf{0.84\ L}$.

c) Concentration $c_2(B|w)$ of carbon dioxide in water:

$$c_2(B|w) = K^{\ominus}_{gd} \times p_2(B|g) = (3.4\times10^{-4}\ \text{mol m}^{-3}\ \text{Pa}^{-1})\times(300\times10^{3}\ \text{Pa}) = 102\ \text{mol m}^{-3}.$$

Amount $n_2(B)$ of carbon dioxide in water:

$$n_2(B) = c_2(B|w)\times V_S = 102\ \text{mol m}^{-3}\times(1\times10^{-3}\ \text{m}^3) = 102\times10^{-3}\ \text{mol} = 102\ \text{mmol}.$$

Volume $V_2(B|g)$ of the gas:

The amount $n_2(B)$ of 102×10^{-3} moles of CO_2 corresponds to a volume $V_2(B|g)$ of the gas of $(102\times10^{-3}\ \text{moles})\times(24.8\times10^{-3}\ \text{m}^3\ \text{mol}^{-1}) = 2.53\times10^{-3}\ \text{m}^3 = 2.53$ L.

Volume $V_O(B|g)$ of gas that "bubbles out" of the liquid:

In order to produce a gas bubble, the pressure inside the bubble has to exceed the atmospheric pressure of about 100 kPa. Therefore, only the volume of CO_2 can "bubble out" that corresponds to the amount of CO_2 dissolved in the pressure range above 100 kPa. In our example we have

$$V_O(B|g) = V_2(B|g) - V_1(B|g) = 2.53\times10^{-3}\ \text{m}^3 - 0.84\times10^{-3}\ \text{m}^3$$

$$= 1.69\times10^{-3}\ \text{m}^3 = \mathbf{1.69\ L}.$$

The remaining CO_2 escapes much more slowly by diffusion without producing bubbles.

1.6.14* Absorption of CO_2 in lime water

	Symbol B	B′	B″	D	D′
Conversion formula:	$Ca^{2+}|w$ +	$2\ OH^-|w$	$+\ CO_2|g \rightleftarrows$	$CaCO_3|s$ +	$H_2O|l$
$\mu^\ominus$/kG:	−553.6	2×(−157.2)	−394.4	−1128.8	−237.1

Drive $\mathcal{A}^\ominus$ of the reaction:

$$\mathcal{A}^\ominus = \{[(-553.6)+2\times(-157.2)+(-394.4)]-[(-1128.8)+(-237.1)]\}\ \text{kG} = \mathbf{+103.5\ kG}.$$

Equilibrium number $\mathcal{K}_{pc}^\ominus$:

$$\mathcal{K}_{pc}^\ominus = \exp\frac{\mathcal{A}^\ominus}{RT^\ominus} = \exp\frac{103.5\times10^3\ \text{G}}{8.314\ \text{G K}^{-1}\times 298\ \text{K}} = \mathbf{1.39\times10^{18}}.$$

Conventional equilibrium constant $K_{pc}^\ominus$:

$K_{pc}^\ominus$ represents a so-called “mixed” equilibrium constant. Therefore, we have to consider both a dimension factor κ_c and a dimension factor κ_p:

$$K_{pc}^\ominus = \mathcal{K}_{pc}^\ominus\times\kappa_c\times\kappa_p, \quad \text{with } \kappa_c = (c^\ominus)^{\nu_c} = (1\ \text{kmol m}^{-3})^{-3} \text{ because of } \nu_c = -1 + (-2) = -3$$

$$\text{and } \kappa_p = (p^\ominus)^{\nu_p} = (100\ \text{kPa})^{-1} \text{ because of } \nu_p = -1.$$

The sum ν_c of the conversion numbers for the dissolved substances includes the conversion numbers of the initial substances $Ca^{2+}|w$ and $OH^-|w$, the sum ν_p of the conversion numbers for the gases only the gas carbon dioxide.

$$K_{pc}^\ominus = (1.39\times10^{18})\times(1\times10^3\ \text{mol m}^{-3})^{-3}\times(100\times10^3\ \text{Pa})^{-1} = \mathbf{1.39\times10^4\ mol^{-3}\ m^9\ Pa^{-1}}.$$

Partial pressure $p_{B''}$ of the carbon dioxide:

For the mass action law we obtain

$$K_{pc}^\ominus = \frac{1}{c_B\times c_{B'}^2\times p_{B''}},$$

since both the precipitate of $Ca(OH)_2$ as a solid as well as water as a solvent do not appear in the mass action law.

We have to take into account that there are twice as many OH^- ions in the calcium hydroxide solution as Ca^{2+} ions:

$$c_B = c, \quad c_{B'} = 2c \quad \text{and therefore}$$

$$K_{pc}^\ominus = \frac{1}{c\times(2c)^2\times p_{B''}} = \frac{1}{4c^3\times p_{B''}}.$$

Solving for $p_{B''}$ results in:

$$p_{B''} = \frac{1}{4c^3 \times K_{pc}^{\ominus}} = \frac{1}{4\times(20\ \text{mol}\,\text{m}^{-3})^3\times(1.39\times10^4\ \text{mol}^{-3}\,\text{m}^9\,\text{Pa}^{-1})} = \mathbf{2.25\times10^{-9}\ Pa}\,.$$

2.6.15 Distribution of iodine

Conversion formula: $I_2|\text{w} \rightleftarrows I_2|\text{Chl}$

$\mu^{\ominus}/\text{kG}$: 16.4 4.2

a) Drive $\mathcal{A}^{\ominus}$ of the reaction:

$$\mathcal{A}^{\ominus} = (16.4-4.2)\ \text{kG} = 12.2\ \text{kG}\,.$$

Equilibrium number $\mathcal{K}_{\text{dd}}^{\ominus}$:

The equilibrium number for the distribution of iodine between the two practically immiscible liquid phases water and chloroform results according to Equation (6.38) in:

$$\mathcal{K}_{\text{dd}}^{\ominus} = \exp\frac{\mathcal{A}^{\ominus}}{RT^{\ominus}} = \exp\frac{12.2\times10^3\ \text{G}}{8.314\ \text{G}\,\text{K}^{-1}\times298\ \text{K}} = 138\,.$$

NERNST's distribution coefficient $K_{\text{dd}}^{\ominus}$:

$$K_{\text{dd}}^{\ominus} = \mathcal{K}_{\text{dd}}^{\ominus}\times\kappa_c \quad \text{with} \quad \kappa_c = (c^{\ominus})^{\nu_c} = (c^{\ominus})^0 = 1 \quad \text{because of} \quad \nu_c = -1+1 = 0\,.$$

$$K_{\text{dd}}^{\ominus} = 138\times1 = 138\,.$$

Since the dimension factor κ_c is equal to 1, the equilibrium number $\mathcal{K}_{\text{dd}}^{\ominus}$ and NERNST's distribution coefficient $K_{\text{dd}}^{\ominus}$ are identical.

Fraction a_{W} of iodine in water:

We start from NERNST's distribution law:

$$K_{\text{dd}}^{\ominus} = \frac{c(\text{B}|\text{Chl})}{c(\text{B}|\text{w})}\,.$$

Since the volumes of the aqueous phase and the chloroform phase should be equal, we can replace the concentration ratio by the amount ratio:

$$K_{\text{dd}}^{\ominus} = \frac{n(\text{B}|\text{Chl})\times V_{\text{S,w}}}{V_{\text{S,Chl}}\times n(\text{B}|\text{w})} = \frac{n(\text{B}|\text{Chl})}{n(\text{B}|\text{w})}\,.$$

Solving for $n(\text{B}|\text{Chl})$ results in:

$$n(\text{B}|\text{Chl}) = K_{\text{dd}}^{\ominus}\times n(\text{B}|\text{w})\,.$$

Eventually, we are looking for the fraction a_{W} of iodine in water:

$$a_W = \frac{n(B|w)}{n_{total}} = \frac{n(B|w)}{n(B|w)+n(B|Chl)} = \frac{n(B|w)}{n(B|w)+K_{dd}^{\ominus}\times n(B|w)} = \frac{1}{1+K_{dd}^{\ominus}}.$$

$$a_W = \frac{1}{1+138} = 0.0072 = \mathbf{0.72\ \%}.$$

Only 0.72 % of the original iodine remains in the aqueous phase.

b) Fraction $a_{W,1}$ of iodine in water after the first extraction:

We can use the same approach as in part a):

$$K_{dd}^{\ominus} = \frac{n(B|Chl)_1 \times V_{S,w}}{V_{S,Chl}\times n(B|w)_1} = \frac{n(B|Chl)_1 \times V_{S,w}}{\frac{1}{2}V_{S,Chl}\times n(B|w)_1} = \frac{2\times n(B|Chl)_1}{n(B|w)_1}$$

$$n(B|Chl)_1 = \tfrac{1}{2}K_{dd}^{\ominus}\times n(B|w)_1$$

$$a_{W,1} = \frac{n(B|w)_1}{n_{total}} = \frac{n(B|w)_1}{n(B|w)_1 + n(B|Chl)_1}$$

$$= \frac{n(B)_1}{n(B|w)_1 + \frac{1}{2}K_{dd}^{\ominus}\times n(B|w)_1} = \frac{1}{1+\frac{1}{2}K_{dd}^{\ominus}}.$$

$$a_{W,1} = \frac{1}{1+\frac{1}{2}\times 138} = 0.0143.$$

Fraction $a_{W,2}$ of iodine in water after the second extraction:

After the first extraction, the total amount of iodine in the aqueous solution in the separating funnel is only $n_{total,1} = a_{W,1}\times n_{total}$. Therefore, we obtain after the second extraction:

$$\frac{n(B|w)_2}{n_{total,1}} = \frac{n(B|w)_2}{a_{W,1}\times n_{total}}.$$

The fraction $a_{W,2}$ of iodine in water after the second extraction (related to the amount n_{total} of iodine in the initial solution) then results in:

$$a_{W,2} = \frac{n(B|w)_2}{n_{total}} = a_{W,1}\times \frac{n(B|w)_2}{n_{total,1}}.$$

The second factor can now be calculated analogously to the procedure in the first part:

$$\frac{n(B|w)_2}{n_{total,1}} = \frac{n(B|w)_2}{n(B|w)_2 + \frac{1}{2}K_{dd}^{\ominus}\times n(B|w)_2} = \frac{1}{1+\frac{1}{2}K_{dd}^{\ominus}} = a_{W,1}$$

We finally obtain for $a_{W,2}$:

$$a_{W,2} = a_{W,1}\times a_{W,1} = 0.0143\times 0.0143 = 0.00020 = \mathbf{0.02\ \%}.$$

It is much more efficient to use two smaller volumes of solvent (multiple extraction) than all the solvent in one large volume (single extraction).

1.6.16* Distribution of iodine for advanced students

Conversion formula: $I_2|w \rightleftarrows I_2|CS_2$

Mass m_x of iodine in the aqueous phase:

The mass of iodine should be determined that remains in the aqueous phase after the extraction with carbon disulfide:

$$K_{dd}^{\ominus} = \frac{c(B|CS_2)}{c(B|w)} = \frac{n(B|CS_2)}{V_{S,CS_2}} \times \frac{V_{S,w}}{n(B|w)}$$

$$= \frac{m(B|CS_2)}{M(B)\times V_{S,CS_2}} \times \frac{M(B)\times V_{S,w}}{m(B|w)} = \frac{m(B|CS_2)}{V(CS_2)} \times \frac{V(H_2O)}{m(B|w)}.$$

Let m_x represent the mass of iodine in the aqueous solution. Then a mass ($m_{total} - m_x$) of iodine remains in the organic phase. Therefore, we have:

$$K_{dd}^{\ominus} = \frac{m_{total} - m_x}{V_{S,CS_2}} \times \frac{V_{S,w}}{m_x}.$$

Solving for m_x yields:

$$K_{dd}^{\ominus} \times V_{S,CS_2} \times m_x = V_{S,w} \times (m_{total} - m_x) = V_{S,w} \times m_{total} - V_{S,w} \times m_x$$

$$[K_{dd}^{\ominus} \times V_{S,CS_2} + V_{S,w}] \times m_x = V_{S,w} \times m_{total}$$

$$m_x = \frac{V_{S,w} \times m_{total}}{K_{dd}^{\ominus} \times V_{S,CS_2} + V_{S,w}}.$$

$$m_x = \frac{(500\times10^{-6}\ m^3)\times(500\times10^{-6}\ kg)}{588\times(50\times10^{-6}\ m^3)+(500\times10^{-6}\ m^3)} = \frac{0.25\times10^{-6}\ m^3\ kg}{0.0299\ m^3}$$

$$= 8.4\times10^{-6}\ kg = \mathbf{8.4\ mg}.$$

Only 8.4 mg of the initial iodine remains in the aqueous phase.

2.6.17 BOUDOUARD reaction

	Symbol B	B′	D
Conversion formula:	C|Graphite +	CO_2|g $\rightleftarrows$	2 CO|g
$\mu^{\ominus}$/ kG:	0	−394.4	2×(−137.2)
α / G K^{-1}:	−5.7	−213.8	2×(−197.7)

a) Drive $\mathcal{A}^{\ominus}$ of the reaction:

$$\mathcal{A}^{\ominus} = \{[0+(-394.4)]-[2\times(-137.2)]\}\ \text{kG} = -120.0\ \text{kG}\,.$$

Equilibrium number $\mathcal{K}_p^{\ominus}$:

$$\mathcal{K}_p^{\ominus} = \exp\frac{\mathcal{A}^{\ominus}}{RT^{\ominus}} = \exp\frac{(-120.0\times10^3\ \text{G})}{8.314\ \text{G}\,\text{K}^{-1}\times298\ \text{K}} = 9.2\times10^{-22}\,.$$

Conventional equilibrium constant $K_p^{\ominus}$:

$$K_p^{\ominus} = \mathcal{K}_p^{\ominus}\times\kappa_p \quad \text{with} \quad \kappa_p = (p^{\ominus})^{\nu_c} = (100\ \text{kPa})^1 \quad \text{because of} \quad \nu_p = -1+2 = 1\,.$$

$$K_p^{\ominus} = (9.2\times10^{-22})\times(100\times10^3\ \text{Pa}) = \mathbf{9.2\times10^{-17}\ Pa}\,.$$

Due to the strongly negative value for the chemical drive and thus the very small value for the equilibrium number (and the conventional equilibrium constant), the equilibrium lies far to the left side. The equilibrium mixture contains (almost) only the reactants.

b) Temperature coefficient α of the reaction:

$$\alpha = \{[-5.7+(-213.8)]-[2\times(-197.7]\}\ \text{G}\,\text{K}^{-1} = 175.9\ \text{G}\,\text{K}^{-1}\,.$$

Equilibrium number $\overset{\circ}{\mathcal{K}}_p$ at 1073 K:

The equilibrium number $\overset{\circ}{\mathcal{K}}_p$ at a temperature of 1073 K can be calculated using Equation (6.40):

$$\overset{\circ}{\mathcal{K}}_p = \exp\frac{\mathcal{A}^{\ominus}+\alpha(T-T^{\ominus})}{RT}\,.$$

$$\overset{\circ}{\mathcal{K}}_p = \exp\frac{(-120.0\times10^3\ \text{G})+175.9\ \text{G}\,\text{K}^{-1}(1073\ \text{K}-298\ \text{K})}{8.314\ \text{G}\,\text{K}^{-1}\times1073\ \text{K}} = 6.2\,.$$

Conventional equilibrium constant $\overset{\circ}{K}_p$ at 1073 K:

$$\overset{\circ}{K}_p = 6.2\times(100\times10^3\ \text{Pa}) = \mathbf{620\ kPa}\,.$$

The equilibrium constant is considerably greater than 1, i.e. the final product carbon monoxide dominates in the equilibrium mixture.

c) Partial pressure p_D of carbon monoxide in the equilibrium mixture:

We start from the mass action law:

$$\overset{\circ}{K}_p = \frac{p_D^2}{p_B}\,.$$

Solving for p_D yields the desired partial pressure of carbon monoxide at equilibrium:

$$p_D = \sqrt{\mathring{K}_p(1073\,\text{K}) \times p_B} = \sqrt{617\,\text{kPa} \times 30\,\text{kPa}} = \mathbf{136\,kPa}.$$

As expected, carbon monoxide dominates in the equilibrium mixture.

d) If the resulting carbon monoxide is continuously removed out of the system, carbon reacts completely with carbon dioxide to carbon monoxide.

2.6.18* BOUDOUARD reaction for advanced students

Symbol	B	B′	D
	C|Graphite +	CO_2|g ⇄	2 CO|g
$p_{i,0}$/kPa:		100	0
p_i:		$p_0 - p_x$	$2p_x$

When a fraction x of carbon dioxide gas decomposes, its initial pressure p_0 is reduced by p_x. However, for every CO_2 molecule that disappears, two CO molecules are produced. Consequently, the partial pressure of carbon monoxide is $2p_x$.

Calculation of p_x at equilibrium:

$$\mathring{K}_p = \frac{p_D^{\nu_D}}{p_{B'}^{|\nu_{B'}|}} = \frac{(2p_x)^2}{p_0 - p_x} = \frac{4p_x^2}{p_0 - p_x}.$$

By rearranging we arrive at the so-called normal form of the quadratic equation, $x^2 + a \times x + b = 0$.

$$4p_x^2 = \mathring{K}_p(p_0 - p_x) = \mathring{K}_p \times p_0 - \mathring{K}_p \times p_x,$$

$$4p_x^2 + \mathring{K}_p \times p_x - \mathring{K}_p \times p_0 = 0,$$

$$p_x^2 + \tfrac{1}{4}\mathring{K}_p \times p_x - \tfrac{1}{4}\mathring{K}_p \times p_0 = 0.$$

The solutions for the normal form of the quadratic equation are:

$$x_{1,2} = -\frac{a}{2} \pm \sqrt{\left(\frac{a}{2}\right)^2 - b}.$$

Since there are no negative pressures, we only need to take the positive square root into consideration in the following calculation:

$$p_x = -\frac{\mathring{K}_p}{8} + \sqrt{\left(\frac{\mathring{K}_p}{8}\right)^2 + \frac{\mathring{K}_p \times p_0}{4}}.$$

Inserting the values results in:

$$p_x = -\frac{81\times10^3\ \text{Pa}}{8} + \sqrt{\left(\frac{81\times10^3\ \text{Pa}}{8}\right)^2 + \frac{(81\times10^3\ \text{Pa})\times(100\times10^3\ \text{Pa})}{4}}$$

$$= -10.125\times10^3\ \text{Pa} + \sqrt{2.128\times10^9\ \text{Pa}^2} = 36\times10^3\ \text{Pa} = 36\ \text{kPa}\,.$$

$$p_{\text{B}'} = p_0 - p_x = 100\ \text{kPa} - 36\ \text{kPa} = \mathbf{64\ kPa}\,.$$

$$p_{\text{D}} = 2p_x = 2\times36\ \text{kPa} = \mathbf{72\ kPa}\,.$$

$$p_{\text{ges}} = p_{\text{B}} + p_{\text{D}} = 64\ \text{kPa} + 72\ \text{kPa} = \mathbf{136\ kPa}\,.$$

2.7 Consequences of Mass Action: Acid-Base Reactions

2.7.1 Proton potential of strong acid-base pairs (I)

a) Proton potential $\mu_{p,1}$ in the hydrochloric acid:

In aqueous solution, the acid of a strongly acidic pair such as hydrochloric acid almost completely loses its protons to the water. If one wants to determine the proton potential of such a strongly acidic pair at arbitrary dilution, it is therefore sufficient to just consider the acid-base pair H_3O^+/H_2O [cf. Eq. (7.3)] (The H_3O^+ concentration is identical to to the HCl concentration due to complete dissociation.):

$$\mu_{p,1} = \mu_p^{\ominus}(H_3O^+/H_2O) + RT^{\ominus}\ln(c_{H_3O^+,1}/c^{\ominus}) .$$

$$\mu_{p,1} = 0\,G + 8.314\,G\,K^{-1} \times 298\,K \times \ln(0.50\,kmol\,m^{-3}/1\,kmol\,m^{-3}) = \mathbf{-1.72\,kG} .$$

The proton potential is thus noticeably lower than the standard value of 0 kG, which is valid for a H_3O^+ concentration of 1 kmol m^{-3} in aqueous solution.

b) Amount of oxonium ions $n_{H_3O^+,2}$ in the hydrochloric acid in the pipette:

$$c_{H_3O^+,1} = \frac{n_{H_3O^+,2}}{V_2} \quad \Rightarrow \quad n_{H_3O^+,2} = c_{H_3O^+,1} \times V_2 .$$

$$n_{H_3O^+,2} = 500\,mol\,m^{-3} \times (50 \times 10^{-6}\,m^3) = 0.025\,mol .$$

Amount of hydroxide ions $n_{OH^-,3}$ in the sodium hydroxide solution in the beaker:

$$c_{OH^-,3} = \frac{n_{OH^-,3}}{V_3} \quad \Rightarrow \quad n_{OH^-,3} = c_{OH^-,3} \times V_3 .$$

$$n_{OH^-,3} = 200\,mol\,m^{-3} \times (50 \times 10^{-6}\,m^3) = 0.010\,mol .$$

Amount of oxonium ions $n_{H_3O^+,3}$ in the mixture:

By mixing the hydrochloric acid with the sodium hydroxide solution, the surplus of oxonium ions is reduced:

$$n_{H_3O^+,3} = n_{H_3O^+,2} - n_{OH^-,3} .$$

$$n_{H_3O^+,3} = 0.025\,mol - 0.010\,mol = 0.015\,mol .$$

Volume V_M of the mixture:

$$V_M = V_2 + V_3 .$$

$$V_M = (50 \times 10^{-6}\,m^3) + (50 \times 10^{-6}\,m^3) = 100 \times 10^{-6}\,m^3 .$$

Concentration of oxonium ions $c_{H_3O^+,3}$ in the mixture:

$$c_{H_3O^+,3} = \frac{n_{H_3O^+,3}}{V_M}.$$

$$c_{H_3O^+,3} = \frac{0.015 \text{ mol}}{100\times10^{-6} \text{ m}^3} = 150 \text{ mol m}^{-3} = \mathbf{0.150 \text{ kmol m}^{-3}}.$$

In a nutshell, the calculation can be presented as follows:

$$c_{H_3O^+,3} = \frac{c_{H_3O^+,1}\times V_2 - c_{OH^-,3}\times V_3}{V_2+V_3}.$$

$$c_{H_3O^+,3} = \frac{500 \text{ mol m}^{-3}\times(50\times10^{-6} \text{ m}^3) - 200 \text{ mol m}^{-3}\times(50\times10^{-6} \text{ m}^3)}{(50+50)\times10^{-6} \text{ m}^3}$$

$$= 150 \text{ mol m}^{-3}.$$

Proton potential $\mu_{p,3}$ in the mixture:

$$\mu_{p,3} = \mu_p^{\ominus}(H_3O^+/H_2O) + RT^{\ominus}\ln(c_{H_3O^+,3}/c^{\ominus}).$$

$$\mu_{p,3} = 0 \text{ G} + 8.314 \text{ G K}^{-1}\times 298 \text{ K}\times\ln(0.150 \text{ kmol m}^{-3}/1 \text{ kmol m}^{-3}) = \mathbf{-4.70 \text{ kG}}.$$

Since the concentration of oxonium ions in the mixture is considerably lower than in the initial hydrochloric acid solution, the proton potential also shows a noticeably lower value compared to that in part a) of the exercise.

2.7.2 Proton potential of strong acid-base pairs (II)

a) Concentration of hydroxide ions $c_{OH^-,1}$ in the sodium hydroxide solution:

$$c_{OH^-,1} = \frac{n_{OH^-,1}}{V_1}.$$

Amount $n_{OH^-,1}$:

Sodium hydroxide completely dissociates in aqueous solution into Na^+ and OH^- ions. Therefore, the amount n_{NaOH} of sodium hydroxide corresponds to the amount n_{OH^-} of hydroxide ions.

$$n_{OH^-,1} = n_{NaOH} = \frac{m_{NaOH}}{M_{NaOH}} = \frac{12.0\times10^{-3} \text{ kg}}{40.0\times10^{-3} \text{ kg mol}^{-1}} = 0.30 \text{ mol}.$$

$$c_{OH^-,1} = \frac{0.30 \text{ mol}}{500\times10^{-6} \text{ m}^3} = 600 \text{ mol m}^{-3} = \mathbf{0.60 \text{ kmol m}^{-3}}.$$

Proton potential $\mu_{p,1}$ in the sodium hydroxide solution:

Determining the proton potential of a dilute strongly alkaline pair follows the same pattern as for a dilute strongly acidic pair, except that in this case the pair H_2O/OH^- (instead of H_3O^+/H_2O) has to be considered [Eq. (7.4)]:

$$\mu_{p,1} = \mu_p^{\ominus}(H_2O/OH^-) - RT^{\ominus}\ln(c_{OH^-}/c^{\ominus}).$$

$$\mu_{p,1} = (-80\times10^3\ G) - 8.314\ G\,K^{-1}\times298\ K\times\ln(0.60\ kmol\,m^{-3}/1\ kmol\,m^{-3}),$$

$$= \mathbf{-78.7\ kG}.$$

The proton potential is noticeably higher than the standard value of –80 kG.

b) Concentration of hydroxide ions $c_{OH^-,3}$ in the mixture:

One proceeds analogously to Solution 2.7.1 b).

$$c_{OH^-,3} = \frac{c_{OH^-,1}\times V_2 - c_{H_3O^+,3}\times V_3}{V_2+V_3}.$$

$$c_{OH^-,3} = \frac{600\ mol\,m^{-3}\times(50\times10^{-6}\ m^3) - 100\ mol\,m^{-3}\times(150\times10^{-6}\ m^3)}{(50+150)\times10^{-6}\ m^3}$$

$$= 75\ mol\,m^{-3} = \mathbf{0.075\ kmol\,m^{-3}}.$$

Proton potential $\mu_{p,3}$ in the mixture:

$$\mu_{p,3} = \mu_p^{\ominus}(H_2O/OH^-) - RT^{\ominus}\ln(c_{OH^-,3}/c^{\ominus}).$$

$$\mu_{p,3} = (-80\times10^3\ G) - 8.314\ G\,K^{-1}\times298\ K\times\ln(0.075\ kmol\,m^{-3}/1\ kmol\,m^{-3})$$

$$= \mathbf{-73.6\ kG}.$$

Compared to the initial alkaline solution, the proton potential has become significantly higher.

2.7.3 Proton potential of weak acid-base pairs (I)

a) Proton potential μ_p in the lactic acid solution:

If one wants to calculate the proton potential of the acid of a weak acid-base pair [such as the lactic acid/lactate pair (HLac/Lac⁻)] in aqueous solution, due to the incomplete proton transfer to the pair H_3O^+/H_2O, both pairs must be taken into account. Assuming that the acid of a weakly acidic pair, dissolved in pure water, is only dissociated to a very small extent, we finally obtain the following relationship for the proton potential [Eq. (7.6)]:

$$\mu_p = \tfrac{1}{2}\times\left[\mu_p^{\ominus}(HLac/Lac^-) + \mu_p^{\ominus}(H_3O^+/H_2O) + RT^{\ominus}\ln(c_{HLac}/c^{\ominus})\right].$$

$$\mu_p = \tfrac{1}{2}\times[(-22\times10^3\ \text{G}) + 0\ \text{G} + 8.314\ \text{G K}^{-1}\times 298\ \text{K}\times\ln(0.30\ \text{kmol m}^{-3}/1\ \text{kmol m}^{-3})]$$

$$= \mathbf{-12.5\ kG}\,.$$

The proton potential is noticeably higher than the standard value of –22 kG. This can be explained by the fact that the lactate concentration is negligibly low compared to the concentration of undissociated lactic acid. (The standard value, however, is valid for a concentration ratio of $c(\text{HLac}):c(\text{Lac}^-)$ of 1:1.)

b) Degree of protonation Θ:

The degree of protonation Θ results according to the "protonation equation" [Eq. (7.14)] in:

$$\Theta = \frac{1}{1+\exp\dfrac{\mu_p^\ominus(\text{HLac/Lac}^-)-\mu_p}{RT^\ominus}}\,.$$

$$\Theta = \frac{1}{1+\exp\dfrac{(-22\times10^3\ \text{G})-(-12.5\times10^3\ \text{G})}{8.314\ \text{G K}^{-1}\times 298\ \text{K}}} = \frac{1}{1+0.0216} = \mathbf{0.979}\,.$$

97.9 % of the amount of acid used exists in protonated form (in the form of lactic acid molecules). Only 2.1 % exists in deprotonated form as lactate ions. Our original assumption that the acid only dissociates to a very small extent, was therefore justified.

2.7.4 Proton potential of weak acid-base pairs (II)

Concentration of cyanide ions c_{CN^-} in the solution:

Since the salt dissociates completely in water, the amount of cyanide ions corresponds to the amount of sodium cyanide.

$$c_{\text{CN}^-} = \frac{n_{\text{NaCN}}}{V_\text{S}} = \frac{m_{\text{NaCN}}/M_{\text{NaCN}}}{V_\text{S}}\,.$$

$$c_{\text{CN}^-} = \frac{(245\times10^{-6}\ \text{kg})\,/\,(49.0\times10^{-3}\ \text{kg mol}^{-1})}{100\times10^{-6}\ \text{m}^{-3}} = 50\ \text{mol m}^{-3} = 0.050\ \text{kmol m}^{-3}\,.$$

Standard proton potential $\mu_p^\ominus$ (HCN/CN⁻) of the weak acid-base pair:

CN^- is the base of the weakly alkaline acid-base pair HCN/CN^-. Therefore, one has to use Equation (7.7) to calculate the proton potential in the solution:

$$\mu_p = \tfrac{1}{2}\times\left[\mu_p^\ominus(\text{HCN/CN}^-)+\mu_p^\ominus(\text{H}_2\text{O/OH}^-)-RT^\ominus\ln(c_{\text{CN}^-}/c^\ominus)\right].$$

Solving for the standard value of the proton potential results in:

$$\mu_p^\ominus(\text{HCN/CN}^-) = 2\mu_p - \mu_p^\ominus(\text{H}_2\text{O/OH}^-) + RT^\ominus\ln(c_{\text{CN}^-}/c^\ominus)\,.$$

$$\mu_p^\ominus(\text{HCN/CN}^-) = 2\times(-62.6\times10^3\ \text{G})-(-80\times10^3\ \text{G})$$
$$+8.314\ \text{G K}^{-1}\times298\ \text{K}\times\ln(0.050\ \text{kmol m}^{-3}/1\ \text{kmol m}^{-3}) = \mathbf{-52.6\ kG}.$$

2.7.5 Titration of a weak acid

a) Proton potential $\mu_{p,0}$ in the benzoic acid solution:

To determine the proton potential caused by the acid of the weakly acidic pair benzoic acid/benzoate (HBenz/Benz⁻) in the initial solution, we use again Equation (7.6):

$$\mu_{p,0} = \tfrac{1}{2}\times\left[\mu_p^\ominus(\text{HBenz/Benz}^-)+\mu_p^\ominus(\text{H}_3\text{O}^+/\text{H}_2\text{O})+RT^\ominus\ln(c_{\text{HBenz}}/c^\ominus)\right].$$

$$\mu_{p,0} = \tfrac{1}{2}\times\left[(-23.9\times10^3\ \text{G})+0\ \text{G}\right.$$
$$\left.+8.314\ \text{G K}^{-1}\times298\ \text{K}\times\ln(0.100\ \text{kmol m}^{-3}/1\ \text{kmol m}^{-3})\right] = \mathbf{-14.8\ kG}.$$

b) Amount of benzoic acid n_{HBenz} in the initial solution:

Assuming that the acid dissociates only to a very small extent, the concentration of the undissociated acid c_{HBenz} can, in first approximation, be equated to the initial concentration $c_{\text{HBenz},0}$.

$$n_{\text{HBenz},0} = c_{\text{HBenz},0}\times V_0 = 100\ \text{mol m}^{-3}\times(100\times10^{-6}\ \text{m}^{-3}) = 0.0100\ \text{mol}.$$

Amount of hydroxide ions $n_{\text{OH}^-,1}$ in the sodium hydroxide solution (titrator) added to the analyte:

$$n_{\text{OH}^-,1} = c_{\text{OH}^-}\times V_{\text{T},1} = 2000\ \text{mol m}^{-3}\times(2.00\times10^{-6}\ \text{m}^3) = 0.0040\ \text{mol}.$$

Proton potential $\mu_{p,1}$ in the titration solution:

The amount $n_{\text{OH}^-,1}$ of OH⁻ ions reacts with the same amount of 0.0040 mole of C_6H_5COOH to $C_6H_5COO^-$, i.e., there is an amount of $n_{\text{Benz}^-,1} = 0.0040$ mol of benzoate ions in the solution. The fraction of the initial amount of benzoic acid in the solution that remains is then $n_{\text{HBenz},1} = 0.0100\ \text{mol} - 0.0040\ \text{mol} = 0.0060$ mol.

We can use the "level equation" [Eq. (7.12)] to calculate the proton potential $\mu_{p,1}$ in the solution:

$$\mu_{p,1} = \mu_p^\ominus(\text{HBenz/Benz}^-)+RT^\ominus\ln\frac{c_{\text{HBenz},1}}{c_{\text{Benz}^-,1}}$$

$$\mu_{p,1} = \mu_p^\ominus(\text{HBenz/Benz}^-)+RT^\ominus\ln\left[\frac{n_{\text{HBenz},1}}{V_\text{S}}\times\frac{V_\text{S}}{n_{\text{Benz}^-,1}}\right].$$

The volume V_S of the solution (in this case $V_0 + V_{\text{T},1}$) cancels out. Therefore, we have:

$$\mu_{\mathrm{p,1}} = \mu_{\mathrm{p}}^{\ominus}(\mathrm{HBenz/Benz^-}) + RT^{\ominus}\ln\frac{n_{\mathrm{HBenz,1}}}{n_{\mathrm{Benz^-,1}}}.$$

$$\mu_{\mathrm{p,1}} = (-23.9\times10^3\ \mathrm{G}) + 8.314\ \mathrm{G\,K^{-1}}\times298\ \mathrm{K}\times\ln\frac{0.0060\ \mathrm{mol}}{0.0040\ \mathrm{mol}} = \mathbf{-22.9\ kG}.$$

As expected, the proton potential was lowered by addition of the base.

c) One proceeds analogously to Solution 2.7.5 b).

Amount of hydroxide ions $n_{\mathrm{OH^-,2}}$ in the added sodium hydroxide solution:

In total, a volume $V_{\mathrm{T,2}} = (2.00 + 1.00)\ \mathrm{mL} = 3.00\ \mathrm{mL}$ of sodium hydroxide solution was added.

$$n_{\mathrm{OH^-,2}} = c_{\mathrm{OH^-}}\times V_{\mathrm{T,2}} = 2000\ \mathrm{mol\,m^{-3}}\times(3.00\times10^{-6}\ \mathrm{m^3}) = 0.0060\ \mathrm{mol}.$$

Proton potential $\mu_{\mathrm{p,2}}$ in the titration solution:

$$n_{\mathrm{Benz^-,2}} = 0.0060\ \mathrm{mol}, \qquad n_{\mathrm{HBenz,2}} = 0.0040\ \mathrm{mol}.$$

$$\mu_{\mathrm{p,2}} = \mu_{\mathrm{p}}^{\ominus}(\mathrm{HBenz/Benz^-}) + RT^{\ominus}\ln\frac{n_{\mathrm{HBenz,2}}}{n_{\mathrm{Benz^-,2}}}.$$

$$\mu_{\mathrm{p,2}} = (-23.9\times10^3\ \mathrm{G}) + 8.314\ \mathrm{G\,K^{-1}}\times298\ \mathrm{K}\times\ln\frac{0.0040\ \mathrm{mol}}{0.0060\ \mathrm{mol}} = \mathbf{-24.9\ kG}.$$

The proton potential has dropped further, but only by about 2 kG. This means that the proton potential has changed only slightly because of the buffering effect of the weakly acidic acid-base pair.

d) One proceeds analogously to Solution 2.1.7.

Conversion formula and basic stoichiometric equation:

Conversion formula: $C_6H_5COOH + NaOH \rightarrow C_6H_5COONa + H_2O$

Basic equation: $$\Delta\xi = \frac{\Delta n_{\mathrm{HBenz}}}{\nu_{\mathrm{HBenz}}} = \frac{\Delta n_{\mathrm{NaOH}}}{\nu_{\mathrm{NaOH}}} \qquad \text{Eq. (1.15)}$$

Volume of sodium hydroxide solution $V_{\mathrm{T,3}}$ required to reach the equivalence point:

Until the equivalence point, the total amount $n_{\mathrm{HBenz,0}}$ of benzoic acid originally present in the solution was consumed. Therefore, the change $\Delta n_{\mathrm{HBenz}}$ ($= 0 - n_{\mathrm{HBenz,0}}$) of the amount of acid results in –0.0100 mol.

The corresponding change of amount of sodium hydroxide can be calculated using the basic stoichiometric equation,

$$\Delta n_{\text{NaOH}} = \frac{\nu_{\text{NaOH}} \times \Delta n_{\text{HBenz}}}{\nu_{\text{HBenz}}} = \frac{(-1)\times(-0.0100\ \text{mol})}{(-1)} = -0.0100\ \text{mol}\,,$$

meaning, because of $\Delta n_{\text{NaOH}} = 0 - n_{\text{NaOH}}$ an amount $n_{\text{NaOH}} = 0.0100$ mol of NaOH was consumed to neutralize the benzoic acid. This amount corresponds to a volume of

$$V_{\text{T,3}} = \frac{n_{\text{NaOH}}}{c_{\text{NaOH}}} = \frac{0.0100\ \text{mol}}{2000\ \text{mol}\,\text{m}^{-3}} = 5.00\times10^{-6}\ \text{m}^3 = \mathbf{5.00\ mL}\,.$$

e) Proton potential $\mu_{\text{p,3}}$ in the titration solution at the equivalence point:

At the equivalence point, instead of the protonated form HBenz, the same amount of the deprotonated form Benz^- is present. As an approximation one can assume that the solution of the "weak base" Benz^- has the same concentration as the original acid ($c_{\text{Benz}^-,3} = c_{\text{HBenz,0}} = 0.100\ \text{kmol}\,\text{m}^{-3}$). The proton potential in the solution can then be calculated according to Equation (7.7):

$$\mu_{\text{p,3}} = \tfrac{1}{2}\times\left[\mu_{\text{p}}^{\ominus}(\text{HBenz/Benz}^-) + \mu_{\text{p}}^{\ominus}(\text{H}_2\text{O/OH}^-) - RT^{\ominus}\ln(c_{\text{Benz}^-,3}/c^{\ominus})\right].$$

$$\mu_{\text{p,3}} = \tfrac{1}{2}\times\left[(-23.9\times10^3\ \text{G}) + (-80\times10^3\ \text{G}) - 8.314\ \text{G}\,\text{K}^{-1}\times298\ \text{K}\times\ln(0.100\ \text{kmol}\,\text{m}^{-3}/1\ \text{kmol}\,\text{m}^{-3})\right] = \mathbf{-49.1\ kG}\,.$$

As expected, the proton potential shows a value noticeably below the neutral value.

For a more accurate calculation, one has to bear in mind that the volume of the solution in the titration flask has increased by the volume $V_{\text{T,3}}$ until the equivalence point. Therefore, we obtain for the concentration $c_{\text{Benz}^-,3}$:

$$c_{\text{Benz}^-,3} = \frac{n_{\text{Benz}^-,3}}{V_0 + V_{\text{T,3}}} = \frac{0.0100\ \text{mol}}{(100.00 + 5.00)\times10^{-6}\ \text{m}^3} = 95\ \text{mol}\,\text{m}^{-3} = 0.095\ \text{kmol}\,\text{m}^{-3}\,.$$

Inserting this value into the above equation results in a $\mu_{\text{p,3}}$ value of –49.0 kG. Apparently, the increase of the amount of water during the process of titration can still be neglected in good approximation, when the concentration difference of analyte to titrator is 1:20 (and more).

2.7.6 Buffer action

a) Proton potential μ_{p} in the buffer solution:

The proton potential μ_{p} in the buffer solution can be calculated with the aid of the "level equation" [Eq. (7.12)]:

$$\mu_{\text{p}} = \mu_{\text{p}}^{\ominus}(\text{HAc/Ac}^-) + RT^{\ominus}\ln\frac{c_{\text{HAc}}}{c_{\text{Ac}^-}}\,.$$

The buffer solution was prepared from equal volumes of acetic acid and sodium acetate solutions. Therefore, the concentrations are both halved:

$$\mu_p = (-27\times10^3\ \text{G}) + 8.314\ \text{G K}^{-1}\times298\ \text{K}\times\ln\frac{0.1\ \text{kmol m}^{-3}}{0.1\ \text{kmol m}^{-3}}$$

$$= -27\times10^{-3}\ \text{G} = \mathbf{-27\ kG}.$$

Since acetic acid and acetate are present in a concentration ratio of 1:1, the proton potential μ_p of the buffer solution corresponds to the standard value $\mu_p^{\ominus}(\text{HAc/Ac}^-)$.

b) Amount of acetate n_{Ac^-} in the buffer solution:

$$n_{\text{Ac}^-} = c_{\text{Ac}^-}\times V_{\text{S,NaAc}}.$$

Since equal volumes of acetic acid and sodium acetate solution were mixed and the total volume of the buffer solution is supposed to be 500 mL, we had to use 250 mL of sodium acetate solution.

$$n_{\text{Ac}^-} = 200\ \text{mol m}^{-3}\times(250\times10^{-6}\ \text{m}^3) = 0.050\ \text{mol}.$$

Amount of acetic acid n_{HAc} in the buffer solution:

The amount of acetic acid n_{HAc} in the buffer solution corresponds to the amount of acetate n_{Ac^-}.

Amount of oxonium ions $n_{\text{H}_3\text{O}^+}$ in the hydrochloric acid:

$$n_{\text{H}_3\text{O}^+} = c_{\text{H}_3\text{O}^+}\times V_{\text{S,HCl}}.$$

$$n_{\text{H}_3\text{O}^+} = (2.00\times10^3\ \text{mol m}^{-3}\times(500\times10^{-9}\ \text{m}^3) = 1.00\times10^{-3}\ \text{mol}.$$

Amount of acetate $n_{\text{Ac}^-,1}$ and acetic acid $n_{\text{HAc},1}$ after addition of hydrochloric acid:

Adding 0.001 mole of HCl to the initial 0.050 mole of acetate reduces the amount of acetate by about 0.001 mole to $n_{\text{Ac}^-,1} = 0.049$ mol according to $\text{Ac}^- + \text{H}_3\text{O}^+ \rightarrow \text{HAc} + \text{H}_2\text{O}$, while the amount of acetic acid has increased by the same amount to $n_{\text{HAc},1} = 0.051$ mol.

Proton potential $\mu_{p,1}$ of the buffer solution after addition of hydrochloric acid:

Applying the "level equation" results in the new proton potential $\mu_{p,1}$ of the buffer solution. Thereby, the concentration ratio can be replaced by the corresponding mole ratio [see Solution 2.7.5 b)].

$$\mu_{p,1} = \mu_p^{\ominus}(\text{HAc/Ac}^-) + RT^{\ominus}\ln\frac{n_{\text{HAc},1}}{n_{\text{Ac}^-,1}}.$$

$$\mu_{p,1} = (-27\times10^3\ \text{G}) + 8.314\ \text{G K}^{-1}\times298\ \text{K}\times\ln\frac{0.051\ \text{mol}}{0.049\ \text{mol}} = \mathbf{-26.9\ kG}.$$

Adding the acid has only changed the proton potential of the buffer solution by 0.1 kG.

c) Concentration of oxonium ions $c_{H_3O^+,2}$ after addition of hydrochloric acid to pure water:

$$c_{H_3O^+,2} = \frac{n_{H_3O^+}}{V_{H_2O}}.$$

$$c_{H_3O^+,2} = \frac{1.00\times10^{-3}\ \text{mol}}{500\times10^{-6}\ \text{m}^3} = 2.00\ \text{mol m}^{-3} = 2.00\times10^{-3}\ \text{kmol m}^{-3}.$$

The very small volume of the added hydrochloric acid can be neglected.

Proton potential $\mu_{p,2}$ of the mixture:

According to Equation (7.3), the proton potential $\mu_{p,2}$ will be

$$\mu_{p,2} = \mu_p^{\ominus}(H_3O^+/H_2O) + RT^{\ominus}\ln(c_{H_3O^+,2}/c^{\ominus}).$$

$$\mu_{p,2} = 0\ \text{G} + 8.314\ \text{G K}^{-1}\times 298\ \text{K}\times\ln(2.00\times10^{-3}\ \text{kmol m}^{-3}/1\ \text{kmol m}^{-3})$$

$$= \mathbf{-15.4\ kG}.$$

By adding the acid, the proton potential has shifted by 24.6 kG in relation to that of pure water with –40 kG (compared to a change of only 0.1 kG ! in the case of the buffer solution).

2.7.7 Buffer capacity

a) Proton potential μ_p in the buffer solution:

To calculate the proton potential μ_p in the buffer solution, we use again the "level equation" [Eq. (7.12)]:

$$\mu_p = \mu_p^{\ominus}(NH_4^+/NH_3) + RT^{\ominus}\ln\frac{c_{NH_4^+}}{c_{NH_3}}.$$

$$\mu_p = (-53\times10^3\ \text{G}) + 8.314\ \text{G K}^{-1}\times 298\ \text{K}\times\ln\frac{0.060\ \text{kmol m}^{-3}}{0.040\ \text{kmol m}^{-3}} = \mathbf{-52.0\ kG}.$$

b) Amount of ammonium ions $n_{NH_4^+}$ in the buffer solution (B):

$$n_{NH_4^+} = c_{NH_4^+}\times V_{S,B} = 60\ \text{mol m}^{-3}\times(50\times10^{-6}\ \text{m}^3) = 3.0\times10^{-3}\ \text{mol} = 3.0\ \text{mmol}.$$

Amount of ammonia n_{NH_3} in the buffer solution (B):

$$n_{NH_3} = c_{NH_3}\times V_{S,B} = 40\ \text{mol m}^{-3}\times(50\times10^{-3}\ \text{m}^3) = 2.0\times10^{-3}\ \text{mol} = 2.0\ \text{mmol}.$$

Maximum amount of hydroxide ions n_{OH^-} that can be added to the buffer:

The buffer solution contains 3.0 mmole of NH_4^+ and 2.0 mmole of NH_3. If one adds n_{OH^-} mmole of sodium hydroxide, the amount of NH_4^+ decreases to (3.0 mmol – n_{OH^-}), while the amount of NH_3 increases to (2.0 mmol + n_{OH^-}). The ratio of acid to corresponding base reaches the value 1:10, if we have:

$$\frac{3.0\ \text{mmol} - n_{OH^-}}{2.0\ \text{mmol} + n_{OH^-}} = 0.1 .$$

Solving for n_{OH^-} results in:

$$n_{OH^-} = 2.5\ \text{mmol} .$$

Consequently, one can add 2.5 mmole of sodium hydroxide to the buffer solution before its buffer capacity is depleted.

Maximum volume of sodium hydroxide solution $V_{S,NaOH}$ that can be added to the buffer:

$$n_{OH^-} = c_{OH^-} \times V_{S,NaOH} \quad \Rightarrow \quad V_{S,NaOH} = \frac{n_{OH^-}}{c_{OH^-}} .$$

$$V_{S,NaOH} = \frac{2.5 \times 10^{-3}\ \text{mol}}{100\ \text{mol}\,\text{m}^{-3}} = 25 \times 10^{-6}\ \text{m}^{-3} = \mathbf{25\ mL} .$$

2.7.8 Indicator

An indicator system obeys the "level equation" [Eq. (7.21)] as well:

$$\mu_p = \mu_p^{\ominus}(\text{HInd/Ind}^-) + RT^{\ominus} \ln \frac{c(\text{HInd})}{c(\text{Ind}^-)} .$$

Proton potential $\mu_{p,1}$ (color change red → yellow):

$$\mu_{p,1} = (-45.1 \times 10^3\ \text{G}) + 8.314\ \text{G}\,\text{K}^{-1} \times 298\ \text{K} \times \ln \frac{30}{1} = -36.7\ \text{kG} .$$

Proton potential $\mu_{p,2}$ (color change yellow → red):

$$\mu_{p,2} = (-45.1 \times 10^3\ \text{G}) + 8.314\ \text{G}\,\text{K}^{-1} \times 298\ \text{K} \times \ln \frac{1}{2} = -46.8\ \text{kG} .$$

Hence, the transition interval of the indicator phenol red lies between a proton potential of **–36.7 kG** and a proton potential of **–46.8 kG.**

2.8 Side Effects of Transformations of Substances

2.8.1 Application of partial molar volumes

Since the molar volumes of water (W) and ethanol (E) in water-ethanol mixtures are plotted in the diagram as a function of the mole fraction x_E of ethanol, we have to calculate first this mole fraction in the homogeneous mixture in question prepared by mixing 50.0 mL of water and 50.0 mL of ethanol.

Masses m_W and m_E of water and ethanol:

$$m_W = \rho_W \times V_W = 997\,\text{kg m}^{-3} \times (50.0 \times 10^{-6}\,\text{m}^3) = 49.9 \times 10^{-3}\,\text{kg}.$$

$$m_E = \rho_E \times V_E = 789\,\text{kg m}^{-3} \times (50.0 \times 10^{-6}\,\text{m}^3) = 39.5 \times 10^{-3}\,\text{kg}.$$

Amounts n_W and n_E of water and ethanol:

$$n_W = \frac{m_W}{M_W} = \frac{49.9 \times 10^{-3}\,\text{kg}}{18.0 \times 10^{-3}\,\text{kg mol}^{-1}} = \mathbf{2.77\,mol}.$$

$$n_E = \frac{m_E}{M_E} = \frac{39.5 \times 10^{-3}\,\text{kg}}{46.0 \times 10^{-3}\,\text{kg mol}^{-1}} = \mathbf{0.86\,mol}.$$

Mole fraction x_E of ethanol:

$$x_E = \frac{n_E}{n_{total}} = \frac{n_E}{n_E + n_W} = \frac{0.86\,\text{mol}}{0.86\,\text{mol} + 2.77\,\text{mol}} = \mathbf{0.24}.$$

Total volume V of the water-ethanol mixture:

The partial molar volumes of water and ethanol can be found in the diagram [However, note the differing scales for water (left) and ethanol (right).]

$V_{m,W} \approx 17.45\,\text{cm}^3\,\text{mol}^{-1}$,

$V_{m,E} \approx 56.25\,\text{cm}^3\,\text{mol}^{-1}$.

The volume of the mixture results from the amounts and volume demands of the components [see Eq. (8.3)]:

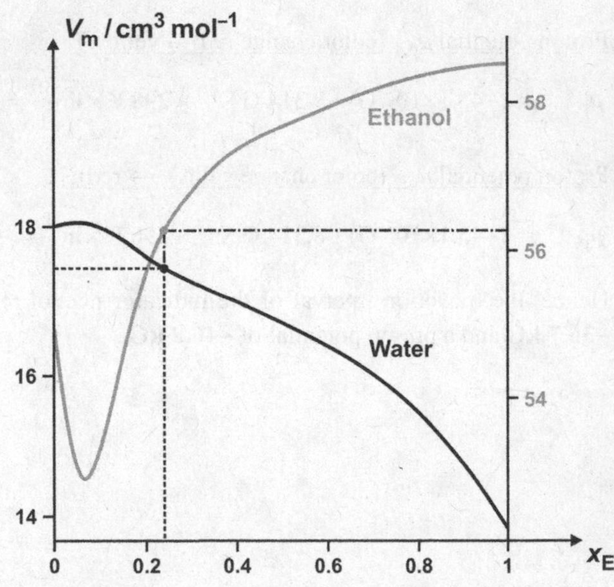

$$V \approx n_W \times V_{m,W} + n_E \times V_{m,E}.$$

$$V \approx 2.77\,\text{mol} \times 17.45\,\text{cm}^3\,\text{mol}^{-1} + 0.86\,\text{mol} \times 56.25\,\text{cm}^3\,\text{mol}^{-1} = 97\,\text{cm}^3 = \mathbf{97 \times 10^{-6}\,m^3}.$$

After mixing the liquids, a "volume shrinkage" of about 3 mL can be observed.

2.8.2 Volume demand of dissolved ions

	Symbol B	D	D′
a) Conversion formula:	$Ca(OH)_2\|s \rightarrow$	$Ca^{2+}\|w +$	$2\ OH^-\|w$
$V_m^\ominus / \text{cm}^3\,\text{mol}^{-1}$:	33.3	−17.7	2×(−5.2)

Molar reaction volume $\Delta_R V^\ominus$:

First, the molar volumes of the substances involved in the reaction have to be known. The molar volume of substance B (solid calcium hydroxide) can be calculated from the data for its volume and amount of substance given in the exercise:

$$V_{m,B}^\ominus = \frac{V_B}{n_B} = \frac{1.00 \times 10^{-6}\,\text{m}^3}{0.030\,\text{mol}} = 33.3 \times 10^{-6}\,\text{m}^3\,\text{mol}^{-1}.$$

The remaining two molar volumes for the substances D ($Ca^{2+}|w$) and D′ ($OH^-|w$) can be taken from the table provided in the text of the exercise.

The molar reaction volume $\Delta_R V^\ominus$ corresponds to the sum of the molar volumes of all reaction partners weighted by the conversion numbers [see Eq. (8.6)]:

$$\Delta_R V^\ominus = \sum_i \nu_i V_{m,i}^\ominus = \nu_B V_{m,B}^\ominus + \nu_D V_{m,D}^\ominus + \nu_{D'} V_{m,D'}^\ominus.$$

$$= [(-1) \times 33.3 + 1 \times (-17.7) + 2 \times (-5.2)] \times 10^{-6}\,\text{m}^3\,\text{mol}^{-1} = \mathbf{-61.4 \times 10^{-6}\,m^3\,mol^{-1}}.$$

Note that the conversion numbers for the reactants are negative, those for the products, however, positive.

Conversion $\Delta\xi$:

0.030 moles of $Ca(OH)_2$ (substance B) are added to water ($n_{B,0} = 0.030$ mol) where it dissolves completely ($n_B = 0$).

$$\Delta\xi = \frac{\Delta n_B}{\nu_B} = \frac{n_B - n_{B,0}}{\nu_B}. \qquad \text{cf. Eq. (1.15)}$$

$$\Delta\xi = \frac{0 - 0.030\,\text{mol}}{-1} = 0.030\,\text{mol}.$$

In the case of the reactant $Ca(OH)_2$, both the change Δn of the amount of substance and the conversion number ν are negative.

Final volume V_{final} (under standard conditions):

The final volume V_{final} under standard conditions is calculated as follows:

$$\Delta V = V_{final} - V_{initial} .$$

Solving for V_{final} results in:

$$V_{final} = V_{initial} + \Delta V .$$

Initial volume $V_{initial}$:

$$V_{initial} = V_W + V_B = (10\times10^{-3}\ m^3) + (1.00\times10^{-6}\ m^3) = 10.001\times10^{-3}\ m^3 .$$

Change in volume ΔV:

The change in volume ΔV is calculated according to Equation (8.6) from the molar reaction volume by multiplying it with the conversion:

$$\Delta V = \Delta_R V^{\ominus} \times \Delta\xi = (-61.4\times10^{-6}\ m^3\ mol^{-1}) \times 0.030\ mol = -1.84\times10^{-6}\ m^3 .$$

$$V_{final} = (10.001\times10^{-3}\ m^3) + (-1.84\times10^{-6})\ m^3 = \mathbf{9.99916\times10^{-3}\ m^3} .$$

The volume thus shrinks when calcium hydroxide is dissolved! This contraction is caused by the H_2O molecules (which are rather loosely packed when in pure water) being concentrated more densely in the hydration shells of the OH^- ions and especially the divalent Ca^{2+} ions.

b)

	Symbol	B		D		D′
Conversion formula:		$H_2O\|l$	$\rightarrow$	$H^+\|w$	+	$OH^-\|w$
$V_m^{\ominus}$ / cm^3 mol^{-1}:		18.1		0.2		−5.2

Molar reaction volume $\Delta_R V^{\ominus}$:

$$\Delta_R V^{\ominus} = [(-1)\times18.1 + 1\times0.2 + 1\times(-5.2)]\times10^{-6}\ m^3\ mol^{-1} = \mathbf{-23.1\times10^{-6}\ m^3} .$$

Conversion $\Delta\xi$:

$$\Delta\xi = \frac{\Delta n_D}{\nu_D} .$$

Change in amount Δn_D:

The acid-base disproportionation of water has to be taken into account (compare Section 7.3 in the textbook "Physical Chemistry from a Different Angle"). Hence, the concentration of H^+ ions (substance D) in equilibrium is:

$$c_D = 1.00\times10^{-7}\ kmol\,m^{-3} = 1.00\times10^{-4}\ mol\,m^{-3} .$$

The change in concentration Δc_D is therefore 1.00×10^{-4} mol m^{-3}, since the water [V_B (= V_w) = 10 L = 10×10^{-3} m^{-3}] is supposed to be in an ion-free state at the beginning.

$$\Delta c_D = \frac{\Delta n_D}{V_B} \Rightarrow \Delta n_D = \Delta c_D \times V_B .$$

$$\Delta n_D = (1.00\times10^{-4}\ \mathrm{mol\,m^{-3}})\times(10\times10^{-3}\ \mathrm{m^{-3}}) = 1.00\times10^{-6}\ \mathrm{mol}.$$

$$\Delta\xi = \frac{1.00\times10^{-6}\ \mathrm{mol}}{1} = 1.00\times10^{-6}\ \mathrm{mol}.$$

Change in volume ΔV (under standard conditions):

$$\Delta V = \Delta_R V^{\ominus} \times \Delta\xi .$$

$$\Delta V = (-23.1\times10^{-6}\ \mathrm{m^3\,mol^{-1}})\times(1.00\times10^{-6}\ \mathrm{mol}) = -23.1\times10^{-12}\ \mathrm{m^3} = \mathbf{-0.023\ mm^3}.$$

2.8.3 Reaction entropy

a) Molar reaction entropy $\Delta_R S^{\ominus}$:

Table A2.1 in the Appendix of the textbook "Physical Chemistry from a Different Angle" lists the temperature coefficients α of the chemical potential of various substances (unit $\mathrm{G\,K^{-1}}$). The molar entropies $S_m^{\ominus}$ can then be determined as follows (using the example of calcium carbide):

$$S_m^{\ominus} = -\alpha = -(-70.3)\ \mathrm{G\,K^{-1}} = 70.3\ \mathrm{G\,K^{-1}} = 70.3\ \mathrm{J\,mol^{-1}\,K^{-1}} = 70.3\ \mathrm{Ct\,mol^{-1}}.$$

The unit G (Gibbs) corresponds to $\mathrm{J\,mol^{-1}}$. Conversely, $\mathrm{J\,K^{-1}}$ corresponds to the unit Ct (Carnot).

	Symbol B	B′	D	D′
Conversion formula:	$CaC_2\|s$ +	$2\ H_2O\|l$ →	$Ca(OH)_2\|s$ +	$C_2H_2\|g$
$S_m^{\ominus}$ / Ct mol^{-1}:	70.3	2×70.0	83.4	200.9

The molar reaction entropy $\Delta_R S^{\ominus}$ corresponds to the sum of the molar entropies of all reaction partners weighted by the conversion numbers [see Eq. (8.13)]:

$$\Delta_R S^{\ominus} = \sum_i \nu_i S_{m,i}^{\ominus} = \nu_B S_{m,B}^{\ominus} + \nu_{B'} S_{m,B'}^{\ominus} + \nu_D S_{m,D}^{\ominus} + \nu_{D'} S_{m,D'}^{\ominus} .$$

$$\Delta_R S^{\ominus} = [(-1)\times70.3 + (-2)\times70.0 + 1\times83.4 + 1\times200.9]\ \mathrm{Ct\,mol^{-1}} = \mathbf{+74.0\ Ct\,mol^{-1}}.$$

The molar reaction entropy is positive because a strongly disordered gas is produced.

b) Conversion $\Delta\xi$:

$$\Delta\xi = \frac{\Delta n_B}{\nu_B} .$$

Change in amount Δn_B:

The change in amount Δn_B of calcium carbide (substance B) can be calculated from its change in mass $\Delta m_B = -8\times10^{-3}$ kg by means of its molar mass $M_B = 64.1\times10^{-3}\ \mathrm{kg\,mol^{-1}}$:

$$\Delta n_B = \frac{\Delta m_B}{M_B} = \frac{-8.0\times10^{-3}\ \text{kg}}{64.1\times10^{-3}\ \text{kg mol}^{-1}} = -0.125\ \text{mol}.$$

$$\Delta\xi = \frac{-0.125\ \text{mol}}{-1} = 0.125\ \text{mol}.$$

Change in entropy ΔS (under standard conditions):

The change in entropy ΔS results from the molar reaction entropy by multiplication with the conversion (see Section 8.5 in the textbook "Physical Chemistry from a Different Angle"):

$$\Delta S = \Delta_R S^\ominus \times \Delta\xi .$$

$$\Delta S = 74.0\ \text{Ct mol}^{-1} \times 0.125\ \text{mol} = \mathbf{9.3\ Ct}.$$

2.8.4 Combustion of natural gas for energy generation

	Symbol	B		B′		D		D′
a) Conversion formula:		$CH_4\vert g$	+	$2\ O_2\vert w$	$\rightarrow$	$CO_2\vert g$	+	$2\ H_2O\vert l$
$\mu^\ominus$/kG:		−50.5		2×0.0		−394.4		2×(−237.1)

Usable energy W^*_{use} :

The following formula can be used to convert the unit kilowatt hour into the SI unit Joule:

$$W = P\times\Delta t = 1\ \text{kW}\times 1\ \text{h} = 1000\ \text{J s}^{-1}\times 3600\ \text{s} = 3.6\times10^6\ \text{J}.$$

Hence, a usable energy W^*_{use} of 3.6 MJ should be released by the combustion of the natural gas (methane).

Chemical drive $\mathcal{A}^\ominus$ of the combustion of natural gas:

$$\mathcal{A}^\ominus = \sum_{\text{initial}} |\nu_i|\mu_i^\ominus - \sum_{\text{final}} \nu_j\mu_j^\ominus . \qquad \text{cf. Eq. (4.3)}$$

$$\mathcal{A}^\ominus = \{[1\times(-50.5)+2\times0.0]-[1\times(-394.4)+2\times(-237.1)]\}\ \text{kG}$$

$$= 818.1\ \text{kG} = 818.1\times10^3\ \text{J mol}^{-1}.$$

Conversion $\Delta\xi$:

The required conversion can be calculated using Equation (8.18),

$$W^*_{\text{use}} = -W_{\rightarrow\xi} = \mathcal{A}^\ominus \times \Delta\xi ,$$

which relates energy conversion and drive. Solving for $\Delta\xi$ results in:

$$\Delta\xi = \frac{W^*_{\text{use}}}{\mathcal{A}^\ominus} = \frac{3.6\times10^6\ \text{J}}{818.1\times10^3\ \text{J mol}^{-1}} = 4.40\ \text{mol}.$$

Change Δn_B of the amount of natural gas:

$$\Delta\xi = \frac{\Delta n_B}{\nu_B} \Rightarrow \Delta n_B = \nu_B \times \Delta\xi .$$

$$\Delta n_B = \frac{4.40\ \text{mol}}{-1} = -4.40\ \text{mol} .$$

Initial amount $n_{B,0}$ of natural gas that is required:

$$\Delta n_B = n_B - n_{B,0} \Rightarrow n_{B,0} = n_B - \Delta n_B .$$

$$n_{B,0} = 0 - (-4.40\ \text{mol}) = \mathbf{4.40\ mol} .$$

b) Initial volume $V_{B,0}$ of natural gas:

The molar volume $V_m^\ominus$ of gases under standard conditions is 24.8 L mol^{-1}. The initial volume $V_{B,0}$ of natural gas can then be calculated according to

$$V_{m,B}^\ominus = \frac{V_{B,0}}{n_{B,0}} \Rightarrow V_{B,0} = n_{B,0} \times V_{m,B}^\ominus .$$

$$V_{B,0} = 4.40\ \text{mol} \times (24.8 \times 10^{-3}\ \text{m}^3\ \text{mol}^{-1}) = 109 \times 10^{-3}\ \text{m}^3 .$$

Since the natural gas is to be burnt completely, the change in volume corresponds to $\mathbf{109 \times 10^{-3}\ m^3}$ = 109 L.

2.8.5 Entropy balance of a reaction

Collection of the necessary data:

	Symbol	B		B′		D
Conversion formula:		$2\ H_2\vert g$	+	$O_2\vert g$	→	$2\ H_2O\vert l$,
$\mu^\ominus$/kG:		2×0.0		0.0		$2 \times (-237.1)$
$S_m^\ominus$ / Ct mol^{-1}:		2×130.7		205.2		2×70.0

a) Molar reaction entropy $\Delta_R S^\ominus$:

$$\Delta_R S^\ominus = [(-2) \times 130.7 + (-1) \times 205.2 + 2 \times 70.0]\ \text{Ct mol}^{-1} = \mathbf{-326.6\ Ct\ mol^{-1}} .$$

The molar reaction entropy is markedly negative, because two (disordered) gases form a relatively ordered liquid.

Conversion $\Delta\xi$:

$$\Delta\xi = \frac{\Delta n_D}{\nu_D} .$$

Change in amount Δn_D of water:

$$\Delta m_D = +1.0\ \text{kg};\ M_D = 18.0\times10^{-3}\ \text{kg mol}^{-1} \Rightarrow$$

$$\Delta n_D = \frac{\Delta m_D}{M_D} = \frac{1.0\ \text{kg}}{18.0\times10^{-3}\ \text{kg mol}^{-1}} = 55.6\ \text{mol}\,.$$

$$\Delta\xi = \frac{55.6\ \text{mol}}{2} = 27.8\ \text{mol}\,.$$

Change in entropy ΔS (under standard conditions):

$$\Delta S = \Delta_R S^{\ominus} \times \Delta\xi\,.$$

$$\Delta S = -326.6\ \text{Ct mol}^{-1} \times 27.8\ \text{mol} = -9080\ \text{Ct} = \mathbf{-9.08\ kCt}\,.$$

b) Chemical drive $\mathcal{A}^{\ominus}$:

$$\mathcal{A}^{\ominus} = \{[2\times0.0+1\times0.0]-[2\times(-237.1)]\}\ \text{kG}$$

$$\mathcal{A}^{\ominus} = 474.2\ \text{kG} = 474.2\times10^3\ \text{G} = \mathbf{474.2\times10^3\ J\,mol^{-1}}\,.$$

Maximum value of usable energy $W^*_{\text{use,max}}$:

In the ideal case of a complete usage of the energy $W^*_{\text{use}} = \eta\times\mathcal{A}\times\Delta\xi$, which the surroundings (index *) receives because of a spontaneous reaction (see Section 8.7 in the textbook "Physical Chemistry from a Different Angle"), meaning for an efficiency of $\eta = 1$, we obtain

$$W^*_{\text{use,max}} = \mathcal{A}^{\ominus}\times\Delta\xi\,.$$

$$W^*_{\text{use,max}} = (474.2\times10^3\ \text{J mol}^{-1})\times27.8\ \text{mol} = 13.18\times10^6\ \text{J} = \mathbf{13.18\ MJ}\,.$$

c) Generated entropy S_g:

As a rule, the energy released during a spontaneous reaction cannot be used completely, but only with an efficiency of $\eta < 1$. The rest of the energy is "burnt" thereby generating entropy [cf. Eq. (8.23)]:

$$S_g = \frac{(1-\eta)\times W^*_{\text{use,max}}}{T^{\ominus}}\,.$$

Entropy S_e exchanged with the environment:

The entropy content S of a system can change in two different ways, by generation (S_g) or by exchange (S_e) of entropy [cf. Eq. (8.20)],

$$\Delta S = S_g + S_e\,,$$

or rearranged,

$S_e \quad = \Delta S - S_g$.

All three cases can be discussed on the basis of the equations above. Therefore, only case 2 ($\eta = 70\ \% = 0.7$) shall be presented in more detail:

$$S_g = \frac{(1-0.7)\times 13.18\times 10^6\ \text{J}}{298\ \text{K}} = 13.3\times 10^3\ \text{J K}^{-1} = 13.3\times 10^3\ \text{Ct} = \mathbf{13.3\ kCt}.$$

$$S_e = (-9.1\times 10^3\ \text{Ct}) - (13.3\times 10^3\ \text{Ct}) = -22.4\times 10^3\ \text{Ct} = \mathbf{-22.4\ kCt}.$$

$$T^{\ominus}\times S_e = 298\ \text{K}\times(-22.4\times 10^3\ \text{J K}^{-1}) = -6.7\times 10^6\ \text{J} = \mathbf{-6.7\ MJ}.$$

The expression $T^{\ominus}\times S_e$ is known as "heat of reaction."

Summary:

	case 1		case 2		case 3	
S_g / kCt:	44.2	,	13.3	,	0	,
S_e / kCt:	−53.3	,	−22.4	,	−9.1	,
$T^{\ominus}\times S_e$ / MJ:	−15.9	,	−6.7	,	−2.7	.

2.8.6 Energy and entropy balance of water evaporation

	Symbol B	D
a) Conversion formula:	$H_2O\|l$	$\rightarrow H_2O\|g,air.$
μ / kG:	−237.1	−238.0

Chemical drive $\mathcal{A}$ for the evaporation of water:

$$\mathcal{A} = \{[1\times(-237.1)] - [1\times(-238.0)]\}\ \text{kG} = 0.9\ \text{kG} = 0.9\times 10^3\ \text{G} = \mathbf{900\ J\ mol^{-1}}.$$

Conversion $\Delta\xi$:

$$\Delta\xi = \frac{\Delta n_B}{\nu_B}.$$

Change in amount Δn_B of (liquid) water:

The change in amount Δn_B of (liquid) water (substance B) can be calculated from its change in volume $\Delta V_B = -1.0\times 10^{-6}\ \text{m}^3$ using its molar volume $V^{\ominus}_{m,B} = 18.1\times 10^{-6}\ \text{m}^3\ \text{mol}^{-1}$ (see Table 8.1 in the textbook "Physical Chemistry from a Different Angle").

$$\Delta n_B = \frac{\Delta V_B}{V^{\ominus}_{m,B}} = \frac{-1.0\times 10^{-6}\ \text{m}^3}{18.1\times 10^{-6}\ \text{m}^3\ \text{mol}^{-1}} = -0.055\ \text{mol}.$$

$$\Delta\xi = \frac{-0.055\ \text{mol}}{-1} = \mathbf{0.055\ mol}.$$

Maximum usable energy $W^*_{\text{use,max}}$:

Analogous to the considerations in Solution 2.8.5 b) we obtain:

$$W^*_{\text{use,max}} = \mathcal{A}\times\Delta\xi = 900\ \text{J mol}^{-1}\times 0.055\ \text{mol} = \mathbf{49.5\ J}.$$

b) Change in height Δh of the weight:

If the usable energy $W^*_{\text{use,max}}$ is used to lift the weight, then it is stored as potential energy W_{pot}:

$$W_{\text{pot}} = W^*_{\text{use,max}}.$$

From the equation for the potential energy,

$$W_{\text{pot}} = m\times g\times\Delta h,$$

we obtain by solving for Δh:

$$\Delta h = \frac{W_{\text{pot}}}{m\times g} = \frac{49.5\ \text{J}}{1\ \text{kg}\times 9.81\ \text{m s}^{-2}} = \frac{49.5\ \text{kg m}^2\,\text{s}^{-2}}{1\ \text{kg}\times 9.81\ \text{m s}^{-2}} = \mathbf{5.0\ m}.$$

c) Generated entropy S_g:

If all released energy remains unused (meaning the efficiciency is $\eta = 0$), Equation (8.23) for the generated entropy results in

$$S_g = \frac{\mathcal{A}\times\Delta\xi}{T} = \frac{900\ \text{J mol}^{-1}\times 0.055\ \text{mol}}{298\ \text{K}} = 0.17\ \text{J K}^{-1} = \mathbf{0.17\ Ct}.$$

d) Chemical potential μ_D of water vapor in air:

We start from the mass action equation 1 [Eq. (6.4)]:

$$\mu_D = \mu_{D,0} + RT\ln\frac{c_D}{c_{D,0}}.$$

Here, μ_D is the chemical potential of water vapor in air. If air with a relative humidity of 100 % is selected as initial state (index 0), then we have in the case of evaporation equilibrium [$\mu_D = \mu_B$ (chemical potential of liquid water)] with $c_D = c_{D,0}$ and thus $RT\ln(c_{D,0}/c_{D,0}) = RT\ln 1 = 0$ finally $\mu_{D,0} = \mu_B$. At a relative humidity of 70 %, one obtains correspondingly $c_D = 0.70\times c_{D,0}$ and consequently:

$$\mu_D = \mu_B + RT\ln\frac{0.70\times c_{D,0}}{c_{D,0}}.$$

At room temperature (25 °C) μ_D is then:

$$\mu_D = (-237.1\times10^3\ \text{G}) + 8.314\ \text{G K}^{-1}\times298\ \text{K}\times\ln 0.70 = \mathbf{-238.0\ kG}.$$

2.8.7* "Working" duck

a) Chemical potential μ_D of water vapor in air:

The value μ_D is calculated analogously to Solution 2.8.6, but instead of a relative humidity of air of 70 % one of 50 % is taken into account:

$$\mu_D = \mu_{D,0} + RT\ln\frac{c_D}{c_{D,0}} = \mu_B + RT\ln\frac{0.50\times c_{D,0}}{c_{D,0}}.$$

At 298 K one obtains:

$$\mu_D = (-237.1\times10^3\ \text{G}) + 8.314\ \text{G K}^{-1}\times298\ \text{K}\times\ln 0.50 = \mathbf{-238.8\ kG}.$$

b) Chemical drive $\mathcal{A}$ for the evaporation of water:

$$\mathcal{A} = \{[1\times(-237.1)] - [1\times(-238.8)]\}\ \text{kG} = 1.7\ \text{kG} = 1.7\times10^3\ \text{G} = 1.7\times10^3\ \text{J mol}^{-1}.$$

Conversion $\Delta\xi$:

$$\Delta\xi = \frac{\Delta n_B}{\nu_B} = \frac{-1\times10^{-3}\ \text{mol}}{-1} = 1\times10^{-3}\ \text{mol}.$$

Maximum usable energy $W^*_{\text{use,max}}$:

$$W^*_{\text{use,max}} = \mathcal{A}\times\Delta\xi = (1.7\times10^3\ \text{J mol}^{-1})\times(1\times10^{-3}\ \text{mol}) = \mathbf{1.7\ kJ}.$$

c) Actually used energy W_{pot}:

$$W_{\text{pot}} = m\times g\times\Delta h = 0.01\ \text{kg}\times9.81\ \text{m s}^{-2}\times0.1\ \text{m} = 1\times10^{-2}\ \text{J}.$$

Efficiency η:

The efficiency η is here the quotient of (actually) used energy and (theoretically) usable energy:

$$\eta = \frac{W_{\text{pot}}}{W^*_{\text{use,max}}}.$$

$$\eta = \frac{1\times10^{-2}\ \text{J}}{1.7\ \text{J}} = 0.006 = \mathbf{0.6\ \%}.$$

Consequently, the efficiency is very low.

2.8.8 "Muscular energy" by oxidation of glucose

	Symbol B	B′	D	D′
a) Conversion formula:	$C_6H_{12}O_6|w$ +	6 $O_2|w$ →	6 $CO_2|w$	+ 6 $H_2O|l$
$\mu^\ominus$/kG:	−917.0	6×16.4	6×(−386.0)	6×(−237,1)

Chemical drive $\mathcal{A}^\ominus$ for the oxidation process:

$$\mathcal{A}^\ominus = \{[1\times(-917.0)+6\times16.4]-[6\times(-386.0)+6\times(-237.1)]\}\ \text{kG}$$

$$= 2920\ \text{kG} = \mathbf{2920\ kJ\ mol^{-1}}.$$

b) Energy W_{min} necessary for the climbing:

$$W_{min} = m\times g\times\Delta h.$$

$$W_{min} = 70\ \text{kg}\times9.81\ \text{m s}^{-2}\times24\ \text{m} = 16480\ \text{kg m}^2\ \text{s}^{-2} = 16480\ \text{N m} = 16480\ \text{J} = \mathbf{16.5\ kJ}.$$

c) Conversion $\Delta\xi$:

$$\Delta\xi = \frac{\Delta n_B}{\nu_B}.$$

Change in amount Δn_B of glucose:

$\Delta m_B = -6\times10^{-3}$ kg; $M_B = 180.0\times10^{-3}$ kg mol^{-1} ⇒

$$\Delta n_B = \frac{\Delta m_B}{M_B} = \frac{-6\times10^{-3}\ \text{kg}}{180.0\times10^{-3}\ \text{kg mol}^{-1}} = -0.033\ \text{mol}.$$

$$\Delta\xi = \frac{-0.033\ \text{mol}}{-1} = 0.033\ \text{mol}.$$

Energy W^*_{use} usable for the climbing:

The usable energy W^*_{use} can be calculated according to:

$$W^*_{use} = \eta\times\mathcal{A}^\ominus\times\Delta\xi = 0.2\times2920\ \text{kJ mol}^{-1}\times0.033\ \text{mol} = \mathbf{19.3\ kJ}.$$

d) Change in mass Δm_B* of the glucose reservoir of the chimney sweeper:

$$\frac{\Delta m_B^*}{\Delta m_B} = \frac{W_{min}}{W^*_{use}}.$$

Solving for Δm_B^* results in:

$$\Delta m_B^* = \Delta m_B\times\frac{W_{min}}{W^*_{use}} = -6\ \text{g}\times\frac{16.5\ \text{kJ}}{19.3\ \text{kJ}} = \mathbf{-5.1\ g}.$$

The chimney sweeper has to eat at least one tablet of glucose if he wants to reach his workplace on the roof of the house.

2.8.9 Fuel Cell

Collection of the necessary data:

	Symbol B	B′	D	D′
Conversion formula:	$C_3H_8\|g$ +	$5\ O_2\|g$	$\rightarrow 3\ CO_2\|g$	$+\ 4\ H_2O\|g$
$\mu^{\ominus}/\text{kG}$:	−23.4	5×0.0	3×(−394.4)	4×(−228.6)
$S_m^{\ominus}/\text{Ct mol}^{-1}$	270.0	5×205.2	3×213.8	4×188.8

a) Chemical drive $\mathcal{A}^{\ominus}$ for the oxidation process:

$$\mathcal{A}^{\ominus} = \{[1\times(-23.4)+5\times 0.0]-[3\times(-394.4)+4\times(-228.6)]\}\ \text{kG}$$

$$= 2074.2\ \text{kG} = \mathbf{2074.2\ kJ\ mol^{-1}}.$$

b) Molar reaction entropy $\Delta_R S^{\ominus}$:

$$\Delta_R S^{\ominus} = [(-1)\times 270.0+(-5)\times 205.2+3\times 213.8+4\times 188.8]\ \text{Ct mol}^{-1} = \mathbf{100.6\ Ct\ mol^{-1}}.$$

c) Usable energy W^*_{use}:

$$W^*_{\text{use}} = \eta\times\mathcal{A}^{\ominus}\times\Delta\xi = 0.8\times(2074.2\times 10^3\ \text{J mol}^{-1})\times 2\ \text{mol} = \mathbf{3.32\times 10^6\ J}.$$

Duration of operation Δt:

$$P = \frac{W^*_{\text{use}}}{\Delta t}.$$

Solving for the duration of operation Δt in question results in:

$$\Delta t = \frac{W^*_{\text{use}}}{P} = \frac{3.32\times 10^6\ \text{J mol}^{-1}}{250\ \text{J s}^{-1}} = 13.3\times 10^3\ \text{s} = \mathbf{3.7\ h}.$$

d) Generated entropy S_g:

$$S_g = \frac{(1-\eta)\times\mathcal{A}^{\ominus}\times\Delta\xi}{T^{\ominus}}.$$

$$S_g = \frac{(1-0.8)\times 2074.2\ \text{kJ mol}^{-1}\times 2\ \text{mol}}{298\ \text{K}} = 2780\ \text{J K}^{-1} = \mathbf{2.78\ kCt}.$$

Exchanged entropy S_e:

$$S_e = \Delta S - S_g = \Delta_R S^{\ominus}\times\Delta\xi - S_g.$$

$$S_e = 100.6\ \text{Ct mol}^{-1}\times 2\ \text{mol} - (2.78\times 10^3\ \text{Ct}) = -2580\ \text{Ct} = \mathbf{-2.58\ kCt}.$$

e) If the fuel cell is short-circuited, the efficiency is $\eta = 0$, i.e. no usable energy is gained anymore.

Generated entropy S'_g :

$$S'_g = \frac{(1-0)\times(2074.2\times10^3\ \text{J mol}^{-1})\times 2\ \text{mol}}{298\ \text{K}} = 13.92\times10^3\ \text{J K}^{-1} = \mathbf{13.92\ kCt}.$$

Exchanged entropy S'_e :

$$S'_e = 100.6\ \text{Ct mol}^{-1}\times 2\ \text{mol} - (13.92\times10^3\ \text{Ct}) = -13.72\times10^3\ \text{Ct} = \mathbf{-13.72\ kCt}.$$

The fuel cell releases a lot of entropy. It would get very hot if one did not let the entropy get away.

2.8.10 Calorimetric measurement of chemical drive

	Symbol	B	B′	D	D′
Conversion formula:		$C_2H_5OH\|l$ +	$3\ O_2\|g$ →	$2\ CO_2\|g$ +	$3\ H_2O\|g$

a) Expended electrical energy W':

The energy expended to electrically heat the calorimeter (including the sample contained therein) can be easily calculated from the voltage U, the current I and the turn-on duration Δt by means of Equation (8.28):

$$W' = U\times I\times\Delta t = 12\ \text{V}\times 1.5\ \text{A}\times 150\ \text{s} = 2700\ \text{J}.$$

Entropy S'_g generated by electric heating to calibrate the calorimeter:

The "burnt" energy, divided by the measured temperature, in this case approximately the standard temperature $T^\ominus$, results in the generated entropy S'_g [cf. Eq. (3.3)]:

$$S'_g = \frac{W'}{T^\ominus} = \frac{2700\ \text{J}}{298\ \text{K}} = 9.1\ \text{J K}^{-1} = 9.1\ \text{Ct}.$$

Entropy $-S_e$ emitted by the reacting substances during the combustion reaction:

The change of temperature ΔT observed due to the reaction is indicative of the entropy $-S_e$ emitted by the sample (see Section 8.8 in the textbook "Physical Chemistry from a Different Angle"):

$$S_e = -\frac{\Delta T}{\Delta T'}S'_g = -\frac{2.49\ \text{K}}{4.92\ \text{K}}\times 9.1\ \text{Ct} = \mathbf{-4.59\ Ct}.$$

b) Conversion $\Delta\xi$:

$$\Delta\xi = \frac{\Delta n_B}{\nu_B} = \frac{-0.001\ \text{mol}}{-1} = 0.001\ \text{mol}.$$

Change in entropy ΔS ($\equiv \Delta S_\ell$):

$$\Delta S \quad = \Delta_R S^\ominus \times \Delta\xi = -138.8\ \text{Ct mol}^{-1} \times 0.001\ \text{mol} = -0.139\ \text{Ct}.$$

Generated entropy S_g:

The entropy S_g generated during the reaction can be calculated by means of the following entropy balance:

$$S_g \quad = \Delta S - S_e = -0.139\ \text{Ct} - (-4.59\ \text{Ct}) = 4.45\ \text{Ct}.$$

Chemical drive $\mathcal{A}^\ominus$:

The generated entropy is related to the drive $\mathcal{A}^\ominus$ of the process by the following formula:

$$S_g \quad = \frac{\mathcal{A}^\ominus \times \Delta\xi}{T^\ominus}.$$

Solving for $\mathcal{A}^\ominus$ results in:

$$\mathcal{A}^\ominus \quad = \frac{T^\ominus \times S_g}{\Delta\xi} = \frac{298\ \text{K} \times 4.45\ \text{Ct}}{0.001\ \text{K}} = 1326 \times 10^3\ \text{J mol}^{-1} = \mathbf{1326\ kG}.$$

If one did not take the latent entropy into consideration, one would obtain for the drive $\mathcal{A}^{\ominus\prime}$:

$$\mathcal{A}^{\ominus\prime} \quad = \frac{T^\ominus \times (-S_e)}{\Delta\xi} = \frac{298\ \text{K} \times 4.59\ \text{Ct}}{0.001\ \text{K}} = 1368 \times 10^3\ \text{J mol}^{-1} = \mathbf{1368\ kG}$$

This would result in a value that is 42 kG and thus 3.2 % too high.

2.9 Coupling

2.9.1 Flip rule

We start with the pressure coefficient β of the chemical potential. It can be formulated as a differential quotient as follows: $\partial\mu/\partial p)_{T,n}$.

1) n is assigned to μ and $-V$ is assigned to p. This means that in the "flipped" differential quotient, n is in the denominator and $-V$ is in the numerator.
2) The sign must change because both n and $-V$ are "position-like" quantities.
3) In the original expression, p and T are unpaired and therefore to be put into the new index.

$$\left(\frac{\partial\mu}{\partial p}\right)_{T,n} \overset{1)}{\times} \left(\frac{\partial(-V)}{\partial n}\right) \xrightarrow{2)} \left(\frac{\partial V}{\partial n}\right) \xrightarrow{3)} \left(\frac{\partial V}{\partial n}\right)_{p,T}.$$

Finally, we obtain

$$\left(\frac{\partial\mu}{\partial p}\right)_{T,n} = \left(\frac{\partial V}{\partial n}\right)_{p,T} = V_{\mathrm{m}},$$

which was to be proven.

2.9.2 Increase in temperature by compression

a) We start with the differential quotient $(\partial V/\partial S)_p$.

   1) $-p$ is assigned to V and T is assigned to S. This means that in the "flipped" differential quotient, $-p$ is in the denominator and T is in the numerator.
   2) The sign must change because both $-p$ and T are "force-like" quantities.
   3) In the original expression, S is unpaired and therefore to be put into the new index.

   $$\left(\frac{\partial V}{\partial S}\right)_{p} \overset{1)}{\times} \left(\frac{\partial T}{\partial(-p)}\right) \xrightarrow{2)} \left(\frac{\partial T}{\partial p}\right) \xrightarrow{3)} \left(\frac{\partial T}{\partial p}\right)_{S}.$$

   Therefore, we finally obtain:

   $$\left(\frac{\partial V}{\partial S}\right)_{p} = \left(\frac{\partial T}{\partial p}\right)_{S}.$$

   The coefficient on the right side expresses that the temperature increases, meaning the object warms up when the pressure on the object increases. In this context, we assume tacitly that no entropy flows into the environment, which could compensate for the effect.

b) Ice water is one of the few exceptions of volume decreasing with an increase of entropy. Therefore, it becomes colder when compressed.

2.9.3 Change in volume of a concrete wall

Volume V of the concrete wall:

$$V = l \times h \times w = 10\ \text{m} \times 3\ \text{m} \times 0.2\ \text{m} = 6\ \text{m}^3 .$$

Change in volume ΔV:

One starts with Equation (9.21) for the volumetric thermal expansion coefficient γ:

$$\gamma = \frac{1}{V}\left(\frac{\partial V}{\partial T}\right)_p .$$

The difference quotient can be used to approximate the differential quotient:

$$\gamma = \frac{1}{V}\frac{\Delta V}{\Delta T} \quad \Rightarrow \Delta V = \gamma \times V \times \Delta T = \gamma \times V \times (T_2 - T_1) .$$

$$\Delta V = (36 \times 10^{-6}\ \text{K}^{-1}) \times 6\ \text{m}^3 \times (308\ \text{K} - 258\ \text{K}) = 0.011\ \text{m}^3 = \mathbf{11000\ cm^3} .$$

2.9.4* Gasoline barrel

Maximum volume of gasoline V_G which may be filled into the barrel:

When the steel barrel and the gasoline inside heat up, both the volume of the gasoline and the volume of the steel barrel increase. The volume of the gasoline at higher temperature results from the volume of the gasoline V_G filled into the barrel plus the increase in volume ΔV_G of the gasoline due to heating. Also in the case of the steel barrel, the increase in volume ΔV_S with increasing temperature has to be added to the initial volume V_S. Thereby, the increase in volume of the hollow barrel is exactly the same as that of a massive object of steel with the same volume. At 50 °C, the barrel should just be completely filled with gasoline, i.e. we have

$$V_G + \Delta V_G = V_S + \Delta V_S .$$

The changes in volume can be calculated again by means of Equation (9.21):

$$V_G + \gamma_G \times V_G \times \Delta T = V_S + \gamma_S \times V_S \times \Delta T .$$

$$V_G(1 + \gamma_G \times \Delta T) = V_S(1 + \gamma_S \times \Delta T) .$$

$$V_G = V_S \times \frac{1 + \gamma_S \times \Delta T}{1 + \gamma_G \times \Delta T} = V_S \times \frac{1 + \gamma_S \times (T_2 - T_1)}{1 + \gamma_G \times (T_2 - T_1)} .$$

$$V_G = (216 \times 10^{-3}\ \text{m}^3) \times \frac{1 + (35 \times 10^{-6}\ \text{K}^{-1}) \times (323\ \text{K} - 293\ \text{K})}{1 + (950 \times 10^{-6}\ \text{K}^{-1}) \times (323\ \text{K} - 293\ \text{K})} = \mathbf{210 \times 10^{-3}\ m^{-3}} .$$

No more than 210 L of gasoline may be filled into the steel barrel.

2.9.5 Compressibility of hydraulic oil

Compressibility χ of hydraulic oil:

According to Equation (9.22), the compressibility χ of a substance (at constant temperature) is:

$$\chi = -\frac{1}{V}\left(\frac{\partial V}{\partial p}\right)_T .$$

The difference quotient can be used to approximate the differential quotient:

$$\chi = -\frac{1}{V}\times\frac{\Delta V}{\Delta p} = -\frac{1}{V}\cdot\frac{(V_2 - V_1)}{(p_2 - p_1)} .$$

$$\chi = -\frac{1}{1.0000\times10^{-3}\ \text{m}^3}\times\frac{(0.9997-1.0000)\times10^{-3}\ \text{m}^3}{(20-10)\times10^5\ \text{Pa}} = \mathbf{3\times10^{-10}\ Pa^{-1}} .$$

2.9.6* Density of water in the depth of the ocean

Relative change $\Delta\rho/\rho$ of the density of water with the depth of the ocean:

We are interested in the relative change $\Delta\rho/\rho$ of the density of water as a function of an increase in pressure by the additional pressure Δp in the depth as a result of the hydrostatic pressure of water.

The density ρ of water at the surface is $\rho = m/V$, the density ρ_d at a depth of 200 m, however, $\rho_d = m/V_d$, where V and V_d are the volumes of a (small) portion of water of mass m, respectively.

The relative change of density then results in (V and V_d can be set approximately equal, since the change in volume is very small):

$$\frac{\Delta\rho}{\rho} = \left(\frac{m}{V_d}-\frac{m}{V}\right)\times\frac{V}{m} = \frac{V-V_d}{V_d\times V}\times V = -\frac{\Delta V}{V_d} \approx -\frac{\Delta V}{V} .$$

Change in volume ΔV of water:

The change in volume ΔV can be calculated by means of Equation (9.22) for the compressibility of a substance (at constant temperature):

$$\chi = -\frac{1}{V}\times\frac{\Delta V}{\Delta p} \quad\Rightarrow\quad \Delta V = -V\times\chi\times\Delta p .$$

Inserting this relationship into the equation for the relative change of density above results in:

$$\frac{\Delta\rho}{\rho} = -\frac{(-V\times\chi\times\Delta p)}{V} = \chi\times\Delta p = (4.6\times10^{-10}\ \text{Pa}^{-1})\times(2\times10^6\ \text{Pa}) = 0.00092 = \mathbf{0.092\ \%} .$$

The density of water at a depth of 200 m has only increased by 0.092 %.

2.9.7 Increase in pressure in a liquid-in-glass thermometer

Increase in pressure Δp in the capillary:

We are interested in the change in pressure Δp caused by heating a substance at constant volume, that is, the quotient $(\partial p/\partial T)_V$. This quotient can be transformed by means of the calculation rules for differential quotients presented in Section 9.4 of the textbook "Physical Chemistry from a Different Angle."

$$\left(\frac{\partial p}{\partial T}\right)_V = -\left(\frac{\partial p}{\partial V}\right)_T\left(\frac{\partial V}{\partial T}\right)_p = -\left(\frac{\partial V}{\partial T}\right)_p\left(\frac{\partial V}{\partial p}\right)_T^{-1} = -\frac{V\gamma}{-V\chi} = \frac{\gamma}{\chi}.$$

A few words of explanation: The initial quotient has the form $(\partial b/\partial b')_a$. The undesired quantity a in the index is inserted into the quotient leading to a negative sign. Subsequently, the first of the two new quotients of the type $(\partial b/\partial a)_{...}$ is inverted. Afterwards, all of the differential quotients already have the desired form. What now remains to be done is to replace the expressions with the usual coefficients [Eqs. (9.21) and (9.22)].

Again, the difference quotient can be used to approximate the differential quotient:

$$\frac{\Delta p}{\Delta T} = \frac{\gamma}{\chi}.$$

Solving for Δp results in:

$$\Delta p = \frac{\gamma}{\chi}\times\Delta T = \frac{1.1\times10^{-3}\ \mathrm{K}^{-1}}{1.5\times10^{-9}\ \mathrm{Pa}^{-1}}\times 5\ \mathrm{K} = \mathbf{37\times10^5\ Pa}\ (= 37\ \mathrm{bar}).$$

2.9.8 Isochoric molar entropy capacity

Isochoric molar entropy capacity $\mathcal{C}_{m,V}$ of ethanol:

The difference between the isobaric molar entropy capacity $\mathcal{C}_{m,p}$ and the isochoric molar entropy capacity $\mathcal{C}_{m,V}$ of a substance results according to Equation (9.25) in:

$$\mathcal{C}_{m,p} - \mathcal{C}_{m,V} = V_m\frac{\gamma^2}{\chi}.$$

The molar volume of a substance can be calculated from its density:

$$\rho = \frac{m}{V} = \frac{M\times n}{V} = \frac{M}{V_m} \quad\Rightarrow\quad V_m = \frac{M}{\rho}.$$

Inserting this relationship into the equation above yields:

$$\mathcal{C}_{m,p} - \mathcal{C}_{m,V} = \frac{M}{\rho}\times\frac{\gamma^2}{\chi} \quad\Rightarrow\quad \mathcal{C}_{m,V} = \mathcal{C}_{m,p} - \frac{M}{\rho}\times\frac{\gamma^2}{\chi}.$$

The molar mass of ethanol is $46.0\times10^{-3}\ \mathrm{kg\,mol^{-1}}$. Thus, we have:

$$\mathcal{C}_{m,V} = 0.370\ \text{Ct mol}^{-1}\,\text{K}^{-1} - \frac{46.0\times10^{-3}\ \text{kg mol}^{-1}}{789\ \text{kg m}^{-3}} \times \frac{(1.4\times10^{-3}\ \text{K}^{-1})^2}{11.2\times10^{-10}\ \text{Pa}^{-1}}$$

$$= 0.370\ \text{Ct mol}^{-1}\,\text{K}^{-1} - 0.102\ \text{Ct mol}^{-1}\,\text{K}^{-1} = \mathbf{0.268\ Ct\ mol^{-1}\,K^{-1}}.$$

In the calculation above, the units in the second summand are converted as follows:

$$\text{mol}^{-1}\,\text{m}^3\,\text{K}^{-2}\,\text{Pa} = \text{mol}^{-1}\,\text{m}^3\,\text{K}^{-2}\,\text{N}\,\text{m}^{-2} = \text{N}\,\text{m}\,\text{K}^{-2}\,\text{mol}^{-1} = \text{J}\,\text{K}^{-1}\,\text{mol}^{-1}\,\text{K}^{-1} = \text{Ct}\,\text{mol}^{-1}\,\text{K}^{-1}.$$

2.9.9* Isentropic compressibility

a) Relationship between isothermal compressibility χ_T and isentropic compressibility χ_S:

We start with the following equation:

$$\chi_S = -\frac{1}{V}\left(\frac{\partial V}{\partial p}\right)_S .$$

With the help of the calculation rules for differential quotients, we obtain:

$$\chi_S = -\frac{1}{V}\left(\frac{\partial V}{\partial p}\right)_S = -\frac{1}{V}\left[\left(\frac{\partial V}{\partial p}\right)_T + \left(\frac{\partial V}{\partial T}\right)_p\left(\frac{\partial T}{\partial p}\right)_S\right] = -\frac{1}{V}\left(\frac{\partial V}{\partial p}\right)_T - \frac{1}{V}\left(\frac{\partial V}{\partial T}\right)_p\left(\frac{\partial T}{\partial p}\right)_S$$

$$= \chi_T - \gamma\times\left(\frac{\partial T}{\partial p}\right)_S = \chi_T - \gamma\times\left[-\left(\frac{\partial T}{\partial S}\right)_p\left(\frac{\partial S}{\partial p}\right)_T\right] = \chi_T - \gamma\times\left[-\left(\frac{\partial S}{\partial T}\right)_p^{-1}\left(\frac{\partial(-V)}{\partial T}\right)_p\right]$$

$$= \chi_T - \gamma\times\left[-\frac{1}{n\mathcal{C}_m}\times -V\gamma\right] = \chi_T - \gamma\times\left[\frac{M}{V\rho\mathcal{C}_m}\times V\gamma\right] = \chi_T - \frac{M\gamma^2}{\rho\mathcal{C}_m}.$$

Explanation to the calculation procedure: First, the quantity S in the index of the differential quotient is replaced by T (rule d). Afterwards, two of the differential quotients can already be substituted by the desired coefficients χ_T and γ [Eqs. (9.22) and (9.21)]. Subsequently, the quantity S is inserted from the index in the remaining differential quotient itself (rule c). The first of the new differential quotients is then inverted (rule a) and the second one is "flipped" (see the example to the flip rule in Section 9.2 of the textbook "Physical Chemistry from a Different Angle"). The inverted differential quotient is replaced by $n\mathcal{C}_m$ according to Equation (3.13), the "flipped" one by $-V\gamma$. Finally, n is replaced according to Equation (1.6) by m/M or $\rho V/M$, respectively (since $m = \rho V$).

b) Isentropic compressibility χ_S of ethanol:

$$\chi_S = \chi_T - \frac{M\gamma^2}{\rho\mathcal{C}_m}.$$

$$\chi_S = (11.2\times10^{-10}\ \text{Pa}^{-1}) - \frac{(46.0\times10^{-3}\ \text{kg mol}^{-1})\times(1.40\times10^{-3}\ \text{K}^{-1})^2}{789\ \text{kg m}^{-3}\times0.370\ \text{Ct mol}^{-1}\,\text{K}^{-1}}$$

$$\chi_S = (11.2\times10^{-10}\ \text{Pa}^{-1}) - (3.1\times10^{-10}\ \text{Pa}^{-1}) = \mathbf{8.1\times10^{-10}\ Pa^{-1}}.$$

The units in the second summand are converted as follows:

$$\mathrm{m^3\,K^{-1}\,Ct^{-1} = m^3\,K^{-1}\,J^{-1}\,K = m^3\,N^{-1}\,m^{-1} = N^{-1}\,m^2 = Pa^{-1}}\,.$$

c) Speed of sound c in ethanol:

$$c = \sqrt{\frac{1}{\chi_S \times \rho}}\,.$$

$$c = \sqrt{\frac{1}{(8.1\times10^{-10}\ \mathrm{Pa^{-1}})\times 789\ \mathrm{kg\,m^{-3}}}} = \mathbf{1250\ m\,s^{-1}}\,.$$

The units in the term below the root are converted as follows:

$$\mathrm{Pa\,kg^{-1}\,m^3 = N\,m^{-2}\,kg^{-1}\,m^3 = kg\,m\,s^{-2}\,kg^{-1}\,m = m^2\,s^{-2}}\,.$$

2.9.10* Compression of ice water

Change in temperature ΔT when ice water is compressed:

We start with the relationship

$$\Delta T = \left(\frac{\partial T}{\partial p}\right)_S \times \Delta p\,.$$

First, the quantity S from the index is inserted (rule c):

$$\Delta T = -\left(\frac{\partial T}{\partial S}\right)_p \times \left(\frac{\partial S}{\partial p}\right)_T \times \Delta p\,.$$

Subsequently, the first differential quotient is inverted (rule a) and the second one is "flipped" (see Section 9.2 Flip rule):

$$\Delta T = \frac{1}{\left(\frac{\partial S}{\partial T}\right)_p} \times \left(\frac{\partial V}{\partial T}\right)_p \times \Delta p\,.$$

If 1 mole of water is considered, the corresponding molar quantities can be used:

$$\Delta T = \frac{\left(\frac{\partial V_\mathrm{m}}{\partial T}\right)_p}{\left(\frac{\partial S_\mathrm{m}}{\partial T}\right)_p} \times \Delta p = \frac{V_\mathrm{m} \times \gamma}{\mathcal{C}_\mathrm{m}} \times \Delta p\,.$$

If the given values are inserted into the equation, one obtains:

$$\Delta T = \frac{(18\times10^{-6}\ \mathrm{m^3\,mol^{-1}})\times(-70\times10^{-6}\ \mathrm{K^{-1}})}{0.28\ \mathrm{Ct\,K^{-1}\,mol^{-1}}} \times (1000\times10^5\ \mathrm{Pa}) = \mathbf{-0.45\ K}\,.$$

2.10 Molecular-Kinetic View of Dilute Gases

2.10.1 Gas cylinder

a) Amount n of nitrogen in the gas cylinder:

Since nitrogen can be treated as ideal in this case, we can use the general gas law [Eq. (10.7)]:

$$p_0 V_0 = nRT_0 .$$

Solving for n results in

$$n = \frac{p_0 V_0}{RT_0} = \frac{(5\times10^6\ \text{Pa})\times(50\times10^{-3}\ \text{m}^3)}{8.314\ \text{G}\,\text{K}^{-1}\times 293\ \text{K}} = \mathbf{102.6\ mol}.$$

Conversion of the units:

$$\frac{\text{Pa}\,\text{m}^3}{\text{G}\,\text{K}^{-1}\,\text{K}} = \frac{\text{N}\,\text{m}^{-2}\,\text{m}^3}{\text{J}\,\text{mol}^{-1}} = \frac{\text{N}\,\text{m}\,\text{m}^3}{\text{N}\,\text{m}\,\text{mol}^{-1}} = \text{mol}.$$

Mass m of nitrogen:

$$m = n\times M = 102.6\ \text{mol}\times(28.0\times10^{-3}\ \text{kg}\,\text{mol}^{-1}) = \mathbf{2.87\ kg}.$$

b) Pressure p_1 of the nitrogen at a temperature ϑ_1 of 45 °C:

We start again from the general gas law and obtain, since the amount of the gas as well as the volume remain constant:

$$p_1 V_0 = nRT_1 .$$

Solving for p_1 results in:

$$p_1 = \frac{nRT_1}{V_0} = \frac{102.6\ \text{mol}\times 8.314\ \text{G}\,\text{K}^{-1}\times 318\ \text{K}}{50\times10^{-3}\ \text{m}^3} = 5.43\times10^6\ \text{Pa} = \mathbf{5.43\ MPa}.$$

The pressure inside the steel cylinder increases to 5.43 MPa.

2.10.2* Weather balloon

Volume V_0 of the weather balloon at sea level:

$$V_0 = \frac{4}{3}\pi r_0^3 = \frac{4}{3}\times 3.142\times(1.5\ \text{m})^3 = 14.1\ \text{m}^3 .$$

Volume V_1 of the weather balloon at an altitude of 10 km:

The formula for the general gas law at sea level is $p_0 V_0 = nRT_0$ und at an altitude of 10 km correspondingly $p_1 V_1 = nRT_1$. Since the amount n of gas in the balloon remains the same, we have:

$$\frac{p_0 V_0}{T_0} = \frac{p_1 V_1}{T_1} .$$

Solving for V_1 results in:

$$V_1 = \frac{p_0 V_0 T_1}{T_0 p_1} = \frac{(100\times10^3 \text{ Pa})\times14.1 \text{ m}^3 \times 223 \text{ K}}{298 \text{ K}\times(30\times10^3 \text{ Pa})} = \mathbf{35.2\ m^3} .$$

Radius r_1 of the weather balloon at an altitude of 10 km:

$$V_1 = \frac{4}{3}\pi r_1^3 .$$

By solving for r_1 we get:

$$r_1 = \sqrt[3]{\frac{3V_1}{4\pi}} = \sqrt[3]{\frac{3\times35.2 \text{ m}^3}{4\times3.142}} = \mathbf{2.03\ m} .$$

The volume and the radius of the weather balloon have increased noticeably.

2.10.3 Potassium superoxide as lifesaver

	Symbol	B		B′		D		D′
Conversion formula:		$4\ KO_2\vert s$	+	$2\ CO_2\vert g$	→	$2\ K_2CO_3\vert s$	+	$3\ O_2\vert g$.

Amount $n_{B',0}$ of carbon dioxide that is present at the beginning:

$$pV_{B'} = n_{B',0}RT .$$

Solving for $n_{B',0}$ results in:

$$n_{B',0} = \frac{pV_{B',0}}{RT} = \frac{(100\times10^3 \text{ Pa})\times(40\times10^{-3} \text{ m}^3)}{8.314 \text{ G K}^{-1}\times283 \text{ K}} = 1.7 \text{ mol} .$$

Change in amount $\Delta n_{B'}$ which is to be achieved by binding the carbon dioxide:

$$\Delta n_{B'} = n_{B'} - n_{B',0} = 0-(1.7 \text{ mol}) = -1.7 \text{ mol} .$$

Since the carbon dioxide is supposed to be completely bound at the end of the reaction, we have $n_{B'} = 0$.

Required change in amount Δn_B of the potassium superoxide:

Due to the basic stoichiometric equation [Eq. (1.15)] we have:

$$\frac{\Delta n_B}{\nu_B} = \frac{\Delta n_{B'}}{\nu_{B'}} \quad\Rightarrow\quad \Delta n_B = \frac{\Delta n_{B'}\times\nu_B}{\nu_{B'}} .$$

$$\Delta n_B = \frac{(-1.7 \text{ mol})\times(-4)}{(-2)} = -3.4 \text{ mol} .$$

Amount $n_{B,0}$ of potassium superoxide that has to be used:

$$\Delta n_B = n_B - n_{B,0} \quad \Rightarrow \quad n_{B,0} = n_B - \Delta n_B .$$

$$n_{B,0} = 0 - (-3.4\ \text{mol}) = 3.4\ \text{mol} .$$

Since the potassium superoxide is also supposed to be completely used up at the end, we have $n_B = 0$.

Mass $m_{B,0}$ of potassium superoxide that has to be used:

$$m_{B,0} = n_{B,0} \times M_B = 3.4\ \text{mol} \times (71.1 \times 10^{-3}\ \text{kg mol}^{-1}) = 0.242\ \text{kg} = \mathbf{242\ g} .$$

2.10.4* Divers' disease

a) Blood volume V_{Bl} of the diver:

$$V_{Bl} = 80\ \text{kg} \times (70 \times 10^{-6}\ \text{m}^3\ \text{kg}^{-1}) = 5.6 \times 10^{-3}\ \text{m}^3 .$$

Partial pressure $p(B|g)_0$ of nitrogen at sea level:

$$p(B|g)_0 = 0.78 \times p_{total,0} = 0.78 \times 100\ \text{kPa} = 78\ \text{kPa} .$$

Amount $n(B|Bl)_0$ of nitrogen dissolved in the diver's blood at sea level:

In order to calculate the amount of nitrogen dissolved in the diver's blood, HENRY's gas law [Eq. (6.37)] is used:

$$\overset{\circ}{K}_{gd} = \frac{c(B|Bl)_0}{p(B|g)_0} = \frac{n(B|Bl)_0}{V_{Bl} \times p(B|g)_0} .$$

Solving for $n(B|Bl)_0$ yields:

$$n(B|Bl)_0 = \overset{\circ}{K}_{gd} \times V_{Bl} \times p(B|g)_0 .$$

$$n(B|Bl)_0 = (5.45 \times 10^{-6}\ \text{mol m}^{-3}\ \text{Pa}^{-1}) \times (5.6 \times 10^{-3}\ \text{m}^3) \times (78 \times 10^3\ \text{Pa})$$
$$= \mathbf{2.38 \times 10^{-3}\ mol} .$$

b) Partial pressure $p(B|g)_1$ of nitrogen at a depth of 20 m:

$$p(B|g)_1 = 0.78 \times p_{total,1} = 0.78 \times 300\ \text{kPa} = 234\ \text{kPa} .$$

Amount $n(B|Bl)_1$ of nitrogen dissolved in the diver's blood at a depth of 20 m:

$$n(B|Bl)_1 = \overset{\circ}{K}_{gd} \times V_{Bl} \times p(B|g)_1 .$$

$$n(B|Bl)_1 = (5.45 \times 10^{-6}\ \text{mol m}^{-3}\ \text{Pa}^{-1}) \times (5.6 \times 10^{-3}\ \text{m}^3) \times (234 \times 10^3\ \text{Pa})$$
$$= \mathbf{7.14 \times 10^{-3}\ mol} .$$

c) Amount $\Delta n(\mathrm{B|g})_2$ of nitrogen released in the diver's bloodstream:

$$\Delta n(\mathrm{B|g})_2 = n(\mathrm{B|Bl})_1 - n(\mathrm{B|Bl})_0 = (7.14\times10^{-3}\ \mathrm{mol}) - (2.38\times10^{-3}\ \mathrm{mol})$$

$$= 4.76\times10^{-3}\ \mathrm{mol}\,.$$

Volume $\Delta V(\mathrm{B|g})_2$ of N_2 gas:

$$p_{\mathrm{total,0}} \times \Delta V(\mathrm{B|g})_2 = \Delta n(\mathrm{B|g})_2 \times R \times T_{\mathrm{Bl}}\,.$$

Solving for $\Delta V(\mathrm{B|g})_2$ results in:

$$\Delta V(\mathrm{B|g})_2 = \frac{\Delta n(\mathrm{B|g})_2 \times R \times T_{\mathrm{Bl}}}{p_{\mathrm{total,0}}}\,.$$

$$\Delta V(\mathrm{B|g})_2 = \frac{(4.76\times10^{-3}\ \mathrm{mol})\times 8.314\ \mathrm{G\,K^{-1}}\times 310\ \mathrm{K}}{100\times10^{3}\ \mathrm{Pa}}$$

$$= 0.12\times10^{-3}\ \mathrm{m^3} = 0.12\ \mathrm{L} = \mathbf{120\ mL}\,.$$

Conversion of the units:

$$\frac{\mathrm{mol\,G\,K^{-1}\,K}}{\mathrm{Pa}} = \frac{\mathrm{mol\,J\,mol^{-1}}}{\mathrm{Pa}} = \frac{\mathrm{N\,m}}{\mathrm{N\,m^{-2}}} = \mathrm{m^3}\,.$$

2.10.5* Volumetric determination of gases

a) Saturation concentration $c(\mathrm{B|w})$ of oxygen in water:

The saturation concentration of oxygen in water is calculated using HENRY's gas law:

$$K_{\mathrm{gd}}^{\ominus} = \frac{c(\mathrm{B|w})}{p(\mathrm{B|g})}\,.$$

Solving for $c(\mathrm{B|w})$ results in:

$$c(\mathrm{B|w}) = K_{\mathrm{gd}}^{\ominus} \times p(\mathrm{B|g}) = (1.3\times10^{-5}\ \mathrm{mol\,m^{-3}\,Pa^{-1}})\times(100\times10^{3}\ \mathrm{Pa}) = \mathbf{1.3\ mol\ m^{-3}}\,.$$

b) Conversion formula: $\mathrm{H_2O|l} \rightleftarrows \mathrm{H_2O|g}$.

$\mu^{\ominus}/\mathrm{kG}$ −237,14 −228,58

Saturation vapor pressure $p_{\mathrm{lg,D}}$ of water:

The saturation vapor pressure of water, i.e. the vapor pressure of water vapor in equilibrium with the liquid under the chosen conditions, can be calculated in the following way [if we proceed as we did in deriving Equation (5.19)]:

$$\mu^{\ominus}(\mathrm{D|l}) = \mu^{\ominus}(\mathrm{D|g}) + RT^{\ominus} \ln \frac{p_{\mathrm{lg,D}}}{p^{\ominus}}\,.$$

We can neglect the pressure dependence of the chemical potential of the liquid because, in comparison to that of a gas, it is lower by three orders of magnitude. Solving for $p_{\mathrm{lg,D}}$ results in:

$$p_{\mathrm{lg,D}} = p^{\ominus} \exp\frac{\mu^{\ominus}(\mathrm{D|l}) - \mu^{\ominus}(\mathrm{D|g})}{RT^{\ominus}}.$$

$$p_{\mathrm{lg,D}} = (100\times10^3\ \mathrm{Pa})\times\exp\frac{(-237.14\times10^3\ \mathrm{G})-(-228.58\times10^3\ \mathrm{G})}{8.314\ \mathrm{G\,K^{-1}}\times298\ \mathrm{K}} = \mathbf{3.16\times10^3\ Pa}.$$

c) Amount $n(\mathrm{B|w})$ of oxygen dissolved in 50 $\mathrm{cm^3}$ of water:

$$c(\mathrm{B|w}) = \frac{n(\mathrm{B|w})}{V_{\mathrm{D}}} \quad\Rightarrow\quad n(\mathrm{B|w}) = c(\mathrm{B|w})\times V_{\mathrm{D}}.$$

$$n(\mathrm{B|w}) = 1.3\ \mathrm{mol\,m^{-3}}\times(50\times10^{-6}\ \mathrm{m^3}) = \mathbf{65\times10^{-6}\ mol}.$$

Volume $V(\mathrm{B|g})$ of oxygen corresponding to the dissolved amount:

The gas volume $V(\mathrm{B|g})$ can be calculated by means of the general gas law [Eq. (10.7)]:

$$p^{\ominus}V(\mathrm{B|g}) = n(\mathrm{B|w})RT^{\ominus} \quad\Rightarrow\quad V(\mathrm{B|g}) = \frac{n(\mathrm{B|w})RT^{\ominus}}{p^{\ominus}}.$$

$$V(\mathrm{B|g}) = \frac{(65\times10^{-6}\ \mathrm{mol})\times8.314\ \mathrm{G\,K^{-1}}\times298\ \mathrm{K}}{100\times10^3\ \mathrm{Pa}} = 1.61\times10^{-6}\ \mathrm{m^3} = \mathbf{1.61\ cm^3}.$$

Because of the solubility of oxygen in water, the volume of gas collected in the eudiometer is reduced by 1.61 $\mathrm{cm^3}$.

d) Total pressure $p_{\mathrm{total,Eu}}$ in the gas-filled part of the eudiometer, if the gas volume is 50 $\mathrm{cm^3}$:

If the gas volume is 50 $\mathrm{cm^3}$, the water levels inside and outside are almost equal according to the figure so that also the pressures inside and outside become equal:

$$p_{\mathrm{total,Eu}} = \mathbf{100\ kPa}.$$

Partial pressure $p(\mathrm{B|g})_{\mathrm{Eu}}$ of the oxygen gas:

According to the law of DALTON, the total pressure p_{total} of a gas mixture is equal to the sum of the partial pressures $p_1, p_2, \ldots$ of all components present [Eq. (6.25)]:

$$p_{\mathrm{ges}} = p_1 + p_2 + \ldots.$$

In the oxygen-water vapor mixture in question, the water vapor exerts the (saturation) partial pressure $p_{\mathrm{lg,D}}$.

$$p_{\mathrm{total,Eu}} = p(\mathrm{B|g})_{\mathrm{Eu}} + p_{\mathrm{lg,D}} \quad\Rightarrow\quad p(\mathrm{B|g})_{\mathrm{Eu}} = p_{\mathrm{total,Eu}} - p_{\mathrm{lg,D}}.$$

$$p(\mathrm{B|g})_{\mathrm{Eu}} = (100\times10^3\ \mathrm{Pa}) - (3.16\times10^3\ \mathrm{Pa}) \approx \mathbf{96.8\ kPa}.$$

e) Contribution $V(\mathrm{D|g})$ of water vapor to the total volume of gas of 50 cm^3:

The partial pressure of a gaseous component corresponds to the pressure that this component would have if it alone occupied the available volume V_{total} . Accordingly, we have $p_1 V_{\text{total}} = n_1 RT$, $p_2 V_{\text{total}} = n_2 RT$, … and finally $p_{\text{total}} V_{\text{total}} = n_{\text{total}} RT$, i.e. we obtain:

$$\frac{p_1}{n_1} = \frac{p_2}{n_2} = \ldots = \frac{p_{\text{total}}}{n_{\text{total}}} .$$

It is also $V_{\text{total}} = V_1 + V_2 + \ldots$ and thus

$$\frac{V_1}{n_1} = \frac{V_2}{n_2} = \ldots = \frac{V_{\text{total}}}{n_{\text{total}}} .$$

Therefore, we have

$$\frac{p_1}{p_{\text{total}}} = \frac{n_1}{n_{\text{total}}} = \frac{V_1}{V_{\text{total}}} \text{ and so on.}$$

The volume $V(\mathrm{D|g})$ of the water vapor can thus be calculated as follows:

$$\frac{p_{\mathrm{lg,D}}}{V(\mathrm{D|g})} = \frac{p_{\mathrm{total,Eu}}}{V_{\mathrm{total,Eu}}} \Rightarrow V(\mathrm{D|g}) = \frac{p_{\mathrm{lg,D}} \times V_{\mathrm{total,Eu}}}{p_{\mathrm{total,Eu}}} .$$

$$V(\mathrm{D|g}) = \frac{(3.16\times10^3\ \mathrm{Pa})\times(50\times10^{-6}\ \mathrm{m^3})}{(100\times10^3\ \mathrm{Pa})} = 1.58\times10^{-6}\ \mathrm{m^3} = \mathbf{1.58\ cm^3} .$$

The gas volume collected in the eudiometer is increased by 1.58 cm^3 due to the water vapor.

1.10.6* Inflating a bicycle tire

Energy W required to inflate the tire:

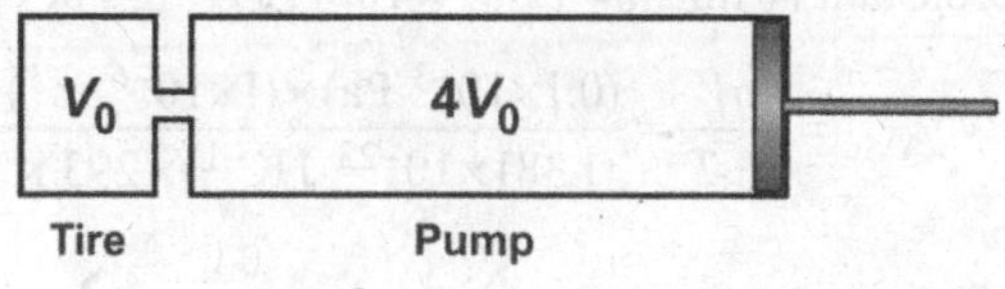

The expenditure of energy W for the compression of an air volume of V_1 = 10 L [volume V_0 of the tire of 2 L and cylinder capacity of the pump of $4V_0$ = 8 L (see figure on the right), filled with air at a pressure of p_1 = 1 bar] to a volume of V_2 = 2 L (tire filled with air at a pressure of p_2 = 5 bar; pump airless because the piston is fully pushed into the cylinder) should be calculated.

If the external pressure were $p_1 = 0$, we would have according to Equation (2.6)

$$\mathrm{d}W = -p\mathrm{d}V .$$

However, since the external pressure p_1 facilitates the pushing of the piston into the cylinder, we have:

$$\mathrm{d}W = -(p - p_1)\mathrm{d}V .$$

Since the pressure p of a given amount of gas at constant temperature increases inversely proportional to its volume V [see Eq. (10.1)], we have

$pV = p_1V_1$ (= const.). and therefore

$$p = \frac{p_1V_1}{V}.$$

Inserting this expression in the above equation results in:

$$W = -\int_{V_1}^{V_2=V_0}(p-p_1)\mathrm{d}V = \int_{V_0}^{V_1}\left(\frac{p_1V_1}{V}-p_1\right)\mathrm{d}V = p_1V_1\int_{V_0}^{V_1}\frac{1}{V}\mathrm{d}V - p_1\int_{V_0}^{V_1}\mathrm{d}V.$$

In this case, the rule of integration was applied that a definite integral changes sign when interchanging the limits of integration.

$$W = p_1V_1\times\ln V\Big|_{V_0}^{V_1} - p_1\times V\Big|_{V_0}^{V_1} = p_1V_1\times\ln\frac{V_1}{V_0} - p_1\times(V_1-V_0) = p_1V_1\left(\ln\frac{V_1}{V_0}-1+\frac{V_0}{V_1}\right).$$

$$W = (100\times10^3\ \mathrm{Pa})\times(10\times10^{-3}\ \mathrm{m}^3)\times\left(\ln\frac{10\times10^{-3}\ \mathrm{m}^3}{2\times10^{-3}\ \mathrm{m}^3}-1+\frac{2\times10^{-3}\ \mathrm{m}^3}{10\times10^{-3}\ \mathrm{m}^3}\right) = \mathbf{809\ J}.$$

2.10.7 Cathode ray tube

Number N of gas molecules in each cm^3 of a cathode ray tube:

We can give the general gas law a slightly different form by means of the relations $n = N\times\tau$ [Gl. (1.2)] und $k_\mathrm{B} = R\times\tau$:

$$pV = N\tau\times\frac{k_\mathrm{B}}{\tau}\times T = Nk_\mathrm{B}T.$$

N is the particle number, τ the elementary amount (of substance) and k_B the BOLTZMANN constant (with $k_\mathrm{B} = 1.381\times10^{-23}\ \mathrm{J\,K^{-1}}$). For the particle number N we thus obtain:

$$N = \frac{pV}{k_\mathrm{B}T} = \frac{(0.1\times10^{-3}\ \mathrm{Pa})\times(1\times10^{-6}\ \mathrm{m}^3)}{(1.381\times10^{-23}\ \mathrm{J\,K^{-1}})\times293\ \mathrm{K}} = \mathbf{2.5\times10^{10}}.$$

2.10.8 Speed and kinetic energy of translation of gas molecules

For the calculation, we need the following relationships, the equation for the root mean square speed of gas molecules [Eq. (10.32)],

$$\sqrt{\overline{v^2}} = \sqrt{3\frac{RT}{M}},$$

and the equation for the average molar kinetic energy [Eq. (10.30)],

$$\overline{W}_\mathrm{kin,m} = \tfrac{1}{2}M\overline{v^2}.$$

a) Root mean square speed of hydrogen molecules at $T_1 = 298$ K and $T_2 = 800$ K:

$$\sqrt{\overline{v_1^2}} = \sqrt{3\frac{RT_1}{M(\mathrm{H_2})}} = \sqrt{3\frac{8.314\ \mathrm{G\,K^{-1}} \times 298\ \mathrm{K}}{2.0\times10^{-3}\ \mathrm{kg\,mol^{-1}}}} = \mathbf{1.93\times10^{-3}\ m\,s^{-1}}.$$

Conversion of the units:

$$\frac{\mathrm{G\,K^{-1}\,K}}{\mathrm{kg\,mol^{-1}}} = \frac{\mathrm{J\,mol^{-1}}}{\mathrm{kg\,mol^{-1}}} = \frac{\mathrm{kg\,m^2\,s^{-2}}}{\mathrm{kg}} = \mathrm{m^2\,s^{-2}}$$

$$\sqrt{\overline{v_2^2}} = \sqrt{3\frac{RT_2}{M(\mathrm{H_2})}} = \sqrt{3\frac{8.314\ \mathrm{G\,K^{-1}} \times 800\ \mathrm{K}}{2.0\times10^{-3}\ \mathrm{kg\,mol^{-1}}}} = \mathbf{3.16\times10^{-3}\ m\,s^{-1}}.$$

Average molar kinetic energy of hydrogen molecules at $T_1 = 298$ K and $T_2 = 800$ K:

$$\overline{W}_{\mathrm{kin,m,1}} = \tfrac{1}{2} M(\mathrm{H_2}) \times \overline{v_1^2} = \tfrac{1}{2}(2.0\times10^{-3}\ \mathrm{kg\,mol^{-1}}) \times (1.93\times10^3\ \mathrm{m\,s^{-1}})^2$$

$$= 3.72\times10^3\ \mathrm{kg\,m^2\,s^{-2}\,mol^{-1}} = 3.72\times10^3\ \mathrm{J\,mol^{-1}} = \mathbf{3.72\ kJ\ mol^{-1}}.$$

$$\overline{W}_{\mathrm{kin,m,2}} = \tfrac{1}{2} M(\mathrm{H_2}) \times \overline{v_2^2}.$$

$$= \tfrac{1}{2}(2.0\times10^{-3}\ \mathrm{kg\,mol^{-1}}) \times (3.16\times10^3\ \mathrm{m\,s^{-1}})^2 = \mathbf{9.98\ kJ\ mol^{-1}}.$$

b) Root mean square speed of oxygen molecules at $T_1 = 298$ K and $T_2 = 800$ K:

$$\sqrt{\overline{v_1^2}} = \sqrt{3\frac{RT_1}{M(\mathrm{O_2})}} = \sqrt{3\frac{8.314\ \mathrm{G\,K^{-1}} \times 298\ \mathrm{K}}{32.0\times10^{-3}\ \mathrm{kg\,mol^{-1}}}} = \mathbf{482\ m\,s^{-1}}.$$

$$\sqrt{\overline{v_2^2}} = \sqrt{3\frac{RT_2}{M(\mathrm{O_2})}} = \sqrt{3\frac{8.314\ \mathrm{G\,K^{-1}} \times 800\ \mathrm{K}}{32.0\times10^{-3}\ \mathrm{kg\,mol^{-1}}}} = \mathbf{790\ m\,s^{-1}}.$$

Average molar kinetic energy of oxygen molecules at $T_1 = 298$ K and $T_2 = 800$ K:

$$\overline{W}_{\mathrm{kin,m,1}} = \tfrac{1}{2} M(\mathrm{O_2}) \times \overline{v_1^2} = \tfrac{1}{2}(32.0\times10^{-3}\ \mathrm{kg\,mol^{-1}}) \times (482\ \mathrm{m\,s^{-1}})^2 = \mathbf{3.72\ kJ\ mol^{-1}}.$$

$$\overline{W}_{\mathrm{kin,m,2}} = \tfrac{1}{2} M(\mathrm{O_2}) \times \overline{v_2^2} = \tfrac{1}{2}(32.0\times10^{-3}\ \mathrm{kg\,mol^{-1}}) \times (790\ \mathrm{m\,s^{-1}})^2 = \mathbf{9.98\ kJ\ mol^{-1}}.$$

Conclusion: The average molar kinetic energy of hydrogen and oxygen molecules is the same at a given temperature. Although the hydrogen molecules are considerably faster than the oxygen molecules, they are also considerably lighter. Equation (10.27) proves that the kinetic energy of translation does not depend from the molar mass:

$$\overline{W}_{\mathrm{kin,m}} = \tfrac{3}{2} RT.$$

2.10.9 Speed of air molecules

Ratio of the root mean square speed of air molecules at different air temperatures:

$$\frac{\sqrt{\overline{v_2^2}}}{\sqrt{\overline{v_1^2}}} = \sqrt{\frac{3RT_2}{M} \times \frac{M}{3RT_1}} = \sqrt{\frac{T_2}{T_1}} = \sqrt{\frac{263\ \text{K}}{308\ \text{K}}} = \mathbf{0.924}.$$

On the winter day, the root mean square speed of the air molecules is 7.6 % lower than on the summer day.

2.10.10* MAXWELL's speed distribution

Fraction of N_2 molecules with a speed between 299.5 m s^{-1} and 300.5 m s^{-1} at 25 °C:

We start with the (corrected) expression

$$\frac{dN(v)}{N} = 4\pi\left(\sqrt{\frac{m}{2\pi k_B T}}\right)^3 \exp\left(-\frac{mv^2}{2k_B T}\right) v^2 dv \qquad \text{Eq. (10.54)}$$

for MAXWELL's speed distribution. Since the speed interval in question is very small, we can write in good approximation (m/k_B was replaced by M/R)

$$\frac{\Delta N(v)}{N} = 4\pi\left(\sqrt{\frac{m}{2\pi k_B T}}\right)^3 \exp\left(-\frac{mv^2}{2k_B T}\right) v^2 \Delta v = 4\pi\left(\sqrt{\frac{M}{2\pi RT}}\right)^3 \exp\left(-\frac{Mv^2}{2RT}\right) v^2 \Delta v$$

and insert for v the value in the middle of the interval, i.e. 300 m s^{-1}:

$$\frac{\Delta N(v)}{N} = 4\times 3.142\left(\sqrt{\frac{28.0\times 10^{-3}\ \text{kg mol}^{-1}}{2\times 3.142\times 8.314\ \text{G K}^{-1}\times 298\ \text{K}}}\right)^3 \times$$

$$\exp\left(-\frac{(28.0\times 10^{-3}\ \text{kg mol}^{-1})\times(300\ \text{m s}^{-1})^2}{2\times 8.314\ \text{G K}^{-1}\times 298\ \text{K}}\right)\times(300\ \text{m s}^{-1})^2\times 1\ \text{m s}^{-1}$$

$$= \mathbf{1.64\times 10^{-3}}.$$

2.11 Substances with Higher Density

2.11.1 VAN DER WAALS equation

a) Pressure p of the carbon dioxide (VAN DER WAALS equation):

The VAN DER WAALS equation [Eq. (11.5)] should be used to calculate the carbon dioxide pressure:

$$p = \frac{nRT}{V - bn} - \frac{an^2}{V^2}.$$

The VAN DER WAALS constants of carbon dioxide result from Table 11.1 in the textbook "Physical Chemistry from a Different Angle": $a = 0.366\ \text{Pa}\,\text{m}^6\,\text{mol}^{-2}$ and $b = 4.3 \times 10^{-5}\ \text{m}^3\,\text{mol}^{-1}$.

Amount n of carbon dioxide:

The amount n of carbon dioxide can be calculated from its mass m using its molar mass $M = 44.0 \times 10^{-3}\ \text{kg}\,\text{mol}^{-1}$:

$$n = \frac{m}{M} = \frac{0.352\ \text{kg}}{44.0 \times 10^{-3}\ \text{kg}\,\text{mol}^{-1}} = 8.00\ \text{mol}.$$

$$p = \frac{8.00\ \text{mol} \times 8.314\ \text{G}\,\text{K}^{-1} \times 300\,\text{K}}{(5.00 \times 10^{-3}\ \text{m}^3) - (4.3 \times 10^{-5}\ \text{m}^3\,\text{mol}^{-1}) \times 8.00\ \text{mol}} - \frac{0.366\ \text{Pa}\,\text{m}^6\,\text{mol}^{-2} \times (8.00\ \text{mol})^2}{(5.00 \times 10^{-3}\ \text{m}^3)^2}$$

$$= (4.29 \times 10^6\ \text{Pa}) - (0.94 \times 10^6\ \text{Pa}) = 3.35 \times 10^6\ \text{Pa} = \mathbf{3.35\ MPa}.$$

b) Pressure p of the carbon dioxide (general gas law):

For comparison, the carbon dioxide should be treated as ideal gas:

$$p = \frac{nRT}{V}. \qquad \text{cf. Eq. (10.7)}$$

$$p = \frac{8.00\ \text{mol} \times 8.314\ \text{G}\,\text{K}^{-1} \times 300\,\text{K}}{5.00 \times 10^{-3}\ \text{m}^3} = 3.99 \times 10^6\ \text{Pa} = \mathbf{3.99\ MPa}.$$

2.11.2 Phase diagram of water

a) Necessary data (listed in Tables A2.1 and 5.2 of the textbook "Physical Chemistry from a Different Angle"):

	$H_2O\vert s$	$H_2O\vert l$	$H_2O\vert g$
$\mu^\ominus/\text{kG}$	−236.55	−237.14	−228.58
$\alpha/\text{G K}^{-1}$	−44.81	−69.95	−188.83
$\beta/\mu\text{G Pa}^{-1}$	19.8	18.1	24.8×10^3

Necessary equations:

Melting:

Conversion formula: $H_2O|s \rightleftarrows H_2O|l$

$$\mathcal{A}_{sl}^\ominus = \mu_s^\ominus - \mu_l^\ominus . \qquad \text{Eq. (4.3)}$$

$$\mathcal{A}_{sl}^\ominus = [(-236.55)-(-237.14)]\,\text{kG} = 0.59\,\text{kG} .$$

$$\alpha_{sl} = \alpha_s - \alpha_l . \qquad \text{Eq. (5.4)}$$

$$\alpha_{sl} = [(-44.81)-(-69.95)]\,\text{G K}^{-1} = 25.14\,\text{G K}^{-1} .$$

$$\beta_{sl} = \beta_s - \beta_l . \qquad \text{Eq. (5.10)}$$

$$\beta_{sl} = (19.8-18.1)\,\mu\text{G Pa}^{-1} = 1.7\,\mu\text{G Pa}^{-1} .$$

General equation for calculating a melting pressure curve:

$$p_{sl} = p^\ominus - \frac{\mathcal{A}_{sl}^\ominus + \alpha_{sl}(T-T^\ominus)}{\beta_{sl}} , \qquad \text{Eq. (11.13)}$$

where the standard conditions have been selected as the initial state.

Solving the equation for T (in this case, p is the independent variable that is given and the corresponding melting temperature T_{sl} is the dependent variable):

$$T_{sl} = T^\ominus - \frac{\mathcal{A}_{sl}^\ominus + \beta_{sl}(p-p^\ominus)}{\alpha_{sl}} .$$

Specific equation for the melting process of ice:

$$T_{sl} = 298.15\,\text{K} - \frac{590\,\text{G} + (1.7\times10^{-6}\,\text{G Pa}^{-1})\times(p-100\times10^3\,\text{Pa})}{25.14\,\text{G K}^{-1}} .$$

Corresponding table:

p/kPa	0	50
T/K	274.688	274.685

The almost vertical melting pressure curve has an extremely small negative slope in the direction of smaller T values. This negative slope is due to the fact that water is one of the very few substances that contracts during melting. As a result, we have $\beta_{sl} = \beta_s - \beta_l = V_{m,s} - V_{m,l} > 0$, whereas β_{sl} is usually almost always negative.

Boiling:

Conversion formula: $H_2O|l \rightleftarrows H_2O|g$

$$\mathcal{A}_{lg}^{\ominus} = [(-237.14) - (-228.58)]\,\text{kG} = -8.56\,\text{kG}\,.$$

$$\alpha_{lg} = [(-69.95) - (-188.83)]\,\text{G}\,\text{K}^{-1} = 118.88\,\text{G}\,\text{K}^{-1}\,.$$

General equation for calculating a boiling pressure curve (vapor pressure curve):

$$p_{lg} = p^{\ominus} \exp\frac{\mathcal{A}_{lg}^{\ominus} + \alpha_{lg}(T - T^{\ominus})}{RT}, \qquad \text{Eq. (11.7)}$$

where again the standard conditions were chosen as initial state.

Specific equation for the boiling process of water:

$$p_{lg} = (100 \times 10^3\ \text{Pa}) \times \exp\frac{-8560\,\text{G} + 118.88\,\text{G}\,\text{K}^{-1}(T - 298.15\,\text{K})}{8.314\,\text{G}\,\text{K}^{-1} \times T}\,.$$

Corresponding table:

T/K	240	260	280	300	320	340	360
p/kPa	0.04	0.23	1.00	3.53	10.63	28.13	66.81

Sublimation:

Conversion formula: $H_2O|s \rightleftarrows H_2O|g$

$$\mathcal{A}_{sg}^{\ominus} = [(-236.55) - (-228.58)]\,\text{kG} = -7.97\,\text{kG}\,.$$

$$\alpha_{sg} = [(-44.81) - (-188.83)]\,\text{G}\,\text{K}^{-1} = 144.02\,\text{G}\,\text{K}^{-1}\,.$$

General equation for calculating a sublimation pressure curve (with $\Delta_{sg}S = \alpha_{sg}$):

$$p_{sg} = p^{\ominus} \exp\frac{\mathcal{A}_{sg}^{\ominus} + \alpha_{sg}(T - T^{\ominus})}{RT}. \qquad \text{Eq. (11.12)}$$

Again, the standard conditions were chosen as the initial state.

Specific equation for the sublimation of ice:

$$p_{sg} = (100 \times 10^3\ \text{Pa}) \times \exp\frac{-7970\,\text{G} + 144.02\,\text{G}\,\text{K}^{-1}(T - 298.15\,\text{K})}{8.314\,\text{G}\,\text{K}^{-1} \times T}\,.$$

Corresponding table:

T/K	240	260	280	300	320	340
p/kPa	0.03	0.20	1.06	4.56	16.32	50.29

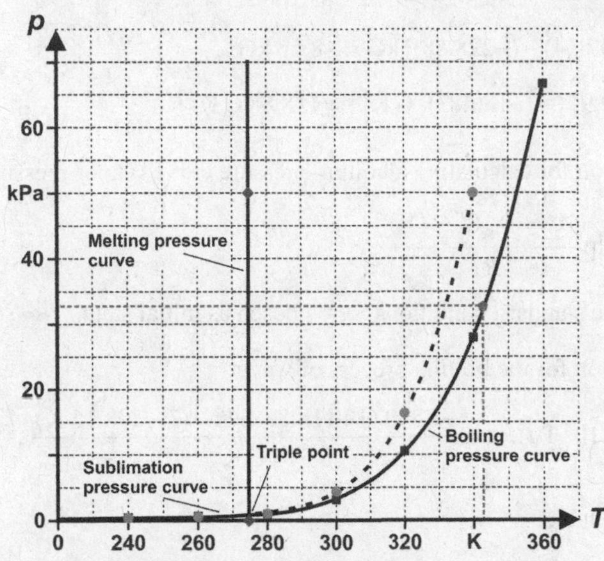

The dashed black line represents the continuation of the sublimation pressure curve.

b) Boiling temperature T_{lg} of water on top of Mount Everest:

One can find a value of about 342 K, i.e. 69 °C, for the boiling temperature of water on top of Mount Everest from the boiling pressure curve above (medium gray pentagon).

2.11.3 Evaporation rate

Vapor pressure p_{lg} of water at differen temperatures:

We insert the particular temperature values in the specific vapor pressure equation for water presented above:

$$p_{\mathrm{lg}} = (100\times10^3\ \mathrm{Pa})\times\exp\frac{-8560\ \mathrm{G}+118.88\ \mathrm{G\,K^{-1}}(T-298\ \mathrm{K})}{8.314\ \mathrm{G\,K^{-1}}\times T}.$$

In this way, we obtain for $\vartheta_1 = 34$ °C (and therefore $T_1 = 307$ K) and $\vartheta_2 = 10$ °C (and therefore $T_2 = 283$ K), respectively:

$p_{\mathrm{lg},1} = 5.32\ \mathrm{kPa}$ and $p_{\mathrm{lg},2} = 1.23\ \mathrm{kPa}$.

Ratio ω_1/ω_2 of evaporation rates:

The higher the temperature, the higher the vapor pressure and therefore the evaporation rate.

$$\frac{\omega_1}{\omega_1} \approx \frac{p_{\mathrm{lg},1}}{p_{\mathrm{lg},2}} = \frac{5.32\,\mathrm{kPa}}{1.23\,\mathrm{kPa}} = \mathbf{4.3}\,.$$

Roughly speaking, water evaporates in summer more than four times faster than in autumn.

2.11.4 Atmosphere in the bathroom

Vapor pressure p_{lg} of water at 38 °C:

Again, we use the specific vapor pressure equation for water [such as in Solutions 2.11.1 a) and 2.11.2]:

$$p_{\mathrm{lg}} = (100\times10^3\ \mathrm{Pa})\times\exp\frac{-8560\,\mathrm{G}+118.88\,\mathrm{G\,K^{-1}}(311\,\mathrm{K}-298\,\mathrm{K})}{8.314\,\mathrm{G\,K^{-1}}\times311\,\mathrm{K}} = 6.63\,\mathrm{kPa}\,.$$

Amount n of water in the air in the bathroom:

As already mentioned, the general gas law is not only valid for pure gases but also for gas mixtures. Therefore, we can write:

$$p_{\mathrm{lg}}V = n\times R\times T\,.$$

$$n = \frac{p_{\mathrm{lg}}V}{RT}\,.$$

Volume V of the bathroom:

The volume of the bathroom results in

$$V = A\times h = 4\,\mathrm{m}^2\times2.5\,\mathrm{m} = 10\,\mathrm{m}^3\,.$$

$$n = \frac{(6.63\times10^3\ \mathrm{Pa})\times10\,\mathrm{m}^3}{8.314\,\mathrm{G\,K^{-1}}\times311\,\mathrm{K}} = \mathbf{25.6\ mol}\,.$$

Mass m of water in the air in the bathroom:

$$m = n\times M = 25.6\,\mathrm{mol}\times(18.0\times10^{-3}\ \mathrm{kg\,mol^{-1}}) = \mathbf{0.46\ kg}\,.$$

2.11.5 Maximum Allowable Concentration (MAC)

Vapor pressure p_{lg} of acetone at 20 °C:

In this case, we start our calculation with Equation (11.10). Instead of choosing an arbitrary initial state, the special case of an equilibrium state, meaning a known boiling temperature such as the standard boiling temperature $T_{\mathrm{lg}}^{\ominus}$, is used here:

$$p_{\text{lg}} = p^{\ominus} \exp \frac{\Delta_{\text{lg}} S_{\text{eq}}^{\ominus} \times (T - T_{\text{lg}}^{\ominus})}{RT} .$$

$\Delta_{\text{lg}} S_{\text{eq}}^{\ominus}$ [$\equiv \Delta_{\text{lg}} S(T_{\text{lg}}^{\ominus})$] is the molar entropy of vaporization at the standard boiling point.

$$p_{\text{lg}} = (100\times10^3 \text{ Pa})\times\exp\frac{88.5 \text{ Ct mol}^{-1}\times(293 \text{ K} - 329 \text{ K})}{8.314 \text{ G K}^{-1}\times 293 \text{ K}} = 27.0\times10^3 \text{ Pa} = 27.0 \text{ kPa} .$$

Amount n of acetone in 1 m^3 of air in the laboratory:

$$n = \frac{p_{\text{lg}} V}{RT} = \frac{(27.0\times10^3 \text{ Pa})\times 1 \text{ m}^3}{8.314 \text{ G K}^{-1}\times 293 \text{ K}} = 11.1 \text{ mol} .$$

Mass m of acetone in 1 m^3 of air in the laboratory:

$$m = n \times M = 11.1 \text{ mol}\times(58.0\times10^{-3} \text{ kg mol}^{-1}) = 0.644 \text{ kg} = \mathbf{644 \text{ g}} .$$

The maximum allowable concentration in workplace air would be exceeded by a factor of 644 g m^{-3} / 1.2 g m^{-3} = 536, i.e. more than 500 times.

2.11.6 Boiling pressure of benzene

Molar entropy of vaporization $\Delta_{\text{lg}} S_{\text{eq}}^{\ominus}$ of benzene:

Since benzene is a nonpolar compound, a molar entropy of vaporization at the standard boiling point of about 88 Ct mol^{-1} can be assumed (PICTET-TROUTON's rule).

Boiling pressure (vapor pressure) p_{lg} of benzene at 60 °C:

We start again from Equation (11.10):

$$p_{\text{lg}} = p^{\ominus} \exp \frac{\Delta_{\text{lg}} S_{\text{eq}}^{\ominus} \times (T - T_{\text{lg}}^{\ominus})}{RT} .$$

$$p_{\text{lg}} \approx (100\times10^3 \text{ Pa})\times\exp\frac{88 \text{ Ct mol}^{-1}\times(333 \text{ K} - 353 \text{ K})}{8.314 \text{ G K}^{-1}\times 333 \text{ K}} = \mathbf{53 \text{ kPa}} .$$

2.11.7 Vaporization of methanol

a) Molar entropy of vaporization $\Delta_{\text{lg}} S_{\text{eq}}^{\ominus}$ of methanol:

To calculate the molar entropy of vaporization, we solve Equation (11.10) for $\Delta_{\text{lg}} S_{\text{eq}}^{\ominus}$:

$$p_{\text{lg}} = p^{\ominus} \exp \frac{\Delta_{\text{lg}} S_{\text{eq}}^{\ominus} \times (T - T_{\text{lg}}^{\ominus})}{RT}$$

$$\ln \frac{p_{\text{lg}}}{p^{\ominus}} = \frac{\Delta_{\text{lg}} S_{\text{eq}}^{\ominus} \times (T - T_{\text{lg}}^{\ominus})}{RT}$$

$$\Delta_{\text{lg}}S_{\text{eq}}^{\ominus} = \frac{RT\times\ln(p_{\text{lg}}/p^{\ominus})}{T-T_{\text{lg}}^{\ominus}} = \frac{8.314\ \text{G}\,\text{K}^{-1}\times 298\ \text{K}\times\ln(16.8\ \text{kPa}/100\ \text{kPa})}{298\ \text{K}-337\ \text{K}}$$

$$\approx \mathbf{113\ Ct\ mol^{-1}}.$$

The literature value is 104.4 Ct mol^{-1} [from: Haynes W M (ed) (2015) CRC Handbook of Chemistry and Physics, 96th edn. CRC Press, Boca Raton].

b) The value determined for the molar entropy of vaporization of methanol is significantly higher than the value of about 88 Ct mol^{-1} according to PICTET-TROUTON's rule. The deviation from this rule can be explained by strong hydrogen bridge bonds between the molecules of methanol (similar to the situation in water); as a result, its liquid phase has a higher degree of order than a nonpolar liquid. Consequently, the increase of "disorder" during the process of vaporization and therefore the entropy of vaporization is higher.

2.11.8* Triple point of 2,2-dimethylpropane

a) Molar entropy of vaporization $\Delta_{\text{lg}}S_{\text{eq}}^{\ominus}$ (at the standard boiling point) of 2,2-di-methyl-propane:

We learn from Equation (11.11) that for the y-intercept b we have:

$$b = \frac{\Delta_{\text{lg}}S_{\text{eq}}^{\ominus}}{R}.$$

The comparison with the empirical formula results in:

$$\frac{\Delta_{\text{lg}}S_{\text{eq}}^{\ominus}}{R} = 10.1945 \quad \text{and thus}$$

$$\Delta_{\text{lg}}S_{\text{eq}}^{\ominus} = 10.1945\times R = 10.1945\times 8.314\ \text{J}\,\text{mol}^{-1}\,\text{K}^{-1} = \mathbf{84.76\ Ct\ mol^{-1}}.$$

Standard boiling point $T_{\text{lg}}^{\ominus}$ of 2,2-dimethylpropane:

For the slope m, we obtain according to Equation (11.11):

$$m = -\frac{\Delta_{\text{lg}}S_{\text{eq}}^{\ominus}\times T_{\text{lg}}^{\ominus}}{R}.$$

Here, the comparison with the empirical equation results in:

$$-\frac{\Delta_{\text{lg}}S_{\text{eq}}^{\ominus}\times T_{\text{lg}}^{\ominus}}{R} = -2877.56\ \text{K} \quad \text{and thus}$$

$$T_{\text{lg}}^{\ominus} = \frac{2877.56\ \text{K}\times R}{\Delta_{\text{lg}}S_{\text{eq}}^{\ominus}} = \frac{2877.56\ \text{K}\times 8.314\ \text{J}\,\text{mol}^{-1}\,\text{K}^{-1}}{84.76\ \text{J}\,\text{K}^{-1}\,\text{mol}^{-1}} = \mathbf{282.26\ K}.$$

Molar entropy of sublimation $\Delta_{sg}S^{\ominus}_{eq}$ (at the standard sublimation point) and standard sublimation point $T^{\ominus}_{sg}$ of 2,2-dimethylpropane:

The procedure corresponds entirely to the preceding calculation:

$$\Delta_{sg}S^{\ominus}_{eq} = 12.9086\times R = 12.9086\times 8.314\ \text{J mol}^{-1}\,\text{K}^{-1} = \mathbf{107.32\ Ct\,mol^{-1}}\,.$$

$$T^{\ominus}_{sg} = \frac{3574.36\ \text{K}\times R}{\Delta_{sg}S^{\ominus}_{eq}} = \frac{3574.36\ \text{K}\times 8.314\ \text{J mol}^{-1}\,\text{K}^{-1}}{107.32\ \text{J K}^{-1}\,\text{mol}^{-1}} = \mathbf{276.90\ K}\,.$$

b) Temperature T_{slg} at the triple point of 2,2-dimethylpropane:

At the triple point, the vapor pressures of the liquid and the solid are the same, i.e., we have

$$-\frac{2877.56\ \text{K}}{T_{slg}} + 10.1945 = -\frac{3574.36\ \text{K}}{T_{slg}} + 12.9086\,.$$

We multiply the equation with T_{slg}, rearrange it,

$$T_{slg}\times(10.1945 - 12.9086) = -3574.36\ \text{K} + 2877.56\ \text{K}\,,$$

and finally obtain

$$T_{slg} = \frac{-696.80\ \text{K}}{-2.7141} = \mathbf{256.73\ K}\,.$$

Vapor pressure p_{slg} at the triple point of 2,2-dimethylpropane:

The vapor pressure at the triple point results from inserting the triple point temperature T_{slg} into one of the two empirical equations:

$$\ln(p_{slg}/p^{\ominus}) = -\frac{2877.56\ \text{K}}{T_{slg}} + 10.1945\,.$$

As a consequence, we obtain:

$$p_{slg} = p^{\ominus}\times\exp\left(-\frac{2877.56\ \text{K}}{T_{slg}} + 10.1945\right).$$

$$p_{slg} = (100\times 10^{3}\ \text{Pa})\times\exp\left(-\frac{2877.56\ \text{K}}{256.73\ \text{K}} + 10.1945\right) = \mathbf{36.3\ kPa}\,.$$

The literature value of the triple point of 2,2-dimethylpropane is 256.58 K and 35.8 kPa [from: Haynes W M (ed) (2015) CRC Handbook of Chemistry and Physics, 96th edn. CRC Press, Boca Raton].

1.11.9* Pressure dependence of the transition temperature

a) We start with the differential quotient $(\partial T/\partial p)_{\mathcal{A},\xi}$. This quotient is transformed using the calculation rules for differential quotients presented in Section 9.4 of the textbook "Physical Chemistry from a Different Angle":

$$\left(\frac{\partial T}{\partial p}\right)_{\mathcal{A},\xi} = -\left(\frac{\partial T}{\partial \mathcal{A}}\right)_{p,\xi}\left(\frac{\partial \mathcal{A}}{\partial p}\right)_{T,\xi} = -\left(\frac{\partial \mathcal{A}}{\partial T}\right)_{p,\xi}^{-1}\left(\frac{\partial \mathcal{A}}{\partial p}\right)_{T,\xi} = \left(\frac{\partial V}{\partial \xi}\right)_{T,p}\left(\frac{\partial S}{\partial \xi}\right)_{T,p}^{-1}.$$

A few words of explanation: The undesired quantity $\mathcal{A}$ in the index is to be inserted into the quotient leading to the negative sign (rule c). The first of the two new quotients is inverted (rule a). Subsequently, the quotients are "flipped" [see Eq. (9.17) and Eq. (9.12)].

In the case of the boiling process, the first expression can be replaced by $\Delta_{\mathrm{lg}}V_{\mathrm{eq}}$, the second, however, by $\Delta_{\mathrm{lg}}S_{\mathrm{eq}}$ (this is analogously valid for other phase transitions such as, for example, s → l). For small pressure changes, the difference quotient can be used to approximate the differential quotient and we obtain:

$$\frac{\Delta T}{\Delta p} = \frac{\Delta_{\mathrm{lg}}V_{\mathrm{eq}}}{\Delta_{\mathrm{lg}}S_{\mathrm{eq}}}.$$

This relationship is equivalent to equation (5.15), because the temperature coefficient α of the chemical drive corresponds in the most general sense to the change in molar entropy $\Delta_{\rightarrow}S$ of the transformation and its pressure coefficient β to the negative change in molar volume $\Delta_{\rightarrow}V$ of the transformation.

b) Change in molar volume $\Delta_{\mathrm{lg}}V_{\mathrm{eq}}$ of vaporization of water (at the boiling point)

Compared to the molar volume of the gas, the molar volume of the liquid can be neglected:

$$\Delta_{\mathrm{lg}}V_{\mathrm{eq}} \approx V_{\mathrm{m,g}} \quad \text{with} \quad V_{\mathrm{m,g}} = \frac{RT}{p}$$

according to the general gas law. In the case of water ($T = 373$ K, $p = 101325$ Pa) we consequently obtain:

$$\Delta_{\mathrm{lg}}V_{\mathrm{eq}} = \frac{8.314\,\mathrm{G\,K^{-1}} \times 373\,\mathrm{K}}{101325\,\mathrm{Pa}} = 0.0306\,\mathrm{m^3\,mol^{-1}}.$$

Conversion of units:

$$\frac{\mathrm{G\,K^{-1}\,K}}{\mathrm{Pa}} = \frac{\mathrm{J\,mol^{-1}}}{\mathrm{N\,m^{-2}}} = \frac{\mathrm{N\,m\,mol^{-1}}}{\mathrm{N\,m^{-2}}} = \mathrm{m^3\,mol^{-1}}.$$

Shift ΔT in the boiling point of water due to a pressure change:

If the pressure is changed from 101325 Pa (= 1 atm) to 100000 Pa (= 1 bar), the corresponding shift in the boiling point of water results in:

$$\Delta T = \frac{\Delta_{\mathrm{lg}}V_{\mathrm{eq}}}{\Delta_{\mathrm{lg}}S_{\mathrm{eq}}} \times \Delta p\,.$$

$$\Delta T = \frac{0.0306\ \mathrm{m^3\,mol^{-1}}}{109.0\ \mathrm{Ct\,mol^{-1}}} \times (10000\ \mathrm{Pa} - 101325\ \mathrm{Pa}) = \mathbf{-0.37\ K}\,.$$

Conversion of units:

$$\frac{\mathrm{m^3\,mol^{-1}\,Pa}}{\mathrm{Ct\,mol^{-1}}} = \frac{\mathrm{m^3\,N\,m^{-2}}}{\mathrm{J\,K^{-1}}} = \frac{\mathrm{N\,m}}{\mathrm{N\,m\,K^{-1}}} = \mathrm{K}\,.$$

In the Celsius scale, this would correspond to a boiling temperature of water at a pressure of 1 bar of 96.63 °C.

2.12 Spreading of Substances

2.12.1 Lowering of vapor pressure

a) Amount n_B of oleic acid:

Taking into account the molar mass M_B [$= M(C_{18}H_{34}O_2)$] of oleic acid with a value of $282.0\times10^{-3}\ \text{kg mol}^{-1}$, we have:

$$n_B = \frac{m_B}{M_B} = \frac{11.3\times10^{-3}\ \text{kg}}{282.0\times10^{-3}\ \text{kg mol}^{-1}} = 0.040\ \text{mol}.$$

Amount n_A of diethyl ether:

Analogously, we obtain with the molar mass M_A [$= M(C_4H_{10}O)$] of diethyl ether with a value of $74.0\times10^{-3}\ \text{kg mol}^{-1}$:

$$n_A = \frac{m_A}{M_A} = \frac{0.100\ \text{kg}}{74.0\times10^{-3}\ \text{kg mol}^{-1}} = 1.35\ \text{mol}.$$

Mole fraction x_B of oleic acid:

$$x_B = \frac{n_B}{n_B + n_A} = \frac{0.040\ \text{mol}}{0.040\ \text{mol} + 1.35\ \text{mol}} = 0.029.$$

Lowering Δp_{lg} of the vapor pressure of diethyl ether by addition of oleic acid:

RAOULT's law [Eq. (12.11)] is used to calculate the lowering of the vapor pressure:

$$\Delta p_{lg} = -x_B \times p^{\bullet}_{lg} = -0.029\times(586\times10^2\ \text{Pa}) = -1700\ \text{Pa} = \mathbf{-1.7\ kPa}.$$

b) Vapor pressure p_{lg} of the solution:

The vapor pressure of the solution results from the vapor pressure of the pure ether reduced by the value for the vapor pressure lowering:

$$p_{lg} = p^{\bullet}_{lg} + \Delta p_{lg} = 58.6\ \text{kPa} - 1.7\ \text{kPa} = \mathbf{56.9\ kPa}.$$

c) Difference in height Δh between the liquid columns in the two legs of the manometer:

The gravitational pressure of the water in the U-tube is $\rho\times g\times\Delta h$. This corresponds to the vapor pressure difference Δp_{lg} in the two washing bottles, i.e. we obtain:

$$\Delta p_{lg} = \rho\times g\times\Delta h \quad\Rightarrow\quad \Delta h = \frac{\Delta p_{lg}}{\rho\times g}.$$

$$\Delta h = \frac{-1700\ \text{Pa}}{998\ \text{kg m}^{-3}\times9.81\ \text{m s}^{-2}} = -0.17\ \text{m} = \mathbf{-17\ cm}.$$

Conversion of units:

$$\frac{\text{Pa}}{\text{kg}\,\text{m}^{-3}\,\text{m}\,\text{s}^{-2}} = \frac{\text{N}\,\text{m}^{-2}}{\text{kg}\,\text{m}^{-2}\,\text{s}^{-2}} = \frac{\text{kg}\,\text{m}\,\text{s}^{-2}}{\text{kg}\,\text{s}^{-2}} = \text{m}\,.$$

2.12.2 "Household chemistry"

a) Lowering of the freezing point ΔT_{sl} of water due to the addition of sugar:

The lowering of the freezing point is calculated using Equation (12.16):

$$\Delta T_{\text{sl}} = -k_{\text{f}} \times b_{\text{B}} = -k_{\text{f}} \times \frac{n_{\text{B}}}{m_{\text{A}}}\,.$$

Table 12.1 in the textbook "Physical Chemistry from a Different Angle" gives a value of 1.86 K kg mol^{-1} for the cryoscopic constant k_{f} of water.

Amount n_{B} of sucrose:

The amount n_{B} of sucrose can be calculated from the mass $m_{\text{B}} = 3 \times 3\ \text{g} = 9\ \text{g}$ of the three sugar cubes by means of the molar mass M_{B} = [$M(C_{12}H_{22}O_{11})$ =] $342.0 \times 10^{-3}\ \text{kg}\,\text{mol}^{-1}$:

$$n_{\text{B}} = \frac{m_{\text{B}}}{M_{\text{B}}} = \frac{9 \times 10^{-3}\ \text{kg}}{342.0 \times 10^{-3}\ \text{kg}\,\text{mol}^{-1}} = 0.026\ \text{mol} = 26\ \text{mmol}\,.$$

Mass m_{A} of the solvent water:

The mass m_{A} of water results from its volume and density according to

$$m_{\text{A}} = \rho_{\text{A}} \times V_{\text{A}} = 1000\ \text{kg}\,\text{m}^{-3} \times (250 \times 10^{-6}\ \text{m}^3) = 0.25\ \text{kg}\,.$$

$$\Delta T_{\text{sl}} = -1.86\ \text{K}\,\text{kg}\,\text{mol}^{-1} \times \frac{26 \times 10^{-3}\ \text{mol}}{0.25\ \text{kg}} \approx -0.19\ \text{K}\,.$$

Freezing point of the sugar solution:

Adding the sugar lowers the freezing point of the solvent water by about 0.2 K. Since the standard freezing point of water is 0 °C, the freezing point of the solution results in **–0.2 °C**.

b) Raising of the boiling point ΔT_{lg} of water due to the addition of table salt:

The addition of the salt results according to Equation (12.17) in a change in the boiling point of water by

$$\Delta T_{\text{lg}} = +2k_{\text{b}} \times b_{\text{B}} = +2k_{\text{b}} \times \frac{n_{\text{B}}}{m_{\text{A}}}\,.$$

The factor 2 is caused by the fact that sodium chloride dissociates in aqueous solution completely into Na^+ and Cl^- ions, meaning twice the number of particles is present. The colligative properties in turn depend solely upon the number of the dissolved particles but not upon their chemical nature.

The ebullioscopic constant k_b of water is +0.51 K kg mol^{-1} according to Table 12.1.

Amount n_B of table salt:

The amount n_B of table salt can be calculated from its mass m_B of 10 g by means of its molar mass $M_B = 58.5\times10^{-3}$ kg mol^{-1}:

$$n_B = \frac{m_B}{M_B} = \frac{10\times10^{-3}\ \text{kg}}{58.5\times10^{-3}\ \text{kg mol}^{-1}} = 0.17\ \text{mol}.$$

$$\Delta T_{lg} = +2\times0.51\ \text{K kg mol}^{-1}\times\frac{0.17\ \text{mol}}{1\ \text{kg}} \approx +0.17\ \text{K}.$$

Boiling point of the "pasta water":

If the boiling point of the water is 100 °C, the "pasta water" will boil consequently at about **100.2 °C**.

2.12.3 Osmotic pressure

a) Amount n_B of urea in 1 L of solution:

The amount n_B of urea in the solution can be determined using VAN'T HOFF's equation [Eq. (12.7)]:

$$p_{osm} = n_B\times\frac{RT}{V}.$$

Solving for n_B results in:

$$n_B = \frac{p_{osm}\times V}{RT} = \frac{(99\times10^3\ \text{Pa})\times(1.00\times10^{-3}\ \text{m}^3)}{8.314\ \text{G K}^{-1}\times298\ \text{K}} = \mathbf{0.040\ mol}.$$

b) Mass m_A of the solvent water in the solution:

$$m_A = \rho_A\times V_A = 1000\ \text{kg m}^{-3}\times(1.00\times10^{-3}\ \text{m}^3) = \mathbf{1.00\ kg}.$$

Amount n_A of water:

$$n_A = \frac{m_A}{M_A} = \frac{1.00\ \text{kg}}{18.0\times10^{-3}\ \text{kg mol}^{-1}} = \mathbf{55.56\ mol}.$$

c) Mole fraction x_B of urea in the solution:

$$x_B = \frac{n_B}{n_B + n_A} = \frac{0.040\ \text{mol}}{0.040\ \text{mol} + 55.56\ \text{mol}} = 7.2\times10^{-4} = \mathbf{7.2\ ‰}.$$

d) Change in chemical potential $\Delta\mu_A$ of the solvent water by addition of the foreign substance urea:

The so-called "colligative lowering of potential" is described by Equation (12.2):

$$\mu_A = \mathring{\mu}_A - RT \times x_B .$$

Consequently, the change in potential $\Delta\mu_A$ results in:

$$\Delta\mu_A = \mu_A - \mathring{\mu}_A = -RT \times x_B = -8.314\ \text{G}\,\text{K}^{-1} \times 298\ \text{K} \times (7.2\times10^{-4}) = \mathbf{-1.78\ G}.$$

Adding the foreign substance lowered the chemical potential of the water by 1.78 G.

e) Lowering of the freezing point ΔT_{sl} of water due to the addition of the foreign substance urea:

In this case, Equation (12.14) is used to calculate the lowering of the freezing point:

$$\Delta T_{sl} = -\frac{RT^{\bullet}_{sl,A} \times x_B}{\Delta_{sl} S^{\bullet}_{eq,A}} .$$

$\Delta_{sl} S^{\bullet}_{eq,A}$ is the molar entropy of fusion of the pure solvent water (at the freezing point); its freezing point $T^{\bullet}_{sl,A}$ is 273 K. Thus, we obtain:

$$\Delta T_{sl} = -\frac{8.314\ \text{G}\,\text{K}^{-1} \times 273\ \text{K} \times (7.2\times10^{-4})}{22.0\ \text{Ct}\,\text{mol}^{-1}} = \mathbf{-0.074\ K}.$$

Conversion of the units:

$$\frac{\text{G}\,\text{K}^{-1}\,\text{K}}{\text{Ct}\,\text{mol}^{-1}} = \frac{\text{J}\,\text{mol}^{-1}}{\text{J}\,\text{K}^{-1}\,\text{mol}^{-1}} = \text{K}$$

The freezing point of the solution is lowered by 0.074 K compared to that of pure water.

f) Amount $n_{B'}$ of magnesium chloride:

Since magnesium chloride in aqueous solution dissociates completely into three particles, a Mg^{2+} ion and two Cl^- ions, and the colligative properties are solely determined by the number of dissolved particles and not their chemical nature, theoretically [neglecting the interactions between the ions; see Solution 2.12.4 f)] one third of the amount of magnesium chloride in the salt solution (compared to the amount of urea) is sufficient to produce the same osmotic pressure as the urea solution.

$$n_{B'} = \tfrac{1}{3} n_B = \tfrac{1}{3} \times 0.040\ \text{mol} = \mathbf{0.013\ mol}.$$

2.12.4 See water

a) Amount n_B of sodium chloride:

$$n_B = \frac{m_B}{M_B}.$$

Mass m_B and molar mass M_B:

One takes 1.000 kg of sea water as starting point. Sea water is suggested to contain a mass fraction w_B of the salt of 3.5 %; this corresponds to 0.035 kg. The molar mass M_B of sodium chloride is $58.5 \times 10^{-3}\ \text{kg mol}^{-1}$.

$$n_B = \frac{0.035\ \text{kg}}{58.5 \times 10^{-3}\ \text{kg mol}^{-1}} = \mathbf{0.60\ mol}.$$

Amount n_A of the solvent water:

$$n_A = \frac{m_A}{M_A}.$$

Mass m_A and molar mass M_A:

The remaining mass of water is 0.965 kg. Its molar mass M_A is $18.0 \times 10^{-3}\ \text{kg mol}^{-1}$.

$$n_A = \frac{0.965\ \text{kg}}{18.0 \times 10^{-3}\ \text{kg mol}^{-1}} = \mathbf{53.6\ mol}.$$

Mole fraction x_B of sodium chloride:

$$x_B = \frac{n_B}{n_A + n_B} = \frac{0.60\ \text{mol}}{53.6\ \text{mol} + 0.60\ \text{mol}} = 0.011.$$

Mole fraction x_F of foreign substance:

Since NaCl dissociates completely into Na^+ and Cl^- ions, we have:

$$x_F = [x_{Na^+} + x_{Cl^-} =]\ 2 \times x_B = 2 \times 0.011 = \mathbf{0.022}.$$

b) Change in chemical potential $\Delta\mu_A$ of water:

When a small amount n_F of foreign substance is dissolved in a liquid A, here water, the chemical potential $\mathring{\mu}_A$ of this liquid will decrease to a new value μ_A [see also Exercise 1.12.3 d)]

$$\mu_A = \mathring{\mu}_A - RT \times x_F \qquad \text{Eq. (12.2)}$$

Rearranging results in:

$$\Delta\mu_A = \mu_A - \mathring{\mu}_A = -RT \times x_F = -8.314\ \text{G K}^{-1} \times 298\ \text{K} \times 0.022 \approx \mathbf{-55\ G}.$$

c) Lowering of vapor pressure Δp_{lg} by dissolving a foreign substance, here sodium chloride, in a pure liquid, here fresh water R:

$$\Delta p_{lg} = -x_F \times \overset{\bullet}{p}_{lg,A} \qquad \text{RAOULT's law}. \qquad \text{Eq. (12.11)}$$

$p_{lg,R}$ $(= \overset{\bullet}{p}_{lg,A})$ is the vapor pressure of fresh water, i.e. pure water.

Vapor pressure $p_{lg,S}$ of sea water:

$$p_{lg,S} = \overset{\bullet}{p}_{lg,A} + \Delta p_{lg} = \overset{\bullet}{p}_{lg,A} - x_F \times \overset{\bullet}{p}_{lg,A} = \overset{\bullet}{p}_{lg,A}(1 - x_F).$$

Ratio of the evaporation rates of sea water, ω_S, and fresh water, ω_R:

The evaporation rate is approximately proportional to the vapor pressure (as we have seen in Exercise 1.11.3), meaning we have:

$$\frac{\omega_S}{\omega_R} \approx \frac{p_{lg,S}}{p_{lg,R}} = \frac{\overset{\bullet}{p}_{lg,A}(1-x_F)}{\overset{\bullet}{p}_{lg,A}} = 1 - x_F.$$

$$\frac{\omega_S}{\omega_R} \approx 1 - 0.022 = \mathbf{0.978}.$$

The evaporation rate of seawater is about 2 % lower than that of freshwater.

d) Lowering of the freezing point ΔT_{sl}:

$$\Delta T_{sl} = -\frac{R\overset{\bullet}{T}_{sl,A} \times x_F}{\Delta_{sl}\overset{\bullet}{S}_{eq,A}}. \qquad \text{Eq. (12.14)}$$

Necessary data:
Freezing point of (pure) water: $\overset{\bullet}{T}_{sl,A} = 273\ \text{K}$,
Molar entropy of fusion of water (at the freezing point): $\Delta_{sl}\overset{\bullet}{S}_{eq,A} = 22.0\ \text{Ct mol}^{-1}$.

$$\Delta T_{sl} = -\frac{8.314\ \text{G K}^{-1} \times 273\ \text{K} \times 0.022}{22.0\ \text{Ct mol}^{-1}} \approx \mathbf{-2.3\ K}.$$

The experimentally determined lowering of the freezing point of sea water with a salinity of 35 g of salt per kg of sea water is −1.922 °C [Millero FJ, Leung WH (1976) The Thermodynamics of Seawater at one Atmosphere. Am J Sci 276:1035-1077].

e) Raising of the boiling point ΔT_{lg}:

$$\Delta T_{lg} = \frac{R\overset{\bullet}{T}_{lg,A} \times x_F}{\Delta_{lg}\overset{\bullet}{S}_{eq,A}}. \qquad \text{Eq. (12.15)}$$

Necessary data:
Boiling point of (pure) water: $T_{lg}^{\bullet} = 373\ \text{K}$,
Molar entropy of vaporization of water (at the boiling point): $\Delta_{lg}S_{eq,A}^{\bullet} = 109.0\ \text{Ct mol}^{-1}$.

$$\Delta T_{lg} = +\frac{8.314\ \text{G K}^{-1} \times 373\ \text{K} \times 0.022}{109.0\ \text{Ct mol}^{-1}} \approx \mathbf{+0.63\ K}.$$

f) Osmotic pressure p_{osm}:

The excess pressure Δp, which has to be applied at the minimum to press sea water through a membrane permeable only to H_2O molecules in order to desalinate it, corresponds to the osmotic pressure p_{osm}:

$$\Delta p = p_{osm} = n_F \frac{RT}{V}. \qquad \text{Eq. (12.8)}$$

Amount n_F:

$$n_F = 2 \times n_B = 2 \times 0.6\ \text{mol} = 1.2\ \text{mol}.$$

Volume V_S of the solution:

$$V_S = \frac{m_S}{\rho_S} = \frac{1.000\ \text{kg}}{1022\ \text{kg m}^{-3}} = 978 \times 10^{-6}\ \text{m}^3.$$

$$p_{osm} = 1.2\ \text{mol} \times \frac{8.314\ \text{G K}^{-1} \times 298\ \text{K}}{978 \times 10^{-6}\ \text{m}^3} = 1.2\ \text{mol} \times \frac{8.314\ \text{N m mol}^{-1}\ \text{K}^{-1} \times 298\ \text{K}}{978 \times 10^{-6}\ \text{m}^3}$$

$$\approx 3.0 \times 10^6\ \text{N m}^{-2} = 3.0 \times 10^6\ \text{Pa} = 3.0\ \text{MPa} = \mathbf{30\ bar}.$$

The experimentally determined osmotic pressure of sea water with a salinity of 35 g/kg is 25.896 bar (at 25 °C) [Millero FJ, Leung WH (1976) The Thermodynamics of Seawater at one Atmosphere. Am J Sci 276:1035-1077]. The quite large derivation from the estimated value of about 30 bar is due to the fact that in the used approach, the interactions between the particles are not taken into account. These interactions are especially strong between charged particles (so-called interionic interactions). A correction factor f may be introduced, for example, to take into account the derivations resulting from the interactions between the particles. Equation (12.8) takes then the following form:

$$p_{osm} = f c_F RT.$$

The correction factor depends on the concentration; for the limiting case $c \to 0$, i.e. a very thin solution, it has a value of 1. In the present case of seawater it is about 0.86. (Analogous correction factors must also be used in the case of the other colligative phenomena such as freezing point depression etc.).

2.12.5 "Frost protection" in the animal world

a) Amount n_B of glycerol:

$$n_B = \frac{m_B}{M_B}.$$

Mass m_B and molar mass M_B:

One takes 0.100 kg of solution as starting point. The mass fraction of glycerol is supposed to be 0.3; this corresponds to 0.030 kg. The molar mass M_B of glycerol is 92.0×10^{-3} kg mol^{-1}.

$$n_B = \frac{0.030\,\text{kg}}{92.0\times10^{-3}\ \text{kg mol}^{-1}} = \mathbf{0.33\ mol}.$$

Amount n_A of water:

$$n_A = \frac{m_A}{M_A}.$$

Mass m_A and molar mass M_A:

The remaining mass of water is 0.070 kg; its molar mass M_A is, as mentioned, 18.0×10^{-3} kg mol^{-1}.

$$n_A = \frac{0.070\ \text{kg}}{18.0\times10^{-3}\ \text{kg mol}^{-1}} = \mathbf{3.89\ mol}.$$

Mole fraction x_F of foreign substance:

$$x_F = \frac{n_B}{n_B + n_A} = \frac{0.326\ \text{mol}}{0.326\ \text{mol} + 3.89\ \text{mol}} = \frac{n_B}{n_B + n_A} = \frac{0.33\ \text{mol}}{0.33\ \text{mol} + 3.90\ \text{mol}} = \mathbf{0{,}0773}.$$

b) Lowering of the freezing point ΔT_{sl} due to the addition of glycerol:

All equations based on mass action are to be applied only with reservation because the content of glycerol in the solution is relatively high. This is also valid for Equation (12.14). However, the lowering of freezing point ΔT_{sl} can at least be estimated with its help:

$$\Delta T_{sl} = -\frac{RT^{\bullet}_{sl,A} \times x_F}{\Delta_{lg} S^{\bullet}_{eq,A}} = -\frac{8.314\ \text{G K}^{-1} \times 273\ \text{K} \times 0.0773}{22.0\ \text{Ct mol}^{-1}} \approx -8\ \text{K}.$$

Freezing point T_{sl} of the hemolymph of the wasp:

The estimated freezing point of the hemolymph of the wasp is **−8 °C**. In fact, the insect can survive even considerably lower temperatures because the hemolymph can be supercooled—i.e. cooled far below the freezing point—without ice crystals forming.

c) Osmotic concentration c_F of the hemolymph at 20 °C:

The osmotic concentration is defined as the amount of all osmotically active particles (here only the glycerol molecules) divided by the volume of solution:

$$c_F = \frac{n_F}{V_S}.$$

Volume V_S of the solution:

$$V_S = \frac{m_S}{\rho_S} = \frac{0.100\ \text{kg}}{1065\ \text{kg m}^3} = 93.9\times10^{-6}\ \text{m}^3.$$

$$c_F = \frac{0.326\ \text{mol}}{93.9\times10^{-6}\ \text{m}^{-3}} = \mathbf{3.47\times10^3\ mol\ m^{-3}}.$$

d) Osmotic pressure of the hemolymph at 20 °C:

Due to the high content of glycerol in the solution, VAN'T HOFF's equation can also only be applied with reservation:

$$p_{osm} = c_F RT. \qquad \text{Eq. (12.8)}$$

$$p_{osm} = 3.47\times10^3\ \text{mol m}^{-3}\times 8.314\ \text{G K}^{-1}\times 293\ \text{K} \approx 8.5\times10^6\ \text{Pa} = 8.5\ \text{MPa} = \mathbf{85\ bar}.$$

2.12.6 "Osmotic power plant"

a) Chemical potential $\mu_{A,R}$ of fresh water at 10 °C (and 100 kPa):

Due to the very low salinity of fresh water (with a mass fraction of the salt of less than 0.1 %), the chemical potential of the water in fresh water (approximately) corresponds to that of pure water. The temperature dependence of the chemical potential is described by Equation (5.2):

$$\mu_{A,R} = \mu_A = \mu_A^\ominus + \alpha(T - T^\ominus) = -237140\ \text{G} - 70\ \text{G K}^{-1}\times(283\ \text{K} - 298\ \text{K}) = \mathbf{-236090\ G}.$$

Chemical potential μ_S of sea water at 10 °C (and 100 kPa):

Taking the "colligative lowering of potential" into account [Eq. (12.2)], the chemical potential $\mu_{A,S}$ of sea water results in:

$$\mu_{A,S} = \mu_{A,R} - RT\times x_F.$$

Mole fraction x_F of foreign substance:

$$x_F = 2\times x_B = 2\times 0.011 = 0.022.$$

$$\mu_{A,S} = -236090\ \text{G} - 8.314\ \text{G K}^{-1}\times 283\ \text{K}\times 0.022 = \mathbf{-236142\ G}.$$

b) Change in pressure Δp:

The change in pressure Δp can be estimated, for example, by Equation (12.4):

$$\mu_{A,S} + \beta \times \Delta p = \mu_{A,R} \, .$$

Solving for Δp results in:

$$\Delta p = \frac{\mu_{A,R} - \mu_{A,S}}{\beta} = \frac{-236090\ \text{G} - (-236142\ \text{G})}{18 \times 10^{-6}\ \text{G}\,\text{Pa}^{-1}} \approx 2.9 \times 10^{6}\ \text{Pa} = 2.9\ \text{MPa} = \mathbf{29\ bar} \, .$$

This change in pressure is a measure for the osmotic pressure p_{osm} of the solution, in this case sea water. The osmotic pressure determined experimentally is 24.544 bar [see Solution 2.12.4.f)].

c) Chemical drive $\mathcal{A}$ for the migration of H_2O from fresh water to sea water:

$$\mathcal{A} = \mu_{A,R} - \mu_{A,S} = -236090\ \text{G} - (-236142\ \text{G}) = \mathbf{52\ G} \, .$$

Usable energy W^*_{use} at an efficiency of $\eta \approx 60\ \%$ at full load:

The usable energy W^*_{use} can be calculated by means of the following equation (see Figure 8.7 in the textbook "Physical Chemistry from a Different Angle"):

$$W^*_{use} = \eta \times \mathcal{A} \times \Delta\xi = 0.6 \times 52\ \text{J}\,\text{mol}^{-1} \times 55500\ \text{mol} = 1.73 \times 10^{6}\ \text{J} = \mathbf{1.73\ MJ} \, .$$

2.12.7 Isotonic saline solution

a) Amount n_F of foreign substance in the saline solution:

We start from VAN'T HOFF's Equation [Eq. (12.8)],

$$p_{osm} = n_F \frac{RT}{V_S} \, ,$$

and solve for n_F,

$$n_F = \frac{p_{osm} \times V_S}{RT} = \frac{(738 \times 10^{3}\ \text{Pa}) \times (500 \times 10^{-6}\ \text{m}^3)}{8.314\ \text{G}\,\text{K}^{-1} \times 310\ \text{K}} = 0.143\ \text{mol} \, .$$

Amount n_B of sodium chloride in the solution:

Since sodium chloride dissociates in aqueous solution completely into Na^+ and Cl^- ions, only half of the amount n_F is necessary:

$$n_B = \tfrac{1}{2} n_F = \tfrac{1}{2} \times 0.143\ \text{mol} = 0.072\ \text{mol} \, .$$

Mass m_B of sodium chloride in the solution:

The mass m_B one need to weigh out can be calculated by means of the molar mass M_B of sodium chloride of $58.5 \times 10^{-3}\ \text{kg}\,\text{mol}^{-1}$:

$$m_B = n_B \times M_B = 0.072\ \text{mol} \times (58.5 \times 10^{-3}\ \text{kg}\,\text{mol}^{-1}) \approx 4.2 \times 10^{-3}\ \text{kg} = \mathbf{4.2\ g} \, .$$

b) Molar concentration c_B of commercial isotonic sodium chloride solution:

$$c_B = \frac{n_B}{V_S} = \frac{m_B}{M_B \times V_S} = \frac{9.0 \times 10^{-3}\ \text{kg}}{(58.5 \times 10^{-3}\ \text{kg mol}^{-1}) \times (1000 \times 10^{-6}\ \text{m}^3)}$$

$$= 154\ \text{mol m}^{-3}.$$

Osmotic concentration c_F of the isotonic sodium chloride solution:

Since sodium chloride dissociates in aqueous solution completely into Na^+ and Cl^- ions, as mentioned, the osmotic concentration c_F results in:

$$c_F = 2 \times c_B = 2 \times 154\ \text{mol m}^{-3} = 308\ \text{mol m}^{-3}.$$

Correction factor f in VAN'T HOFF's equation:

$$p_{osm} = f c_F RT \quad \Rightarrow$$

$$f = \frac{p_{osm}}{c_F RT} = \frac{738 \times 10^3\ \text{Pa}}{308\ \text{mol m}^{-3} \times 8.314\ \text{G K}^{-1} \times 310\ \text{K}} = \mathbf{0.93}.$$

Conversion of the units:

$$\frac{\text{Pa}}{\text{mol m}^{-3}\ \text{G K}^{-1}\ \text{K}} = \frac{\text{N m}^{-2}}{\text{mol m}^{-3}\ \text{J mol}^{-1}} = \frac{\text{N}}{\text{m}^{-1}\ \text{N m}} = 1.$$

2.12.8 Determining molar masses by cryoscopy

a) Molar mass M_B of the unknown substance:

We start from Equation (12.16) for the lowering of the freezing point:

$$\Delta T_{sl} = -k_f \times \frac{n_B}{m_A} = -k_f \times \frac{m_B}{m_A \times M_B}.$$

Solving for the desired molar mass M_B results in [the cryoscopic constant k_f of camphor has a value of 37.8 K kg mol^{-1} (Table 12.1)]:

$$M_B = -\frac{k_f \times m_B}{m_A \times \Delta T_{sl}} = -\frac{37.8\ \text{K kg mol}^{-1} \times (40 \times 10^{-6}\ \text{kg})}{(10,0 \times 10^{-3}\ \text{kg}) \times (-0.92\ \text{K})} = \mathbf{164 \times 10^{-3}\ kg\ mol^{-1}}.$$

b) Molecular formula of the unknown substance:

The molar mass of the formula unit C_5H_6O is 82×10^{-3} kg mol^{-1}. The number N of formula units in the molecular formula results in:

$$N = \frac{\text{molar mass of the unknown substance}}{\text{molar mass of the formula unit}} = \frac{164 \times 10^{-3}\ \text{kg mol}^{-1}}{82 \times 10^{-3}\ \text{kg mol}^{-1}} = 2.$$

Consequently, the molecular formula of the unknown substance is $C_{10}H_{12}O_2$. For example, it may be eugenol, the main constituent in the essential oil extracted from the clove plant. Eugenol is used, inter alia, as a local antiseptic and analgesic in dentistry.

2.12.9 Determining molar masses by osmometry

Molar mass M_B of the enzyme catalase:

Starting point is VAN'T HOFF's equation [Eq. (12.7)]:

$$p_{osm} = n_B \times \frac{RT}{V_S} = \frac{m_B}{M_B} \times \frac{RT}{V_S} .$$

Solving for M_B results in:

$$M_B = \frac{m_B \times RT}{p_{osm} \times V_S} .$$

$$M_B = \frac{(10.0\times10^{-3}\ \text{kg})\times 8.314\ \text{G K}^{-1}\times 300\ \text{K}}{104\ \text{Pa}\times(1.00\times10^{-3}\ \text{m}^3)} \approx \mathbf{240\ kg\,mol^{-1}} .$$

As macromolecular substance, the enzyme catalase has a high molar mass of about 240 kg mol^{-1}.

2.13 Homogeneous and Heterogeneous Mixtures

2.13.1 Ideal liquid mixture

a) Mole fractions x_A and x_B of benzene (component A) and toluene (component B):

$$x_A = \frac{n_A}{n_A + n_B} = \frac{0.8\ \text{mol}}{0.8\ \text{mol} + 1.2\ \text{mol}} = 0.4\,.$$

For a binary mixture (such as in this example), we have:

$$x_B = 1 - x_A = 0.6\,.$$

Chemical drive $\mathcal{A}_{mix}$ for the mixing process:

The chemical drive $\mathcal{A}_{mix}$ for the mixing of two indifferent components A and B, such as benzene and toluene, is given by the following equation:

$$\mathcal{A}_{mix} = -RT(x_A \times \ln x_A + x_B \times \ln x_B)\,.$$

$$\mathcal{A}_{mix} = -8.314\ \text{G K}^{-1} \times 298\ \text{K} \times (0.4 \times \ln 0.4 + 0.6 \times \ln 0.6) = 1670\ \text{G} = \mathbf{1.67\ kG}\,.$$

Molar entropy of mixing $\Delta_{mix}S$:

The molar entropy of mixing $\Delta_{mix}S$ can be calculated as follows [cf. Eq. (13.16)]:

$$\Delta_{mix}S = -R(x_A \times \ln x_A + x_B \times \ln x_B) = \frac{\mathcal{A}_{mix}}{T}\,.$$

$$\Delta_{mix}S = \frac{1670\ \text{J mol}^{-1}}{298\ \text{K}} = 5.60\ \text{J K}^{-1}\ \text{mol}^{-1} = \mathbf{5.60\ Ct\ mol^{-1}}\,.$$

Molar volume of mixing $\Delta_{mix}V$:

In an ideal mixture, such as in this case, we have [(Eq. (13.15)]:

$$\Delta_{mix}V = \mathbf{0.}$$

b) Mole fraction x_A at which the maximum possible entropy of the mixture occurs:

In order to determine the mole ratio in question, we have to look for the extreme value of the function given by Equation (13.16). For this purpose, we have to take the derivative of this function with respect to x_A and to determine the x_A value for which the derivative is equal to zero. In a binary mixture, we have $x_B = 1 - x_A$; therefore, we can write:

$$\Delta_{mix}S = -R(x_A \times \ln x_A + x_B \times \ln x_B) = -R[x_A \times \ln x_A + (1 - x_A) \times \ln(1 - x_A)]\,.$$

Concerning the derivation, we have to consider that we obtain in the case of the natural logarithm function:

$$y' = \frac{1}{x} \quad \text{for } x > 0\,. \qquad \text{Eq. (A1.8)}$$

In addition, the product rule [Eq. (A1.11)] as well as the chain rule [Eq. (A1.13)] have to be used:

$$\frac{\mathrm{d}\Delta_{\mathrm{mix}}S}{\mathrm{d}x_{\mathrm{A}}} = -R\left[1\times\ln x_{\mathrm{A}} + x_{\mathrm{A}}\times\frac{1}{x_{\mathrm{A}}} + (-1)\times\ln(1-x_{\mathrm{A}}) + (1-x_{\mathrm{A}})\times\frac{1}{1-x_{\mathrm{A}}}\times(-1)\right]$$

$$= -R\left[\ln x_{\mathrm{A}} + 1 - \ln(1-x_{\mathrm{A}}) - 1\right] = -R\left[\ln x_{\mathrm{A}} - \ln(1-x_{\mathrm{A}})\right] = -R\ln\frac{x_{\mathrm{A}}}{1-x_{\mathrm{A}}}.$$

Subsequently, we set the derivative equal to zero and obtain:

$$0 = -R\ln\frac{x_{\mathrm{A}}}{1-x_{\mathrm{A}}} \quad \text{and therefore} \quad \mathrm{e}^0 = 1 = \frac{x_{\mathrm{A}}}{1-x_{\mathrm{A}}} \quad \text{and finally} \quad x_{\mathrm{A}} = \tfrac{1}{2}.$$

The maximum possible entropy of mixing, $\Delta_{\mathrm{mix}}S$, thus occurs when the mole fractions of both components are the same.

This result can be illustrated by Figure 13.7 in the textbook "Physical Chemistry from a Different Angle." Due to $\Delta_{\mathrm{mix}}S = \mathcal{A}_{\mathrm{mix}}/T$, at constant temperature the molar entropy of mixing $\Delta_{\mathrm{mix}}S$ becomes maximum at the same amount x_{A} at which it is also $\mathcal{A}_{\mathrm{mix}}$, meaning at $x_{\mathrm{A}} = x_{\mathrm{B}} = \tfrac{1}{2}$, as can be seen in the figure.

2.13.2 Chemical potential of a homogeneous mixture

a) Chemical potential μ_{M} of a homogeneous mixture:

Relevant equation:

$$\mu_{\mathrm{M}} = x_{\mathrm{A}}\times(\mathring{\mu}_{\mathrm{A}} + RT\ln x_{\mathrm{A}}) + x_{\mathrm{B}}\times(\mathring{\mu}_{\mathrm{B}} + RT\ln x_{\mathrm{B}})$$

Since both oxygen (O_2, component A) and nitrogen (N_2, component B) are elements, the basic value of the chemical potential under standard conditions (298 K, 100 kPa) is zero in both cases, meaning $\mathring{\mu}_{\mathrm{A}} = \mathring{\mu}_{\mathrm{B}} = 0$.

$$\mu_{\mathrm{M}} = x_{\mathrm{A}}\times RT\ln x_{\mathrm{A}} + x_{\mathrm{B}}\times RT\ln x_{\mathrm{B}} = RT[x_{\mathrm{A}}\times\ln x_{\mathrm{A}} + x_{\mathrm{B}}\times\ln x_{\mathrm{B}}]$$

$$= RT[(1-x_{\mathrm{B}})\times\ln(1-x_{\mathrm{B}}) + x_{\mathrm{B}}\times\ln x_{\mathrm{B}}].$$

Corresponding table:

	0	0.01	0.02	0.05	0.10	0.20	0.30	0.40	0.50
μ_{M} / kG	0	−0.14	−0.24	−0.49	−0.81	−1.24	−1.51	−1.67	−1.72

The second branch of the curve is symmetrical to the first. As can be seen from the values for $x_{\mathrm{B}} < 0.10$, the $\mu(x_{\mathrm{B}})$ curve is very steep at the beginning (and at the end) of the range of values. At $x_{\mathrm{B}} = 0$ as well as $x_{\mathrm{B}} = 1$, the slope is finally $\pm\infty$, i.e. the curve has vertical tangents, a fact that is not easy to recognize in the figure.

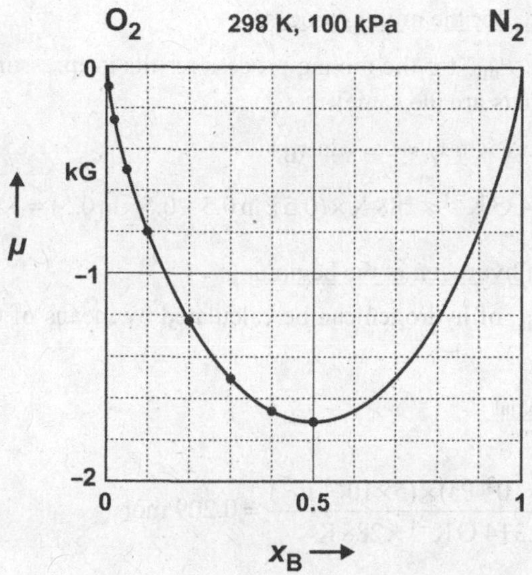

b) μ_M value of air:

Simplifying, air can be considered as a gas mixture with a volume fraction of 79 % of nitrogen and 21 % of oxygen. Assuming ideal gas behavior, the volume fraction corresponds to the mole fraction [$V \sim n$ at constant T (here 298 K) and p (here 100 kPa)] [cf. Eq. (10.5)]. Therefore, the mole fraction x_B of nitrogen is 0.79.

$$\mu_M = RT^{\ominus}[(1-x_B)\times\ln(1-x_B)+x_B\times\ln x_B].$$

$$\mu_M = 8.314\,\mathrm{G\,K^{-1}}\times 298\,\mathrm{K}\times[(1-0.79)\times\ln(1-0.79)+0.79\times\ln 0.79]$$
$$= -1270\,\mathrm{G} = \mathbf{-1.27\,kG}.$$

2.13.3 Mixing of ideal gases

a) We can formulate the general conversion formula for the mixing process as follows,

$$x_A\mathrm{A}+x_B\mathrm{B}\rightarrow \mathrm{A}_{x_A}\mathrm{B}_{x_B},$$

where A stands for hydrogen and B for nitrogen.

Since both gases are at the same pressure and mixed in the ratio 1:1, the mole fraction of hydrogen is $x_A = 0.5$ and that of nitrogen $x_B = 0.5$. Therefore, the conversion formula for the current mixing process is:

$$\tfrac{1}{2}\mathrm{A}+\tfrac{1}{2}\mathrm{B}\rightarrow \mathrm{A}_{0.5}\mathrm{B}_{0.5}.$$

Chemical drive $\mathcal{A}_{mix}$ for the mixing process:

The chemical drive $\mathcal{A}_{mix}$ for the mixing process results in (pressure and temperature of all reactants and products are the same):

$$\mathcal{A}_{mix} = -RT(x_A \times \ln x_A + x_B \times \ln x_B).$$

$$\mathcal{A}_{mix} = -8.314\,\text{G}\,\text{K}^{-1} \times 288\,\text{K} \times (0.5 \times \ln 0.5 + 0.5 \times \ln 0.5) = +1660\,\text{G} = \mathbf{1.66\,kG}.$$

Amount $n_{A,initial}$ of hydrogen at the beginning:

The amount $n_{A,initial}$ of hydrogen can be calculated by means of the general gas law [Eq. (10.7)]:

$$n_{A,initial} = \frac{p_A V_{initial}}{RT}.$$

$$n_{A,initial} = \frac{(100 \times 10^3\,\text{Pa}) \times (5 \times 10^{-3}\,\text{m}^3)}{8.314\,\text{G}\,\text{K}^{-1} \times 288\,\text{K}} = 0.209\,\text{mol}.$$

Conversion $\Delta\xi$:

The conversion $\Delta\xi$ is then [cf. Eq. (1.14)]:

$$\Delta\xi = \frac{\Delta n_A}{\nu_A} = \frac{n_{A,final} - n_{A,initial}}{\nu_A}.$$

$$\Delta\xi = \frac{0 - 0.209\,\text{mol}}{-0.5} = 0.418\,\text{mol}.$$

Energy W_f released during the mixing process:

The energy W_f can be determined by means of Equation (8.18):

$$W_f = -W_{\rightarrow\xi} = \mathcal{A}_{mix} \times \Delta\xi.$$

$$W_f = 1660\,\text{J}\,\text{mol}^{-1} \times 0.418\,\text{mol} = \mathbf{694\,J}.$$

b*) Amount $n_{A,initial}$ of hydrogen and amount $n_{B,initial}$ of nitrogen:

In the present case, the pressures of the two components are different: $p_A = 100$ kPa [such as in part a)] and $p_B = 300$ kPa. The amount $n_{A,initial}$ of hydrogen has remained the same, the amount $n_{B,initial}$ of nitrogen results in:

$$n_{B,initial} = \frac{p_B V_{initial}}{RT} = \frac{(300 \times 10^3\,\text{Pa}) \times (5 \times 10^{-3}\,\text{m}^3)}{8.314\,\text{G}\,\text{K}^{-1} \times 288\,\text{K}} = 0.627\,\text{mol}.$$

Mole fraction x_A of hydrogen and mole fraction x_B of nitrogen:

$$x_A = \frac{n_{A,initial}}{n_{A,initial} + n_{B,initial}} = \frac{0.209\,\text{mol}}{0.209\,\text{mol} + 0.627\,\text{mol}} = 0.25.$$

$$x_B \quad = 1 - x_A = 1 - 0.25 = 0.75 .$$

Therefore, the conversion formula of the mixing process is:

$$\tfrac{1}{4}A + \tfrac{3}{4}B \rightarrow A_{0.25}B_{0.75} .$$

Pressure p_M of the gas mixture:

$$p_M \quad = \frac{n_{total}RT}{V_{final}} = \frac{(0.209\ \text{mol} + 0.627\ \text{mol}) \times 8.314\ \text{G K}^{-1} \times 288\ \text{K}}{10 \times 10^{-3}\ \text{m}^3} = 200\ \text{kPa} .$$

Chemical drive $\mathcal{A}_{mix}$ for the mixing process:

Since the pressures p are different for all substances involved (pure A, pure B, mixture M), we have to use the following equation:

$$\mathcal{A}_{mix} \quad = x_A \mathring{\mu}_A(p_A) + x_B \mathring{\mu}_B(p_B) - \mu_M(p_M) .$$

We can write for the chemical potential $\mu_M(p_M)$ of the homogeneous mixture:

$$\mu_M(p_M) = x_A \mathring{\mu}_A(p_M) + x_B \mathring{\mu}_B(p_M) + RT(x_A \times \ln x_A + x_B \times \ln x_B) .$$

Based on the mass action equation 2 [Eq. (6.24)], one obtains for the term $\mathring{\mu}_A(p_M)$:

$$\mathring{\mu}_A(p_M) = \mathring{\mu}_A(p_A) + RT \ln \frac{p_M}{p_A} .$$

The same applies to $\mathring{\mu}_B(p_M)$.

Inserting the expressions into the equation above for the chemical potential of the homogeneous mixture results in:

$$\mu_M(p_M) = x_A \mathring{\mu}_A(p_A) + RTx_A \times \ln \frac{p_M}{p_A} + x_B \mathring{\mu}_B(p_B) + RTx_B \times \ln \frac{p_M}{p_B}$$

$$+RT(x_A \times \ln x_A + x_B \times \ln x_B)$$

$$= x_A \mathring{\mu}_A(p_A) + x_B \mathring{\mu}_B(p_B) + RT\left[x_A \times \ln\left(x_A \times \frac{p_M}{p_A}\right) + x_B \times \ln\left(x_B \times \frac{p_M}{p_B}\right)\right] .$$

If one inserts this expression for $\mu_M(p_M)$ into the equation for the chemical drive $\mathcal{A}_{mix}$ of the mixing process, only the following expression remains:

$$\mathcal{A}_{mix} \quad = -RT\left[x_A \times \ln\left(x_A \times \frac{p_M}{p_A}\right) + x_B \times \ln\left(x_B \times \frac{p_M}{p_B}\right)\right] .$$

$$\mathcal{A}_{mix} \quad = -8.314\ \text{G K}^{-1} \times 288\ \text{K} \times$$

$$\left[0.25 \times \ln\left(0.25 \times \frac{200 \times 10^3\ \text{Pa}}{100 \times 10^3\ \text{Pa}}\right) + 0.75 \times \ln\left(0.75 \times \frac{200 \times 10^3\ \text{Pa}}{300 \times 10^3\ \text{Pa}}\right)\right]$$

$$\mathcal{A}_{mix} \quad = -8.314\ \text{G K}^{-1} \times 288\ \text{K} \times (0.25 \times \ln 0.5 + 0.75 \times \ln 0.5) = +1600\ \text{G} = \mathbf{+1.60\ kG} .$$

Conversion $\Delta\xi$:

$$\Delta\xi = \frac{n_{A,final} - n_{A,initial}}{\nu_A} = \frac{0 - 0.209 \text{ mol}}{-0.25} = 0.836 \text{ mol}.$$

Energy W_f released during the mixing process:

$$W_f = \mathcal{A}_{mix} \times \Delta\xi = 1660 \text{ J mol}^{-1} \times 0.836 \text{ mol} = \mathbf{1388\ J}.$$

2.13.4 Real mixture

a) The factor $a = 0.49RT$ in the formula of the extra potential $\overset{+}{\mu}_G(x_B)$ and therefore the extra potential itself is positive. By addition of the extra potential, the curve of the chemical potential μ_M of an ideal mixture in dependence on composition forming a "drooping belly" is deformed in upward direction. However, the factor a is smaller than $2RT$ meaning the two components A and B are lowly compatible [see Section 13.5 (Real mixtures) in the textbook "Physical Chemistry from a Different Angle"].

b) Mole fraction x_A of component A and mole fraction x_B of component B:

Because the components A and B should be mixed in the amount-of-substance ratio 1:4 the mole fraction of the first component is

$$x_A = \frac{n_A}{n_A + n_B} = \frac{1}{1+4} = \frac{1}{5} = 0.2$$

and that of the second one is $x_B = 4/5 = 0.8$. Correspondingly, the conversion formula is:

$$\tfrac{1}{5}\text{A} + \tfrac{4}{5}\text{B} \rightarrow \text{A}_{0.2}\text{B}_{0.8}.$$

Chemical drive $\mathcal{A}_{mix}$ for the mixing process:

The chemical drive $\mathcal{A}_{mix}$ for the mixing process results again in:

$$\mathcal{A}_{mix} = x_A \overset{\bullet}{\mu}_A + x_B \overset{\bullet}{\mu}_B - \mu_M.$$

By inserting the formula for the chemical potential of the real mixture (see again Section 13.5), one obtains:

$$\mathcal{A}_{mix} = x_A \overset{\bullet}{\mu}_A + x_B \overset{\bullet}{\mu}_B - \left[x_A \overset{\bullet}{\mu}_A + x_B \overset{\bullet}{\mu}_B + RT(x_A \times \ln x_A + x_B \times \ln x_B) + \underbrace{0.49RT \times x_A \times x_B}_{\overset{+}{\mu}_M}\right]$$

$$= -RT\left[x_A \times \ln x_A + x_B \times \ln x_B + 0.49 \times x_B \times (1 - x_B)\right].$$

$$\mathcal{A}_{mix} = -8.314 \text{ G K}^{-1} \times 303 \text{ K} \times \left[0.2 \times \ln 0.2 + 0.8 \times \ln 0.8 + 0.49 \times 0.2 \times 0.8\right]$$

$= 1063\ \text{G} = \mathbf{1.063\ kG}$.

Molar entropy of mixing $\Delta_{\text{mix}}S$:

$$\Delta_{\text{mix}}S = \frac{\mathcal{A}_{\text{mix}}}{T} = \frac{1063\ \text{J mol}^{-1}}{303\ \text{K}} = 3.51\ \text{J K}^{-1}\ \text{mol}^{-1} = \mathbf{3.51\ Ct\ mol^{-1}}.$$

c) Conversion $\Delta\xi$:

The conversion $\Delta\xi$ results in:

$$\Delta\xi = \frac{\Delta n_{\text{A}}}{\nu_{\text{A}}} = \frac{n_{\text{A,final}} - n_{\text{A,initial}}}{\nu_{\text{A}}} = \frac{0 - 1\ \text{mol}}{-0.2} = 5\ \text{mol}.$$

Energy W_{f} released by the mixing process:

Again, the energy W_{f} can be determined by means of Equation (8.18):

$$W_{\text{f}} = \mathcal{A}_{\text{mix}} \times \Delta\xi = 1063\ \text{J mol}^{-1} \times 5\ \text{mol} = 5320\ \text{J} = \mathbf{5.32\ kJ}.$$

2.13.5 Boiling equilibrium in the butane-pentane system

First, we plot the double tangent for better orientation (see the figure below). The two points of contact limit the so-called miscibility gap meaning a two-phase region where a vapor g of composition $x_{\text{B}}^{\text{g}} = 0.5$ and a liquid l of composition $x_{\text{B}}^{\text{l}} = 0.8$ coexist.

	liquid: x^{l}	x_{B}^{l}	vapor: x^{g}	x_{B}^{g}
a)	0 %,	– %	100 %,	0 %
b)	0 %,	– %	100 %,	20 %
c)	0 %,	80 %	100 %,	50 %
d)	33 %,	80 %	67 %,	50 %
e)	57 %	80 %	43 %,	50 %
f)	100 %,	90 %	0 %,	– %

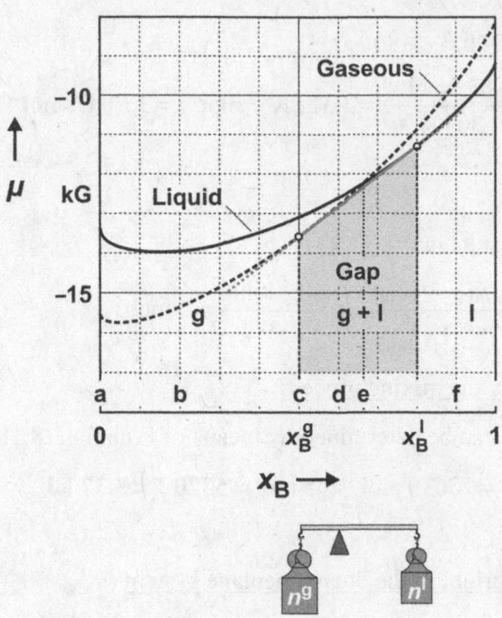

Explanation:

a) Only the pure phase A (butane) exists as vapor.

b) Only a vapor phase with the composition $x_B = 0.2$ exists.

c) The composition $x_B = 0.5$ represents one of the limits of the "miscibility gap." At this point, just only a vapor phase exists.

d) The composition $x_B = 0.6$ lies in the range of the "gap" meaning a vapor of the composition $x_B^g = 0.5$ coexists with a liquid of the composition $x_B^l = 0.8$. The ratio of amounts of gaseous and liquid phases is given by the "lever rule" [Eq. (13.14)]:

$$\frac{n^g}{n^l} = \frac{x_B^l - x_B^{\blacktriangle}}{x_B^{\blacktriangle} - x_B^g} = \frac{0.8-0.6}{0.6-0.5} = \frac{2}{1}.$$

In equilibrium, a fraction of $x^g = n^g/(n^g + n^g) = 2/(2+1) = 0.67 = 67\ \%$ is present as vapor with a composition of $x_B^g = 0.5$ and a fraction of $x^l = 33\ \%$ is present as liquid with the composition $x_B^l = 0.8$.

e) The composition $x_B = 0.67$ also lies in the range of the "gap." But the ratio (calculated analogously to part a) shifts in favor of the liquid.

f) At this point outside the "gap," only a liquid with the mole fraction $x_B = 0.9$ can be found.

2.14 Binary Systems

2.14.1 Miscibility diagram

a) In the following miscibility diagram, the phase boundary line for the phase poorer in phenol was plotted in dark gray and that for the phase richer in phenol in medium gray. The two-phase region was shaded in light gray.

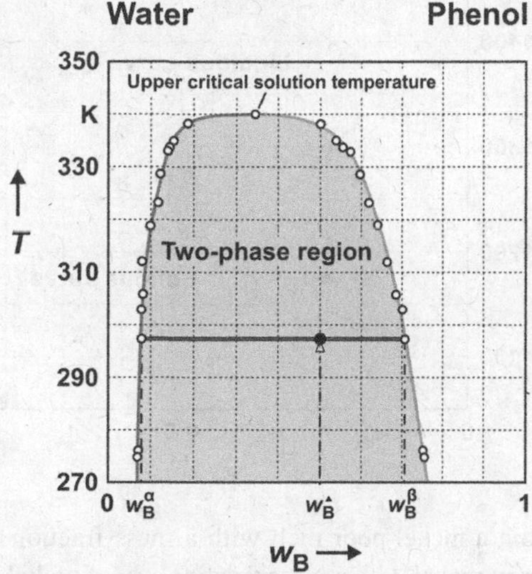

b) If a mixture of 5 g of water and 5 g of phenol (mass fraction $w_B = 0.5$) is prepared at 298 K, two separate liquid mixed phases occur. The mass fraction of phenol is **0.08** for the phenol-poor phase and **0.71** for the phenol-rich phase.

The mass ratio of phenol-poor phase (α) to phenol-rich phase (β) at the given temperature can be calculated quite analogously to the ratio of amounts by means of the lever rule [(cf. Eq. (13.14)] (see the lever indicated in the figure):

$$\frac{m^\alpha}{m^\beta} = \frac{w_B^\beta - w_B^\blacktriangle}{w_B^\blacktriangle - w_B^\alpha} = \frac{0.71-0.5}{0.5-0.08} = 0.5\left(=\frac{0.5}{1}\right).$$

The fraction of the phenol-poor phase (α) in the mixture results in $m^\alpha/(m^\alpha+m^\beta) = 0.5/(0.5+1) =$ **33 %**, that of the phenol-rich phase correspondingly in **67 %**.

c) Heating the sample above approx. **338 K** results in a single phase.

2.14.2 Liquid-solid phase diagram of copper and nickel

a) see figure under c). The solidus curve (melting curve) was plotted in dark gray, the liquidus curve (freezing curve) in medium gray; the two-phase region was shaded in light gray.

b) The metals copper and nickel are indifferent to each other, i.e. they are infinitely soluble in each other both in the liquid as well as the solid states.

c)

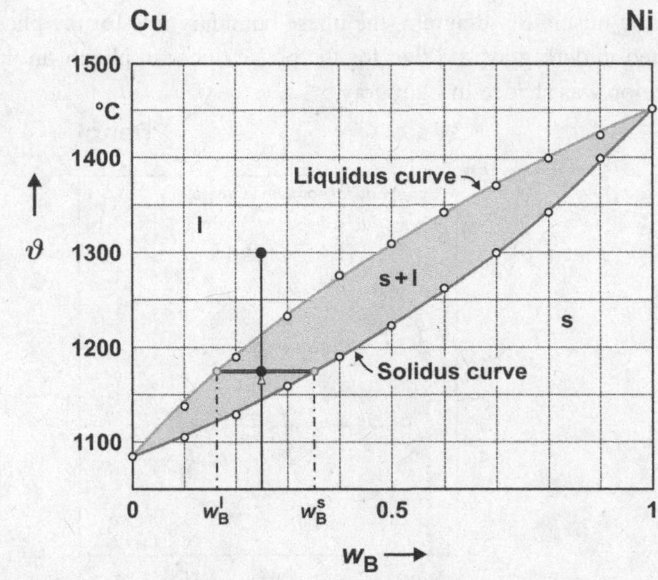

At 1175 °C, we obtain a nickel-poor melt with a mass fraction w_B^l of nickel of 0.17 and mixed crystals richer in nickel with a mass fraction w_B^s of nickel of 0.35.

d) The mass ratio can be calculated again by means of the lever rule:

$$\frac{m^l}{m^s} = \frac{w_B^s - w_B^\blacktriangle}{w_B^\blacktriangle - w_B^l} = \frac{0.35 - 0.25}{0.25 - 0.17} = 1.25\,.$$

Then, the fraction of the melt results in $1.25/(1.25 + 1) =$ **56 %**, that of the mixed crystals correspondingly in **44 %**.

e) Since we assumed the mass of the melt to be 10 kg and since the fraction of mixed crystals at 1175 °C is 44 %, we have **4.4 kg** of crystallites at this temperature.

2.14.3 Liquid-solid phase diagram of bismuth and cadmium

a) see figure under c). The solidus curve was again plotted in dark gray, the liquidus curve in medium gray; the two-phase regions were shaded in light gray.

b) The metals bismuth and cadmium are completely incompatible, i.e. they are (almost) insoluble in each other in the solid state.

c) The mole fraction of bismuth in the melt of the partially molten bismuth-cadmium alloy is supposed to be 0.3; therefore, the mole fraction x_B of cadmium is $1 - 0.3 = 0.7$. We can deduce from the figure that the temperature in question is about 210 °C (starting point of the tie-line). (Almost) pure crystals of cadmium (Phase β; $x_B^\beta = 1$) have separated from the melt (end point of the tie-line).

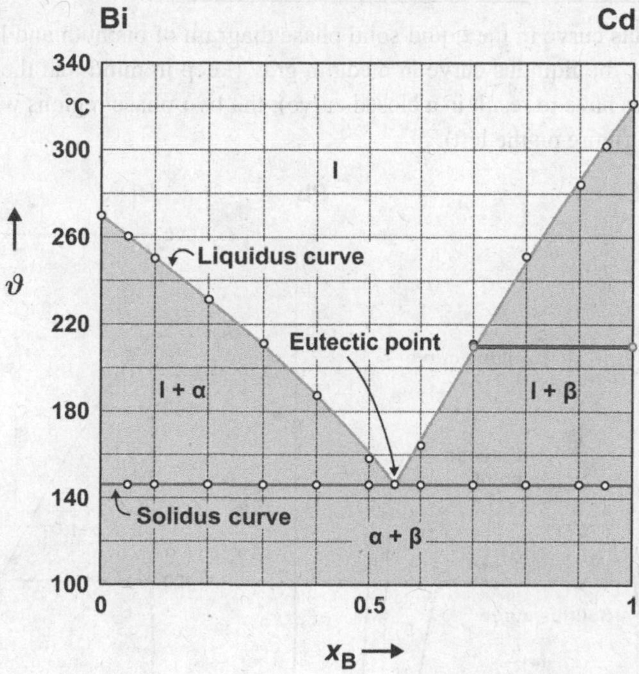

d) Since we supposed that the total amount of liquid and solid phase is 20 moles and that 10 moles of pure cadmium crystals (Phase β) have separated from the melt at about 210 °C, there still has to be 10 moles of melt present. In order to determine the composition $x_B^\blacktriangle$ of the starting mixture, we use the lever rule:

$$\frac{n^l}{n^\beta} = \frac{x_B^\beta - x_B^\blacktriangle}{x_B^\blacktriangle - x_B^l} .$$

Inserting the given values and solving for $x_B^\blacktriangle$ results in:

$$\frac{10 \text{ mol}}{10 \text{ mol}} = \frac{1 - x_B^\blacktriangle}{x_B^\blacktriangle - 0.7}$$

$$x_B^\blacktriangle - 0.7 = 1 - x_B^\blacktriangle$$

$$x_B^\blacktriangle = \frac{1.7}{2} = 0.85 .$$

The mole fraction of cadmium in the starting mixture is **0.85**.

e) Since the total amount of liquid and solid phase is 20 moles and the mole fraction of bismuth in the starting mixture is $1 - 0.85 = 0.15$, the amount of bismuth crystallites after complete solidification of the alloy is 3 moles.

2.14.4* Liquid-solid phase diagram of bismuth and lead

a) b) The solidus curve in the liquid-solid phase diagram of bismuth and lead was plotted in dark gray, the liquidus curve in medium gray (keep in mind that the solidus and liquidus curves have to result in a closed curve); the two-phase regions were shaded in light gray (see figure on the left).

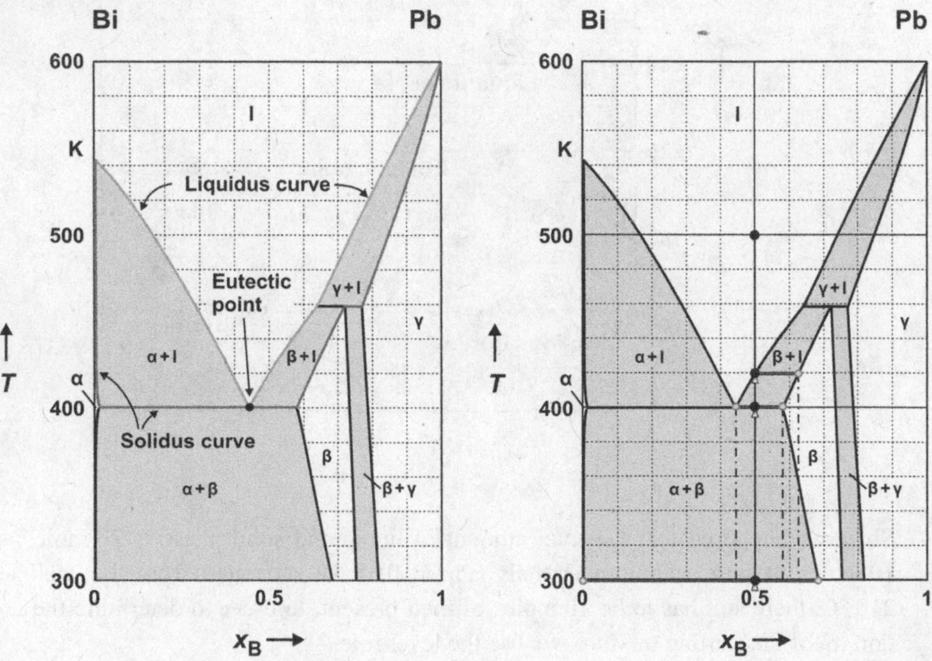

c) Table of the mole fractions x of the phases in question (first number) and corresponding lead content x_B (second number):

T/K	Phase "l"	Phase "α"	Phase "β"	Phase "γ"
500	100 %, 50 %	– %, – %	– %, – %	– %, – %
420	100 %, 50 %	– %, – %	0 %, 63 %	– %, – %
401	57 %, 44 %	– %, – %	43 %, 58 %	– %, – %
300	– %, – %	26 %, 0 %	74 %, 68 %	– %, – %

Comments:

- At a temperature of 500 K, we have only a melt with a lead content of 50 %.
- If the temperature is lowered to 420 K, the boundary line of the two-phase region (β+1) is just reached. The melt 1 with a lead content of 50 % is in equilibrium with β mixed crystals with a lead content of 63 %.
- Further cooling causes more and more solid substance to crystallize. Thereby, the concentration of the melt changes according to the solidus curve and the concentration of the precipitated β mixed crystals according to the liquidus curve. Eventually, at 401 K, the horizontal line is almost reached. The ratio of melt to β mixed crystals at this temperature is given by the lever rule (see the lever indicated in the figure above on the right):

$$\frac{n^{\mathrm{l}}}{n^{\beta}} = \frac{x_{\mathrm{B}}^{\beta} - x_{\mathrm{B}}^{\blacktriangle}}{x_{\mathrm{B}}^{\blacktriangle} - x_{\mathrm{B}}^{\mathrm{l}}} = \frac{58-50}{50-44} = \frac{8}{6}.$$

The fraction of melt then results in 8/(8+6) = 57 %, that of β mixed crystals in 43 %.

After only a slight further cooling down to 400 K (more precisely 399 K), the system reaches definitively the horizontal line. The entire residual melt has to solidify at this so-called eutectic temperature; thereby α mixed crystals and β mixed crystals precipitate simultaneously. These crystallites are extremely fine-grained (so-called eutectic structure) whereas the primary crystals of the β phase precipitated during the previous cooling process are much larger.

- Further cooling alters the composition of the mixed crystals. The α mixed crystals become continuously richer in bismuth and the β mixed crystals become continuously richer in lead. At 300 K, finally, (almost) pure bismuth is in equilibrium with β mixed crystals with a lead content of 68 %. The corresponding relative amounts of both phases are again given by the lever rule:

$$\frac{n^{\alpha}}{n^{\beta}} = \frac{x_{\mathrm{B}}^{\beta} - x_{\mathrm{B}}^{\blacktriangle}}{x_{\mathrm{B}}^{\blacktriangle} - x_{\mathrm{B}}^{\alpha}} = \frac{68-50}{50-0} = \frac{18}{50}.$$

The fraction of α mixed crystals (pure bismuth) results in 18/(18 + 50) = 26 %, that of β mixed crystals in 74 %.

In reality, such changes of composition in the solid state require a very long time because the diffusion of atoms in solids is extremely slow.

1.14.5 Liquid mixed phase and corresponding mixed vapor

a) Amount n_{A} of ethanol in the liquid mixed phase:

$$n_{\mathrm{A}} = \frac{m_{\mathrm{A}}}{M_{\mathrm{A}}} = \frac{0.050\ \mathrm{kg}}{46.0\times10^{-3}\ \mathrm{kg\,mol^{-1}}} = 1.09\ \mathrm{mol}.$$

Amount n_B of methanol in the liquid mixed phase:

$$n_B = \frac{m_B}{M_B} = \frac{0.050\ \text{kg}}{32.0\times10^{-3}\ \text{kg mol}^{-1}} = 1.56\ \text{mol}.$$

Mole fraction x_A^l of ethanol in the liquid mixed phase:

$$x_A^l = \frac{n_A}{n_A + n_B} = \frac{1.09\ \text{mol}}{1.09\ \text{mol} + 1.56\ \text{mol}} = \mathbf{0.41}.$$

Mole fraction x_B^l of methanol in the liquid mixed phase:

$$x_B^l = 1 - x_A^l = 1 - 0.41 = \mathbf{0.59}.$$

b) Partial pressure p_A of ethanol in the mixed vapor:

The partial pressure of a component i in a mixed vapor is equal to the product of the vapor pressure of the pure component and its mole fraction x_i^l in the liquid mixed phase [RAOULT's law; Eq. (14.1)]:

$$p_A = x_A^l \times p_A^\bullet = 0.41\times(5.8\times10^3\ \text{Pa}) = 2.38\times10^3\ \text{Pa} = \mathbf{2.38\ kPa}.$$

Partial pressure p_B of methanol in the mixed vapor:

$$p_B = x_B^l \times p_B^\bullet = 0.59\times(12.9\times10^3\ \text{Pa}) = 7.61\times10^3\ \text{Pa} = \mathbf{7.61\ kPa}.$$

Total vapor pressure p_{total} above the mixture:

$$p_{total} = p_A + p_B = (2.38 + 7.61)\times10^3\ \text{Pa} = 9.99\times10^3\ \text{Pa} = \mathbf{9.99\ kPa}.$$

c) Mole fraction x_A^g of ethanol in the mixed vapor:

The mole fraction x_A^g of ethanol in the mixed vapor results according to Equation (14.7) in:

$$x_A^g = \frac{p_A}{p_{total}} = \frac{2.38\times10^3\ \text{Pa}}{9.99\times10^3\ \text{Pa}} = \mathbf{0.24}.$$

Mole fraction x_B^g of methanol in the mixed vapor:

$$x_B^g = \frac{p_B}{p_{total}} = \frac{7.61\times10^3\ \text{Pa}}{9.99\times10^3\ \text{Pa}} = \mathbf{0.76}.$$

As expected, the vapor is enriched with the more volatile component, here methanol.

2.14.6 Vapor pressure diagram of m-xylene and benzene

a) Mole fraction x_B^g of benzene in the mixed vapor and total vapor pressure above the liquid mixed phase:

$$x_B^g = \frac{p_B}{p_{total}} = \frac{x_B^l \times p_B^\bullet}{x_B^l \times p_B^\bullet + x_A^l \times p_A^\bullet} = \frac{x_B^l \times p_B^\bullet}{x_B^l \times p_B^\bullet + (1 - x_B^l) \times p_A^\bullet}.$$

The mole fraction x_B^g of benzene in the mixed vapor above a liquid mixed phase with a mole fraction $x_B^l = 0.05$ of benzene results in:

$$x_B^g = \frac{0.05 \times (10 \times 10^3 \text{ Pa})}{0.05 \times (10 \times 10^3 \text{ Pa}) + (1 - 0.05) \times (0.83 \cdot 10^3 \text{ Pa})} = \mathbf{0.39}.$$

The total vapor pressure above the liquid mixed phase is

$$p_{total} = x_B^l \times p_B^\bullet + (1 - x_B^l) \times p_A^\bullet.$$

$$p_{total} = 0.05 \times (10 \times 10^3 \text{ Pa}) + (1 - 0.05) \times (0.83 \times 10^3 \text{ Pa}) = 1.29 \times 10^3 \text{ Pa} = \mathbf{1.29\ kPa}.$$

The remaining values are calculated in an analogous way.

x_B^l	x_B^g	p_{total} / kPa
0.05	0.39	1.29
0.10	0.57	1.75
0.25	0.80	3.12
0.50	0.92	5.42
0.75	0.97	7.71

b) see figure under d). The boiling point curve was plotted in dark gray and the dew point curve in medium gray; the two-phase region was shaded in light gray.

c) Mole fraction x_B^l of benzene in the liquid mixed phase:

$$x_B^l = \frac{n_B}{n_B + n_A} = \frac{2 \text{ mol}}{2 \text{ mol} + 1 \text{ mol}} = 0.67.$$

The mixture begins to boil at about **7.0 kPa** (starting point of tie-line 1 in the diagram).

d) As can be seen from the diagram (end point of tie-line 1), the mole fraction of benzene in the mixed vapor is **0.96** and therefore that of m-xylene **0.04**.

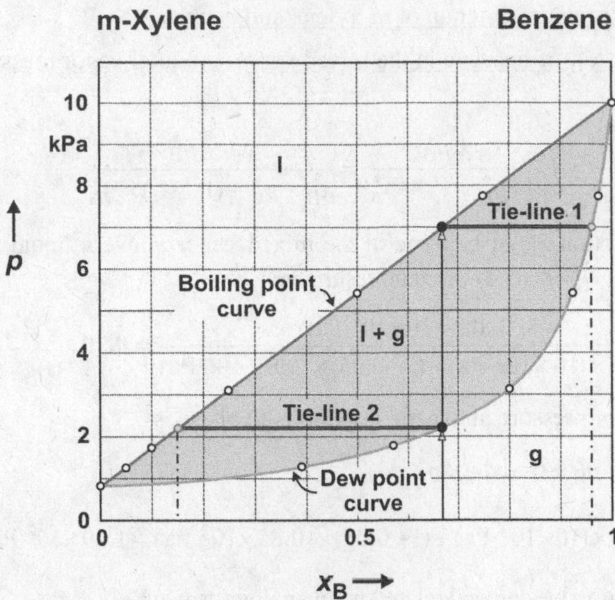

e) The mole fraction of benzene in the last remaining droplets of the liquid is **0.14** and therefore that of m-xylene **0.86** (starting point of tie-line 2). The corresponding vapor pressure is **2.1 kPa**.

1.14.7 Boiling temperature diagram of toluene and benzene as well as distillation

a) In order to construct the boiling point curve, we have to calculate the composition of the liquid phase at the given temperatures considering the condition that the sum of the partial pressures of the components toluene $p_A(T)$ and benzene $p_B(T)$ corresponds to the total pressure $p_{total} = 100$ kPa:

$$p_{total} \quad = p_A(T) + p_B(T)\,.$$

Since the components of the homogeneous mixture are indifferent to each other, we can use RAOULT's law [Eq. (14.3)] over the entire composition range and obtain:

$$p_A(T) \;=\; x_A^l(T) \times p_A^\bullet(T) \qquad \text{and} \qquad p_B(T) \;=\; x_B^l(T) \times p_B^\bullet(T)\,.$$

Inserting both expressions into the equation for the total pressure with regard to the relationship

$$x_A^l(T) \;=\; 1 - x_B^l(T)$$

finally results in

$$p_{total} \quad = [1 - x_B^l(T)] \times p_A^\bullet(T) + x_B^l(T) \times p_B^\bullet(T) = p_A^\bullet(T) + x_B^l(T) \times [p_B^\bullet(T) - p_A^\bullet(T)]\,.$$

By solving for $x_B^l(T)$, we obtain:

$$x_B^l(T) = \frac{p_{total} - p_A^{\bullet}(T)}{p_B^{\bullet}(T) - p_A^{\bullet}(T)}.$$

The dew point curve represents the composition of the vapor phase, which is in equilibrium with the corresponding liquid phase at the given temperature [Eq. (14.8)]:

$$x_B^g(T) = \frac{p_B^{\bullet}(T)}{p_{total}} \times x_B^l(T).$$

As an example, we calculate the values for x_B^l and x_B^g at a temperature of 358 K:

$$x_B^l(358\text{ K}) = \frac{100\text{ kPa} - 46\text{ kPa}}{116\text{ kPa} - 46\text{ kPa}} = \mathbf{0.77}.$$

$$x_B^g(358\text{ K}) = \frac{116\text{ kPa}}{100\text{ kPa}} \cdot 0{,}77 = \mathbf{0.89}.$$

The remaining data are summarized in the following table.

T/K	x_B^l	x_B^g
353	1.00	1.00
358	0.77	0.89
363	0.58	0.77
368	0.40	0.62
373	0.25	0.45
378	0.12	0.25
384	0.00	0.00

b) see figure under c). Again, the boiling point curve was plotted in dark gray and the dew point curve in medium gray; the two-phase region was shaded in light gray.

c) Mole fraction $x_B^{\blacktriangle}$ of benzene in the mixture:

The mass ratio of toluene to benzene is supposed to be 1:1. If one starts, for example, from 1.0 kg of the mixture, this mixture contains 0.5 kg of benzene and 0.5 kg of toluene. The mole fraction $x_B^{\blacktriangle}$ of benzene is defined as

$$x_B^{\blacktriangle} = \frac{n_B}{n_A + n_B}.$$

Since the amount n can be expressed by $n = m/M$, we obtain:

$$x_B^{\blacktriangle} = \frac{m_B/M_B}{m_A/M_A + m_B/M_B}.$$

The molar mass of benzene is 78.0×10^{-3} kg mol^{-1}, that of toluene is 92.0×10^{-3} kg mol^{-1}. Thus, we finally have:

$$x_B^{\blacktriangle} = \frac{0.5\text{ kg}/78.0\times10^{-3}\text{ kg mol}^{-1}}{0.5\text{ kg}/92.0\times10^{-3}\text{ kg mol}^{-1} + 0.5\text{ kg}/78.0\times10^{-3}\text{ kg mol}^{-1}}$$

$$= \frac{6.41\text{ mol}}{5.43\text{ mol} + 6.41\text{ mol}} = \mathbf{0.54}.$$

If the mixture is heated to 365 K at constant pressure, the two-phase region is reached and the system consists of a liquid and a gaseous phase with different compositions, which are in equilibrium with each other. The mole fractions of benzene in both phases can be read from the boiling temperature diagram:

$x_B^l \approx \mathbf{0.50}$, $x_B^g \approx \mathbf{0.72}$.

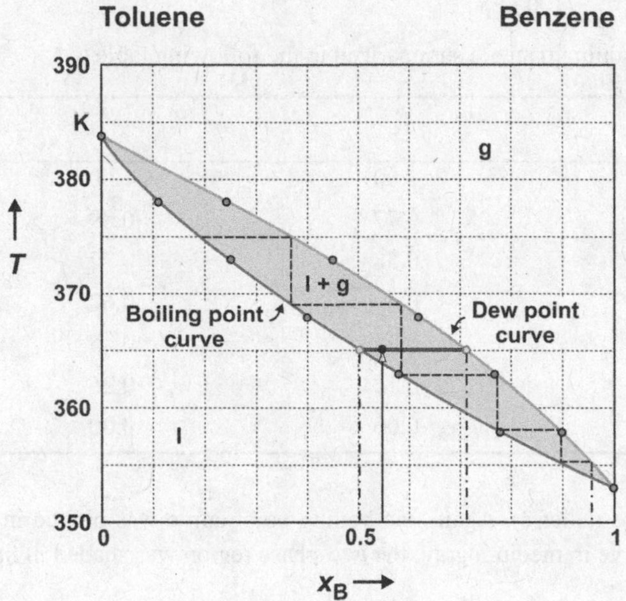

d) If the five theoretical plates are plotted as "steps" into the boiling temperature diagram, each step consisting of a horizontal part (for the boiling process) and a vertical part (for the condensation process), the composition of the distillate can be determined:

$x_B^l \approx \mathbf{0.96}$.

This corresponds to a purity of 96 % related to the amount of substance.

2.14.8 Boiling temperature diagram with azeotropic maximum

a) The boiling point curve in the following boiling temperature diagram was plotted in dark gray, the dew point curve, however, in medium gray; the two-phase regions were shaded in light gray.

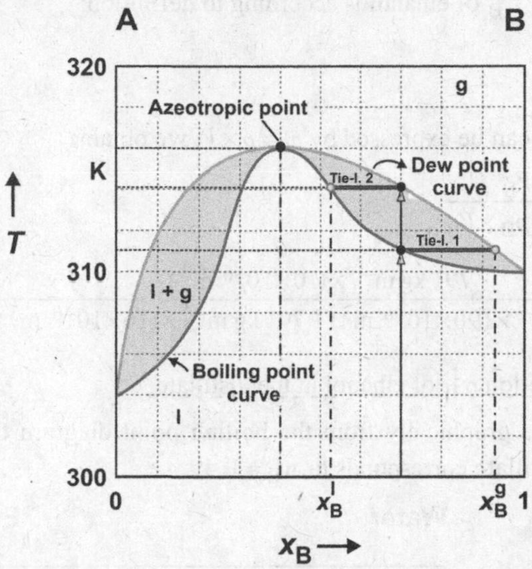

b) A liquid mixture with a mole fraction x_B^l of B of 0.7 boils at a temperature of **311 K**, easy to read from the diagram. The fraction of B in the resulting vapor is x_B^g = **0.93** [end point of tie-line 1].

c) A vapor with a mole fraction x_B^g of B of 0.7 condenses at a temperature of **314 K**. The fraction of B in the resulting condensate is x_B^l = **0.52** [starting point of tie-line 2].

d) The azeotropic mixture boils at **316 K**. At the azeotropic point, liquid and vapor have the same composition. Therefore, we have: $x_B^l = x_B^g$ = **0.40**.

e) The two liquids show a boiling point maximum and thus a vapor pressure minimum. However, a negative deviation of the vapor pressure curves from RAOULT's law is characteristic for a "highly compatible" behavior of the components in the mixture.

2.14.9 Distillation of mixtures of water and ethanol

a) see the boiling point diagram below

b) Mass fraction w_B^1 of ethanol in the starting mixture:

The volume concentration σ_B of the mixture of water and ethanol is supposed to be 10 % of ethanol. If we assume, for example, a mixture of 100 cm³, we have approximately 10 cm³ of ethanol and 90 cm³ of water (see the comment in the text of Exercise 1.14.9). The mass fraction w_B^1 of ethanol is according to definition:

$$w_B^1 = \frac{m_B}{m_A + m_B}.$$

Since the mass m can be expressed by $m = \rho \times V$, we obtain:

$$w_B^1 = \frac{\rho_B \times V_B}{\rho_A \times V_A + \rho_B \times V_B}.$$

$$w_B^1 = \frac{791\ \text{kg m}^{-3} \times (10 \times 10^{-6}\ \text{m}^3)}{998\ \text{kg m}^{-3} \times (90 \times 10^{-6}\ \text{m}^3) + 791\ \text{kg m}^{-3} \times (10 \times 10^{-6}\ \text{m}^3)} \approx 0.08.$$

Volume concentration σ_B^1 of ethanol in the distillate:

We can determine graphically from the boiling point diagram that the mass fraction of ethanol in the distillate corresponds to $w_B^1 \approx 0.49$.

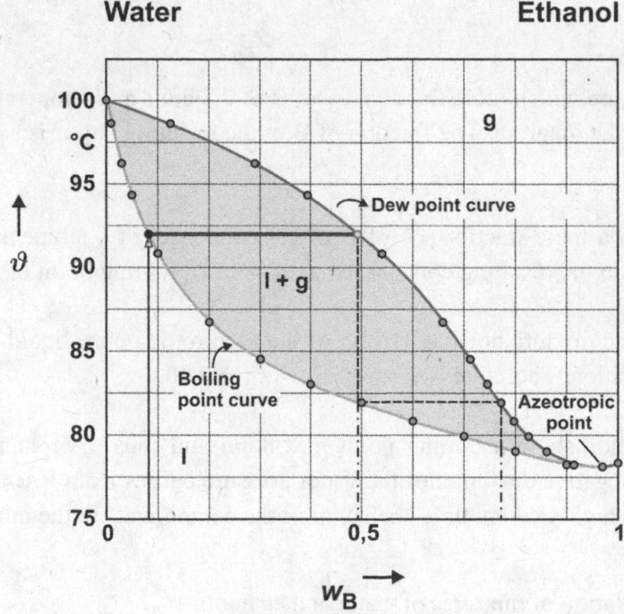

The conversion into the volume concentration is carried out analogously to the above calculation (only that one now assumes, for example, 100 g of the mixture):

$$\sigma_B^1 = \frac{V_B}{V_A + V_B} = \frac{m_B/\rho_B}{m_A/\rho_A + m_B/\rho_B}.$$

$$\sigma_B^1 = \frac{0.049\ \text{kg}/791\ \text{kg m}^{-3}}{0.051\ \text{kg}/998\ \text{kg m}^{-3} + 0.049\ \text{kg}/791\ \text{kg m}^{-3}} = \mathbf{0.55}.$$

The distillate thus has an "alcoholic" content of about 55 "vol%." This corresponds approximately to the "alcoholic" content of some types of rum (e.g. "Pott Rum"). These rums are already suitable to prepare a "Feuerzangenbowle" (a traditional German alcoholic drink for which a rum-soaked sugarloaf is set on fire and drips into mulled wine) (rums with a significantly lower "alcoholic" content do not burn). But also the liqueur Green Chartreuse contains 55 "vol%" of alcohol.

c) Mass fraction w_B^1 of ethanol in the distillate:

The distillate should have a volume concentration σ_B of 80 % of ethanol. For the calculation, we proceed exactly as in part b) of this exercise:

$$w_B^1 = \frac{\rho_B \times V_B}{\rho_A \times V_A + \rho_B \times V_B}.$$

$$w_B^1 = \frac{791\ \text{kg m}^{-3} \times (80 \times 10^{-6}\ \text{m}^3)}{998\ \text{kg m}^{-3} \times (20 \times 10^{-6}\ \text{m}^3) + 791\ \text{kg m}^{-3} \times (80 \times 10^{-6}\ \text{m}^3)} = 0.76.$$

As can be seen from the diagram, two theoretical plates are sufficient to obtain a distillate with a volume concentration of "alcohol" of at least 80 %.

2.15 Interfacial Phenomena

2.15.1 Surface energy

Energy $W_{\to A}$ required to create the new surface area:

In order to calculate the energy $W_{\to A}$ required to create the surface area A, Equation (15.2) is used:

$$W_{\to A} = \sigma \times A = \sigma \times l \times s\,.$$

l is the total width of the surface on the front and the back of the liquid film, i.e. we have l = 6 cm.

$$W_{\to A} = (30\times10^{-3}\ \text{N m}^{-1})\times(6\times10^{-2}\ \text{m})\times(4\times10^{-2}\ \text{m}) = 72\times10^{-6}\ \text{N m} = \mathbf{72\ \mu J}\,.$$

2.15.2* Atomization

Volume V_{dr} of a spherical water droplet:

$$V_{\text{dr}} = \frac{4}{3}\pi r_{\text{dr}}^3 = \frac{4}{3}\times3.142\times(0.5\times10^{-6}\ \text{m})^3 = 5.24\times10^{-19}\ \text{m}^3\,.$$

Number N_{dr} of droplets produced from the given volume of water:

The number N_{dr} of water droplets is obtained by dividing the total volume V by the droplet volume V_{dr}:

$$N_{\text{dr}} = \frac{V}{V_{\text{dr}}} = \frac{1.00\times10^{-3}\ \text{m}^3}{5.24\times10^{-19}\ \text{m}^3} = 1.91\times10^{15}\,.$$

Surface A_{dr} of a droplet:

$$A_{\text{dr}} = 4\pi r_{\text{dr}}^2 = 4\times3.142\times(0.5\times10^{-6}\ \text{m})^2 = 3.14\times10^{-12}\ \text{m}^2\,.$$

Surface A of all newly produced droplets together:

$$A = N_{\text{dr}}\times A_{\text{dr}} = (1.91\times10^{15})\times(3.14\times10^{-12}\ \text{m}^2) = 6000\ \text{m}^2\,.$$

Energy $W_{\to A}$ necessary to create the new surface area:

The value for the surface tension of water at 25 °C can be found in Table 15.1 of the textbook "Physical Chemistry from a Different Angle": $\sigma(H_2O) = 72.0\ \text{mN m}^{-1}$.

$$W_{\to A} = \sigma\times A = (72.0\times10^{-3}\ \text{N m}^{-1})\times6000\ \text{m}^2 = 432\ \text{N m} = \mathbf{0.43\ kJ}\,.$$

2.15.3 Capillary pressure

Excess pressure p_σ in a spherical water droplet:

The excess pressure in a spherical water droplet can be calculated by means of Equation (15.6):

$$p_\sigma = \frac{2\sigma}{r}.$$

The surface tension σ of water at 283 K is 74.2 mN m^{-1} (see Table 15.2):

$$p_\sigma = \frac{2\times(74.2\times10^{-3}\ \mathrm{N\,m^{-1}})}{250\times10^{-9}\ \mathrm{m}} = 594\times10^{3}\ \mathrm{N\,m^{-2}} = \mathbf{594\ kPa}.$$

2.15.4* Floating drops and bubbles

a) a$_1$) $\sigma_1 = \sigma_2 = \sigma_3$, a$_2$) $\sigma_1 > \sigma_2 + \sigma_3$, a$_3$) $\sigma_2 > \sigma_1 + \sigma_3$, a$_4$) $\sigma_3 > \sigma_1 + \sigma_3$

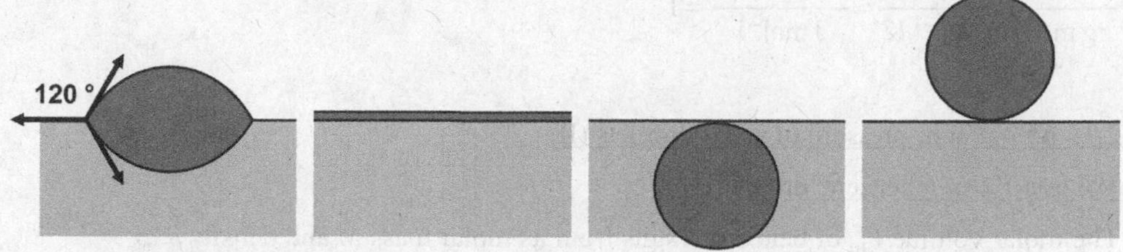

b) b$_1$) To calculate the excess pressure in the small air bubble (see left sketch on the bottom), Equation (15.6) is used again:

$$p_\sigma = \frac{2\sigma}{r} = \frac{2\times(73.0\times10^{-3}\ \mathrm{N\,m^{-1}})}{0.1\times10^{-3}\ \mathrm{m}} = 1460\ \mathrm{N\,m^{-2}} = \mathbf{1460\ Pa}.$$

b$_2$) In the case of the large air bubble (approximately a hemisphere), two interfaces occur (see right sketch on the bottom). Therefore, Equation (15.5) is used here:

$$p_\sigma = \frac{4\sigma}{r} = \frac{4\times(73.0\times10^{-3}\ \mathrm{N\,m^{-1}})}{10\times10^{-3}\ \mathrm{m}} = 29\ \mathrm{N\,m^{-2}} = \mathbf{29\ Pa}.$$

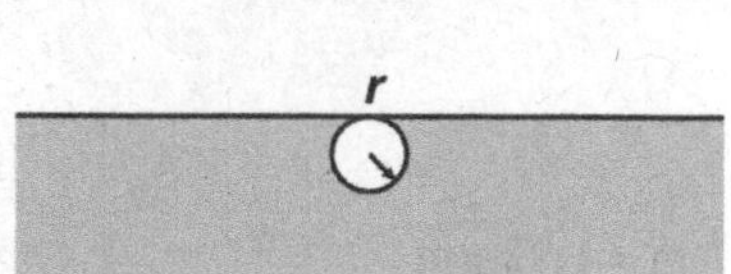

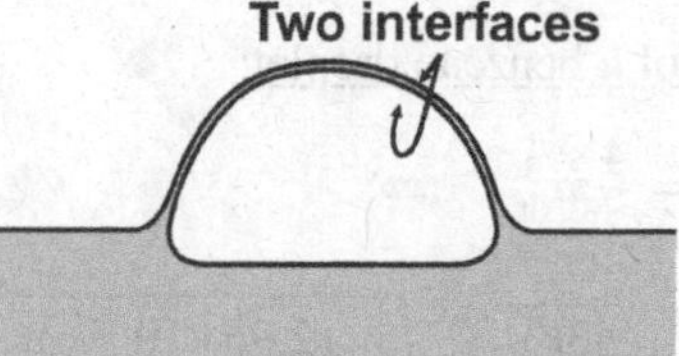

2.15.5 Vapor pressure of small droplets (I)

Vapor pressure $p_{\mathrm{lg},r}$ of a spherical water droplet:

The vapor pressure of the water droplet can be determined by means of the KELVIN equation [Eq. (15.7)],

$$p_{\mathrm{lg},r} = p_{\mathrm{lg},r=\infty} \exp\frac{2\sigma V_\mathrm{m}}{rRT} = p_{\mathrm{lg},r=\infty} \exp\frac{2\sigma M}{\rho rRT},$$

where the molar volume V_m of the substance in question can be expressed by its density ρ: $V_\mathrm{m} = M/\rho$. Inserting this expression and taking into account the molar mass of water with a value of $18{,}0\times 10^{-3}\ \mathrm{kg\,mol^{-1}}$ results in:

$$p_{\mathrm{lg},r} = 3167\ \mathrm{Pa}\times\exp\frac{2\times(72.0\times10^{-3}\ \mathrm{N\,m^{-1}})\times(18.0\times10^{-3}\ \mathrm{kg\,mol^{-1}})}{997\ \mathrm{kg\,m^{-3}}\times(5\times10^{-9}\ \mathrm{m})\times 8.314\ \mathrm{G\,K^{-1}}\times 298\ \mathrm{K}} = \mathbf{3907\ Pa}.$$

Conversion of units for the exponential expression:

$$\frac{\mathrm{N\,m^{-1}\,kg\,mol^{-1}}}{\mathrm{kg\,m^{-3}\,m\,G\,K^{-1}\,K}} = \frac{\mathrm{N\,m\ mol^{-1}}}{\mathrm{J\,mol^{-1}}} = 1.$$

2.15.6* Vapor pressure of small droplets (II)

Volume V_dr of a benzene droplet:

The molar volume V_m of benzene results from its molar mass M and density ρ to

$$V_\mathrm{m} = \frac{M}{\rho}.$$

By dividing the molar volume by the AVAGADRO constant N_A (which represents the number of particles in a portion of substance of one mole), and multiplying by the number of molecules N_dr in one droplet, we obtain the volume V_dr of a benzene droplet:

$$V_\mathrm{dr} = \frac{V_\mathrm{m}\times N_\mathrm{dr}}{N_\mathrm{A}} = \frac{M\times N_\mathrm{dr}}{\rho\times N_\mathrm{A}}.$$

$$V_\mathrm{dr} = \frac{(78.0\times10^{-3}\ \mathrm{kg\,mol^{-1}})\times 200}{(876\ \mathrm{kg\,m^{-3}})\times(6.022\times10^{23}\ \mathrm{mol^{-1}})} = \mathbf{2.96\times10^{-26}\ m^3}.$$

Radius r_dr of a benzene droplet:

$$V_\mathrm{dr} = \frac{4}{3}\pi r_\mathrm{dr}^3 \quad\Rightarrow$$

$$r_\mathrm{dr} = \sqrt[3]{\frac{3V_\mathrm{dr}}{4\pi}} = \sqrt[3]{\frac{3\times(2.96\times10^{-26}\ \mathrm{m^3})}{4\times 3.142}} = 1.92\times10^{-9}\ \mathrm{m} = \mathbf{1.92\ nm}.$$

Ratio $p_{\mathrm{lg},r}/p_{\mathrm{lg},r=\infty}$:

The ratio $p_{\mathrm{lg},r}/p_{\mathrm{lg},r=\infty}$ results according to the KELVIN equation [Eq. (15.7)] in:

$$\frac{p_{\mathrm{lg},r}}{p_{\mathrm{lg},r=\infty}} = \exp\frac{2\sigma V_{\mathrm{m}}}{r_{\mathrm{dr}}RT} = \exp\frac{2\sigma M}{\rho r_{\mathrm{dr}}RT}.$$

$$\frac{p_{\mathrm{lg},r}}{p_{\mathrm{lg},r=\infty}} = \exp\frac{2\times(28.2\times10^{-3}\ \mathrm{N\,m^{-1}})\times78.0\times10^{-3}\ \mathrm{kg\,mol^{-1}})}{876\ \mathrm{kg\,m^{-3}}\times(1.92\times10^{-9}\ \mathrm{m})\times8.314\ \mathrm{G\,K^{-1}}\times298\ \mathrm{K}} = \mathbf{2.9}.$$

The vapor pressure of the small benzene droplets is thus almost three times higher than that of the bulk liquid.

2.15.7 Determination of the surface tension

Surface tension σ of an ethanol-water mixture:

The surface tension of the given ethanol-water mixture can be determined using Equation (15.8):

$$h = \frac{2\sigma}{\rho r_{\mathrm{c}} g} \quad \Rightarrow$$

$$\sigma = \frac{\rho r_{\mathrm{c}} g h}{2} = \frac{955\ \mathrm{kg\,m^{-3}}\times(0.200\times10^{-3}\ \mathrm{m})\times9.81\ \mathrm{m\,s^{-2}}\times(3.58\times10^{-2}\ \mathrm{m})}{2}$$

$$= 0.0335\ \mathrm{N\,m^{-1}} = \mathbf{33.5\ mN\,m^{-1}}.$$

2.15.8* Capillary action

a) Capillary rise h_1 at 0 °C in a capillary with a radius $r_{\mathrm{c},1}$ of 0.01 mm:

The capillary rise of water at 0 °C can be calculated according to Equation (15.8):

$$h_1(0\ ^\circ\mathrm{C}) = \frac{2\sigma(0\ ^\circ\mathrm{C})}{\rho(0\ ^\circ\mathrm{C}) r_{\mathrm{c},1} g}.$$

$$h_1(0\ ^\circ\mathrm{C}) = \frac{2\times(76\times10^{-3}\ \mathrm{N\,m^{-1}})}{1000\ \mathrm{kg\,m^{-3}}\times(0.1\times10^{-3}\ \mathrm{m})\times9.81\ \mathrm{m\,s^{-2}}} = 0.155\ \mathrm{m} = 155\ \mathrm{mm}.$$

Capillary rise h_1 at 100 °C in a capillary with a radius $r_{\mathrm{c},1}$ of 0.01 mm:

$$h_1(100\ ^\circ\mathrm{C}) = \frac{2\sigma(100\ ^\circ\mathrm{C})}{\rho(100\ ^\circ\mathrm{C}) r_{\mathrm{c},1} g}.$$

$$h_1(100\ ^\circ\mathrm{C}) = \frac{2\times(59\times10^{-3}\ \mathrm{N\,m^{-1}})}{958\ \mathrm{kg\,m^{-3}}\times(0.1\times10^{-3}\ \mathrm{m})\times9.81\ \mathrm{m\,s^{-2}}} = 0.126\ \mathrm{m} = 126\ \mathrm{mm}.$$

Capillary rise h_2 at 0 °C in a capillary with a radius $r_{c,2}$ of 0.02 mm:

$$h_2(0\,°\text{C}) = \frac{2\sigma(0\,°\text{C})}{\rho(0\,°\text{C})r_{c,2}g}.$$

$$h_2(0\,°\text{C}) = \frac{2\times(76\times10^{-3}\ \text{N}\,\text{m}^{-1})}{1000\ \text{kg}\,\text{m}^{-3}\times(0.2\times10^{-3}\ \text{m})\times9.81\ \text{m}\,\text{s}^{-2}} = 0.077\ \text{m} = 77\ \text{mm}.$$

Capillary rise h_2 at 100 °C in a capillary with a radius $r_{c,2}$ of 0.02 mm:

$$h_2(100\,°\text{C}) = \frac{2\sigma(100\,°\text{C})}{\rho(100\,°\text{C})r_{c,2}g}.$$

$$h_2(100\,°\text{C}) = \frac{2\times(59\times10^{-3}\ \text{N}\,\text{m}^{-1})}{958\ \text{kg}\,\text{m}^{-3}\times(0.2\times10^{-3}\ \text{m})\times9.81\ \text{m}\,\text{s}^{-2}} = 0.063\ \text{m} = 63\ \text{mm}.$$

Capillary rises h at 0 °C and 100 °C in the capillary shown in the figure:
The thin part of the capillary until a height h_0 of 70 mm is completely filled at 0 °C as well as at 100 °C. In the part above, the water can rise up to **77 mm** at 0 °C, but at 100 °C it cannot exceed 70 mm [because $h_1(100\,°\text{C})$ = 126 mm > h_0, but $h_2(100\,°\text{C})$ = 63 mm < h_0].

b) Radius r of curvature of the water surface at 0 °C and 100 °C:

In the case of a contact angle θ of ≈ 0 °, the radius of curvature is just equal to the capillary radius r_c. By solving Equation (15.8) for r we obtain

$$r = \frac{2\sigma}{\rho h g}$$

and thus at 0 °C

$$r(0\,°\text{C}) = \frac{2\sigma(0\,°\text{C})}{\rho(0\,°\text{C})h(0\,°\text{C})g}$$

$$r(0\,°\text{C}) = \frac{2\times(76\times10^{-3}\ \text{N}\,\text{m}^{-1})}{1000\ \text{kg}\,\text{m}^{-3}\times(77\times10^{-3}\ \text{m})\times9.81\ \text{m}\,\text{s}^{-2}} = 0.20\times10^{-3}\ \text{m} = \mathbf{0.20\ mm},$$

which, as expected, corresponds to the tube radius in the upper area. At 100 °C, however, we have

$$r(100\,°\text{C}) = \frac{2\sigma(100\,°\text{C})}{\rho(100\,°\text{C})h(100\,°\text{C})g}$$

$$r(100\,°\text{C}) = \frac{2\times(59\times10^{-3}\ \text{N}\,\text{m}^{-1})}{958\ \text{kg}\,\text{m}^{-3}\times(70\times10^{-3}\ \text{m})\times9.81\ \text{m}\,\text{s}^{-2}} = 0.18\times10^{-3}\ \text{m} = \mathbf{0.18\ mm}.$$

c) The pressure depends only on the height h, but not on the tube diameter d. Compared to the external air pressure, a negative pressure prevails in the capillary at $h > 0$.

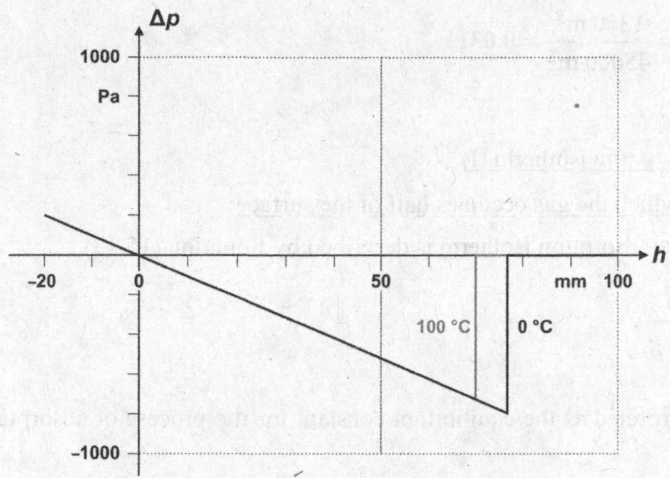

2.15.9 Fractional coverage

Number N of the nitrogen molecules:

We start from the general gas law [Eq. (10.7)]:

$$pV = nRT\,.$$

According to Equation (1.2), the amount of substance n is

$$n = N\tau\,,$$

where N represents the number of particles in the portion of substance in question and τ the elementary amount (of substance). Inserting the expression in the general gas law results in:

$$pV = N\tau RT \quad \Rightarrow$$

$$N = \frac{pV}{\tau RT} = \frac{(490\times10^{3}\ \text{Pa})\times(46.0\times10^{-6}\ \text{m}^3)}{(1.66\times10^{-24}\ \text{mol})\times 8.314\ \text{G}\,\text{K}^{-1}\times 190\ \text{K}} = 8.6\times10^{21}\,.$$

Surface area A_N covered by nitrogen molecules:

$$A_N = N\times A_M\,,$$

where A_M represents the area occupied by one nitrogen molecule:

$$A_N = (8.6\times10^{21})\times(0.16\times10^{-18}\ \text{m}^2) = 1380\ \text{m}^2\,.$$

Surface area A_{ac} of the activated charcoal:

$$A_{ac} = m_{ac}\times A_{sp,ac} = (50.0\times10^{-3}\ \text{kg})\times(900\times10^{3}\ \text{m}^2\,\text{kg}^{-1}) = 45000\ \text{m}^2\,.$$

Fractional coverage Θ:

The fractional coverage represents the fraction of the surface that is covered [cf. Eq. (15.9)]:

$$\Theta = \frac{A_N}{A_{ac}} = \frac{1380\ m^2}{45000\ m^2} = \mathbf{0.031}\,.$$

2.15.10 LANGMUIR isotherm (I)

Pressure p at which the gas occupies half of the surface:

The LANGMUIR adsorption isotherm is described by Equation (15.13),

$$\Theta = \frac{\overset{\circ}{K}\times p}{1+\overset{\circ}{K}\times p}\,.$$

$\overset{\circ}{K}$ can be interpreted as the equilibrium constant for the process of absorption. Solving for p yields:

$$p = \frac{\Theta}{\overset{\circ}{K}(1-\Theta)}\,.$$

If half of the surface is occupied by the gas, the fractional coverage Θ is 0.5.

$$p = \frac{0.5}{(0.65\times10^3\ Pa^{-1})\times(1-0.5)} = 1540\ Pa = \mathbf{1.54\ kPa}\,.$$

2.15.11* LANGMUIR isotherm (II)

a) Determination of the constant $\overset{\circ}{K}$ and the mass m_{mono}:

We start again from the LANGMUIR adsorption isotherm [Eq. (15.13)]:

$$\Theta = \frac{\overset{\circ}{K}\times p}{1+\overset{\circ}{K}\times p}\,.$$

For the fractional coverage Θ, we have [Eq. (15.9)]:

$$\Theta = \frac{m}{m_{mono}}\,.$$

The two expressions can be combined to

$$\frac{m}{m_{mono}} = \frac{\overset{\circ}{K}\times p}{1+\overset{\circ}{K}\times p}\,.$$

Therefore, we obtain for the first pair of measured values (p_1, m_1)

$$\frac{m_1}{m_{mono}} = \frac{\mathring{K}\times p_1}{1+\mathring{K}\times p_1}$$

and for the second pair (p_2, m_2)

$$\frac{m_2}{m_{mono}} = \frac{\mathring{K}\times p_2}{1+\mathring{K}\times p_2}.$$

Division of the first by the second equation results in:

$$\frac{m_1}{m_2} = \frac{p_1}{p_2}\times\frac{1+\mathring{K}\times p_2}{1+\mathring{K}\times p_1} = \frac{p_1+\mathring{K}\times p_1\times p_2}{p_2+\mathring{K}\times p_1\times p_2}.$$

Solving for $\mathring{K}$ yields:

$$m_1(p_2+\mathring{K}\times p_1\times p_2) = m_2(p_1+\mathring{K}\times p_1\times p_2)$$

$$m_1\times p_2+m_1\times\mathring{K}\times p_1\times p_2 = m_2\times p_1+m_2\times\mathring{K}\times p_1\times p_2$$

$$\mathring{K}\times p_1\times p_2(m_1-m_2) = m_2\times p_1-m_1\times p_2$$

$$\mathring{K} = \frac{m_2\times p_1-m_1\times p_2}{p_1\times p_2\times(m_1-m_2)}.$$

$$\mathring{K} = \frac{(51.34\times10^{-6}\text{ kg})\times(400\times10^2\text{ Pa})-(31.45\times10^{-6}\text{ kg})\times(800\times10^2\text{ Pa})}{(400\times10^2\text{ Pa})\times(800\times10^2\text{ Pa})\times[(31.45\times10^{-6}\text{ kg})-(51.34\times10^{-6}\text{ kg})]}$$

$$= \frac{-0,4624\text{ Pa kg}}{-63650\text{ Pa}^2\text{ kg}} = \mathbf{7.27\times10^{-6}\ Pa^{-1}}.$$

Subsequently, we return, for example, to the equation for the first pair of measured values and solve for m_{mono}:

$$m_{mono} = \frac{m_1(1+\mathring{K}\times p_1)}{\mathring{K}\times p_1}.$$

$$m_{mono} = \frac{(31.45\times10^{-6}\text{ kg})\times[1+(7.27\times10^{-6}\text{ Pa}^{-1})\times(400\times10^2\text{ Pa})]}{(7.27\times10^{-6}\text{ Pa}^{-1})\times(400\cdot10^2\text{ Pa})},$$

$$= 140.0\times10^{-6}\text{ kg} = \mathbf{140.0\ mg}.$$

b) Fractional coverage Θ_1:

$$\Theta_1 = \frac{m_1}{m_{mono}} = \frac{31.45\times10^{-6}\text{ kg}}{140.0\times10^{-6}\text{ kg}} = \mathbf{0.225}.$$

Fractional coverage Θ_2:

$$\Theta_2 = \frac{m_2}{m_{\text{mono}}} = \frac{51.34\times10^{-6}\ \text{kg}}{140.0\times10^{-6}\ \text{kg}} = \mathbf{0.367}\,.$$

2.16 Basic Principles of Kinetics

2.16.1 Conversion rate and rate density

Total conversion $\Delta\xi$:

The total conversion can be calculated by means of the "basic stoichiometric equation" [Eq. (1.14)] as well as the defining equation for the molar mass [Eq. (1.5) or (1.6)]:

$$\Delta\xi = \frac{\Delta n_B}{\nu_B} = \frac{\Delta m_B}{M_B \times \nu_B}.$$

Since the entire fuel mass m_0 is to be converted, we can also write:

$$\Delta\xi = \frac{m_B - m_{B,0}}{M_B \times \nu_B} = \frac{0 - m_{B,0}}{M_B \times \nu_B} = \frac{-m_{B,0}}{M_B \times \nu_B}.$$

Candle:

In the case of a candle, paraffin [formula ≈ (CH_2)] serves as fuel.

$$\Delta\xi = \frac{-1\times10^{-2}\ \text{kg}}{(14.0\times10^{-3}\ \text{kg mol}^{-1})\times(-2)} = \mathbf{0.36\ mol}.$$

Sun:

In the sun, however, hydrogen is converted.

$$\Delta\xi = \frac{-2\times10^{30}\ \text{kg}}{(1.0\times10^{-3}\ \text{kg mol}^{-1})\times(-4)} = \mathbf{500\times10^{30}\ mol}.$$

Conversion rate ω:

The conversion rate ω corresponds here to the quotient of total conversion $\Delta\xi$ and burning duration Δt [cf. Eq. (16.2)]:

$$\omega = \frac{\Delta\xi}{\Delta t}.$$

Candle:

$$\omega = \frac{0.36\ \text{mol}}{3600\ \text{s}} = \mathbf{100\times10^{-6}\ mol\ s^{-1}}.$$

Sun:

$$\omega = \frac{500\times10^{30}\ \text{mol}}{5\times10^{18}\ \text{s}} = \mathbf{100\times10^{12}\ mol\ s^{-1}}.$$

Rate density r:

The rate density r results in [cf. Eq. (16.6)]:

$$r = \frac{\omega}{V}.$$

Candle:

$$r = \frac{100\times10^{-6}\ \mathrm{mol\,s^{-1}}}{1\times10^{-6}\ \mathrm{m^3}} = \mathbf{100\ mol\,m^{-3}\,s^{-1}}.$$

Sun:

$$r = \frac{100\times10^{12}\ \mathrm{mol\,s^{-1}}}{1\times10^{27}\ \mathrm{m^3}} = \mathbf{100\times10^{-15}\ mol\,m^{-3}\,s^{-1}}.$$

Heating power P:

$$P = \mathcal{A}\times\omega.$$

Candle:

$$P = (1.2\times10^6\ \mathrm{J\,mol^{-1}})\times(100\times10^{-6}\ \mathrm{mol\,s^{-1}}) = \mathbf{120\ W}.$$

Sun:

$$P = (3\times10^{12}\ \mathrm{J\,mol^{-1}})\times(100\times10^{12}\ \mathrm{mol\,s^{-1}}) = \mathbf{300\times10^{24}\ W}.$$

Power density φ:

$$\varphi = \mathcal{A}\times r.$$

Candle:

$$\varphi = (1.2\times10^6\ \mathrm{J\,mol^{-1}})\times100\ \mathrm{mol\,m^{-3}\,s^{-1}} = \mathbf{120\times10^6\ W\,m^{-3}}.$$

Sun:

$$\varphi = (3\times10^{12}\ \mathrm{J\,mol^{-1}})\times(100\cdot10^{-15}\ \mathrm{mol\,m^{-3}\,s^{-1}}) = \mathbf{0.3\ W\,m^{-3}}.$$

2.16.2 Rate law

a) Order of the reaction with respect to the individual reactants B, B′ and B″:

We expect a rate law in the form of a power function [cf. Eq. (16.8)],

$$r = k\times c_{\mathrm{B}}^{b}\times c_{\mathrm{B'}}^{b'}\times c_{\mathrm{B''}}^{b''}.$$

We call the exponents b, b' and b'' the order of the reaction with respect to the individual reactants B, B′ and B″. Usually, small integers such as 1 and 2 appear as exponents.

In order to determine the order of the reaction with respect to a particular substance, we have to compare two rows of the table in which the substance in question occurs in different concentrations while the concentrations of the other substances remain the same.

For example, when the concentration of the reactant B is increased from 0.2 to 0.3 kmol m^{-3}, meaning by a factor of 1.5 (at constant concentrations of B′ and B″), the rate density rises from 3.2 to 4.8 mol m^{-3} s^{-1}, meaning also by a factor of 1.5 (as the comparison of the first and the second row of the table shows). Therefore, the reaction is first order with respect to B.

However, when the concentration of the reactant B′ is increased from 0.2 to 0.3 kmol m^{-3}, meaning also by a factor of 1.5, the rate density increases from 3.2 to 7.2 mol m^{-3} s^{-1}, meaning by a factor of 2.25 (comparison of the first and the third row). This value corresponds to $(1.5)^2$. Consequently, the reaction is second order with respect to B′.

Finally, when the concentration of the reactant B″ is doubled from 0.2 to 0.4 kmol m^{-3}, the rate density also doubles (comparison of the second and the fourth row). Therefore, the reaction is first order with respect to B″.

Rate law:

Eventually, we obtain the following rate law:

$$r = k \times c_{\mathrm{B}}^1 \times c_{\mathrm{B'}}^2 \times c_{\mathrm{B''}}^1 = k \times c_{\mathrm{B}} \times c_{\mathrm{B'}}^2 \times c_{\mathrm{B''}}.$$

The overall order of the reaction is $1+2+1=4$.

b) Rate coefficient k:

In order to calculate the rate coefficient k, we can use, for example, the data in the first row of the table:

$$k = \frac{r}{c_{\mathrm{B}} \times c_{\mathrm{B'}}^2 \times c_{\mathrm{B''}}}.$$

$$k = \frac{3.2\,\mathrm{mol\,m^{-3}\,s^{-1}}}{(0.2\times10^3\,\mathrm{mol\,m^{-3}})\times(0.2\times10^3\,\mathrm{mol\,m^{-3}})^2\times(0.2\times10^3\,\mathrm{mol\,m^{-3}})}$$

$$= \mathbf{2\times10^{-9}\,m^9\,mol^{-3}\,s^{-1}}.$$

2.16.3* Rate law for advanced students

a) Change in concentration $\dot{c}_{\mathrm{B'},0}$:

The (differential) change in concentration $\dot{c}_{\mathrm{B'},0} = \mathrm{d}c_{\mathrm{B'}}/\mathrm{d}t$ at $t = 0$ corresponds to the (in this case negative) slope of the so-called "initial tangent" to the $c_{\mathrm{B'}}(t)$ curve in question (see the figure below).

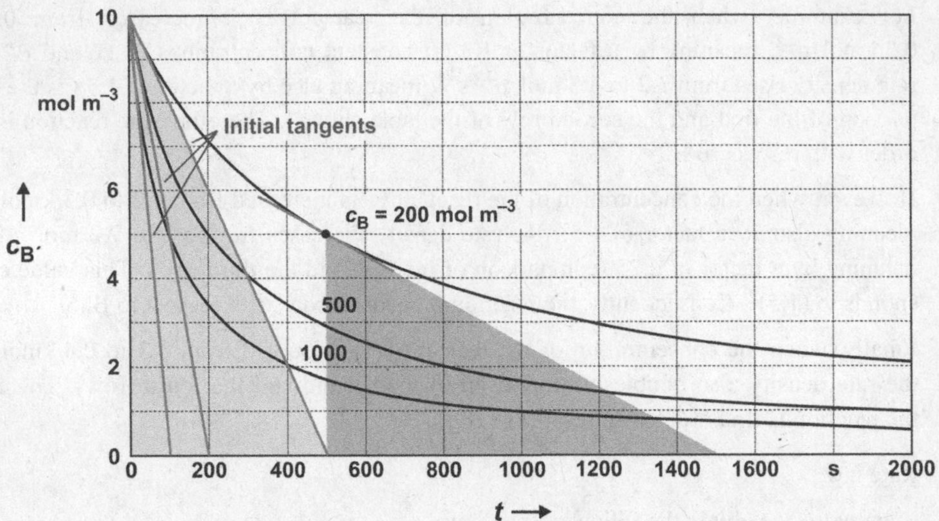

As an example, we determine $\dot{c}_{B',0}$ at a concentration $c_B = 200\ \text{mol m}^{-3}$. The slope $m_{T,0}$ of the corresponding "initial tangent" results from the ratio of "height" $\Delta c_{B'}$ to "basic line" Δt of a suitably selected "slope triangle" (highlighted in light gray) (here approximately formed by the intercepts of the tangents drawn at the initial point of the curve):

$$m_{T,0} = \frac{c_{B',1} - c_{B',0}}{t_1 - t_0} = \frac{0 - c_{B',0}}{t_1 - 0} = \dot{c}_{B',0}\,.$$

While the section on the axis of ordinates is fixed, the section on the axis of abscissas has to be determined graphically; according to the figure, the point of intersection of the tangent with the t axis lies at $t_1 \approx 500$ s (the graphical approach for the determination of t_1 is associated with a certain uncertainty, which is very roughly about 10 %):

$$\dot{c}_{B',0} = \frac{0\ \text{mol m}^{-3} - 10\ \text{mol m}^{-3}}{500\ \text{s} - 0\ \text{s}} = \mathbf{-0.020\ mol\ m^{-3}\ s^{-1}}\,.$$

The other two values are determined in the same way.

Initial rate density r_0:

Equation (16.7) is used to determine the rate density r:

$$r = \frac{1}{\nu_{B'}} \times \frac{dc_{B'}}{dt} = \frac{1}{\nu_{B'}} \times \dot{c}_{B'}\,.$$

According to the conversion formula, the conversion number $\nu_{B'}$ of the reactant B′ is −1, i.e., we obtain for r_0, the rate density at time $t = 0$ (so-called "initial rate density"):

$$r_0 = \frac{1}{-1} \times (-0.020\ \text{mol m}^{-3}\ \text{s}^{-1}) = \mathbf{+0.020\ mol\ m^{-3}\ s^{-1}}\,.$$

$c_B/\text{mol m}^{-3}$	200	500	1000
$\dot{c}_{B'}/\text{mol m}^{-3}\,\text{s}^{-1}$	**−0.020**	**−0.050**	**−0.100**
$r_0/\text{mol m}^{-3}\,\text{s}^{-1}$	**+0.020**	**+0.050**	**+0.100**

b) Rate density r:

As an example, we select the determination of the rate density r at the concentrations c_B = 200 mol m^{-3} and $c_{B'}$ = 5 mol m^{-3}. First, we mark the concentration of reactant B′ on the corresponding $c_{B'}(t)$ curve (figure above; black dot) and construct subsequently the tangent to the curve at this point. We can determine the slope m_T of the tangent by means of the corresponding slope triangle (highlighted in medium gray):

$$m_T = \frac{c_{B',2} - c_{B',1}}{t_2 - t_1}.$$

$$m_T = \frac{0\text{ mol m}^{-3} - 5\text{ mol m}^{-3}}{1500\text{ s} - 500\text{ s}} = -0.005\text{ mol m}^{-3}\,\text{s}^{-1}.$$

(Here, too, the graphical approach for the determination of the t values is associated with the mentioned uncertainty of roughly 10 %.)

The rate density r results in

$$r = \mathbf{+0.005\ mol\ m^{-3}\ s^{-1}}.$$

The determination of all other values follows the same scheme. The values in the table represent ideal values, which may vary a little bit depending on the individual skill to construct a tangent line.

$c_B/\text{mol m}^{-3}$	200	500	1000
$c_{B'}/\text{mol m}^{-3}$ = 2	–	0.002	0.004
5	0.005	0.0125	0.025
10	0.020	0.050	0.100

c) Rate law:

We expect a rate law of the type

$$r = k \times c_B^b \times c_{B'}^{b'},$$

where the exponents b and b' are small integers such as 1 and 2.

A look at the results in part b) of the exercise shows that the rate density r increases linearly with increasing concentration of reactant B (comparison of the values in one row). Therefore, the reaction is first order with respect to B.

However, the rate density r increases as the square of increasing concentration of reactant B′ (comparison of the values in one column). If one increases for example the concentration of reactant B′ from 5 to 10 kmol m^{-3} (at a constant concentration of B of 1000 mol m^{-3}), meaning by a factor of 2, the rate density will increase from 0.025 to $0.100 \text{ mol m}^{-3} \text{ s}^{-1}$, meaning by a factor of 4 [= $(2)^2$]. Consequently, the reaction is second order with respect to B′.

Eventually, we obtain the following rate law:

$$r = k \times c_{\mathrm{B}} \times c_{\mathrm{B'}}^2 .$$

Rate coefficient k:

The rate coefficient k can be estimated using data from the table above:

$$k = \frac{r}{c_{\mathrm{B}} \times c_{\mathrm{B'}}^2} = \frac{0.100 \text{ mol m}^{-3} \text{ s}^{-1}}{1000 \text{ mol m}^{-3} \times (10 \text{ mol m}^{-3})^2} = \mathbf{1.0 \times 10^{-6} \text{ m}^6 \text{ mol}^{-2} \text{ s}^{-1}} .$$

2.16.4 Decomposition of dinitrogen pentoxide

a) Chemical drive $\mathcal{A}_{\mathrm{sg}}^{\ominus}$ for the sublimation of dinitrogen pentoxide:

The chemical potentials of the substances in question under standard conditions can be found in Table A2.1 in the Appendix of the textbook "Physical Chemistry from a Different Angle."

$$\begin{array}{lll} & N_2O_5|s \rightarrow & N_2O_5|g \\ \mu^{\ominus}/\text{kG:} & 113.9 & 115.1 \end{array}$$

$$\mathcal{A}_{\mathrm{sg}}^{\ominus} = \sum_{\text{initial}} |\nu_i| \mu_i - \sum_{\text{final}} \nu_j \mu_j = (113.9 - 115.1) \text{ kG} = \mathbf{-1.2 \text{ kG}} .$$

Since dinitrogen pentoxide sublimes just above room temperature, we expect a negative chemical drive under standard conditions.

Chemical drive $\mathcal{A}_{\mathrm{D}}^{\ominus}$ for the decomposition of dinitrogen pentoxide:

$$\begin{array}{llll} & N_2O_5|s \rightarrow & 2\ NO_2|g + & \tfrac{1}{2}\ O_2|g \\ \mu^{\ominus}/\text{kG:} & 113.9 & 2 \times 52.3 & \tfrac{1}{2} \times 0 \end{array}$$

$$\mathcal{A}_{\mathrm{D}}^{\ominus} = [113.9 - (2 \times 104.6 + \tfrac{1}{2} \times 0)] \text{ kG} = \mathbf{+9.3 \text{ kG}} .$$

The chemical drive for the process is positive, i.e. dinitrogen pentoxide decomposes even at room temperature into NO_2 and O_2.

b) Plot $c_B = f(t)$:

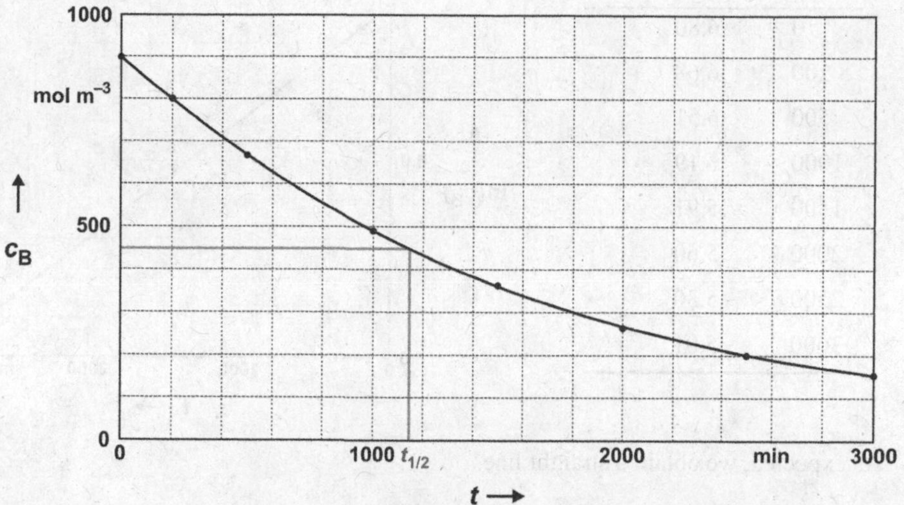

In order to figure out if the data can be represented by an exponential function, we can use the criterion presented in Appendix A1.1 of the textbook "Physical Chemistry from a Different Angle": We say that y is exponentially dependent upon x when an increase of x by a fixed amount a causes an increase of y by a fixed factor β,
in short: $y = f(x)$ is exponential if: $x \rightarrow x + a \Rightarrow y \rightarrow y \times \beta$.

Inserting the values given in the table results in (with $a = 500$ und $\beta \approx 0.74$):

$t \rightarrow t + 500 \quad \Rightarrow c_B \rightarrow c_B \times 0.74$

$t \rightarrow t + 1000 \Rightarrow c_B \rightarrow c_B \times 0.54 \, [\approx (0.74)^2]$

$t \rightarrow t + 1500 \Rightarrow c_B \rightarrow c_B \times 0.41 \, [\approx (0.74)^3] \ldots$

Indeed, the data can be represented by an exponential function, meaning the reaction is first order with respect to dinitrogen pentoxide and also the overall order is 1.

c) Plot $\ln\{c_B\} = f(t)$:

The logarithmic relation (16.13), for example, is suitable for verifying the reaction order found in part b),

$$\ln\{c_B\} = \ln\{c_{B,0}\} - kt .$$

If we do have a first-order reaction, we will obtain a straight line when $\ln\{c_B\}$ is plotted as a function of t.

t / min	$\ln\{c_B\}$
0	6.80
200	6.68
500	6.51
1000	6.19
1500	5.91
2000	5.60
2500	5.30
3000	5.01

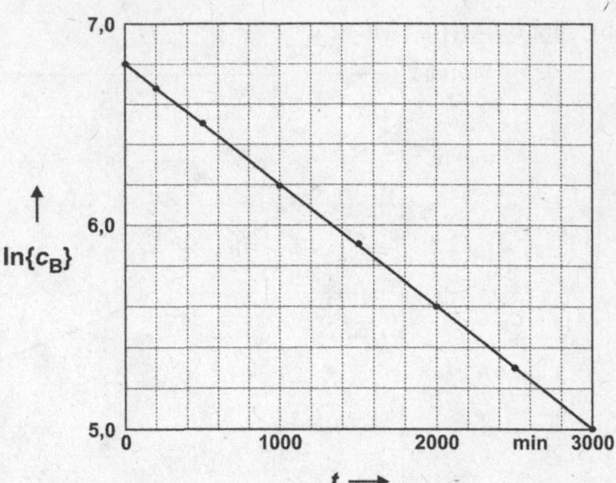

As expected, we obtain a straight line.

d) Rate coefficient k:

The (negative) slope m of the straight line, which can be determined either graphically with the help of a slope triangle or computationally by means of linear regression is -6.0×10^{-4} min^{-1}. Consequently, the rate coefficient k is 6.0×10^{-4} min^{-1} = $\mathbf{1.0\times10^{-5}}$ **s^{-1}**.

Half-life $t_{1/2}$:

The half-life of the first-order reaction in question is according to Equation (16.15)

$$t_{1/2} = \frac{\ln 2}{k} = \frac{0.6931}{1.0\times10^{-5}\ \text{s}^{-1}} = \mathbf{6.9\times10^{4}\ s}\ (= 1150\ \text{min}^{-1}).$$

This result corresponds quite well with the value that can be determined graphically with the help of the plot in part b).

2.16.5 Dissociation of ethane

a) Ratio $n_B/n_{B,0}$ after 2 hours:

The ratio of amount n_B of ethane after a certain time (in this case 2 hours) to the initial amount $n_{B,0}$ can be calculated using Equation (16.12):

$$\ln\frac{c_B}{c_{B,0}} = -kt \quad \Rightarrow$$

$$\frac{c_B}{c_{B,0}} = \frac{n_B}{n_{B,0}} = \exp(-kt) = \exp[(-5.5\times10^{-4}\ \text{s})\times 7200\ \text{s}] = \mathbf{0.019}.$$

After two hours, only 1.9 % of the initial amount is still present meaning 98.1 % of the ethane has decomposed.

b) Half-life $t_{1/2}$ of ethane:

$$t_{1/2} = \frac{\ln 2}{k} = \frac{0.6931}{5.5\times10^{-4}\ \mathrm{s}^{-1}} = \mathbf{1260\ s}.$$

2.16.6 Decomposition of dibenzoyl peroxide

Rate coefficient k of the reaction:

We start for example from Equation (16.12),

$$\ln\frac{c_{\mathrm{B},0}}{c_{\mathrm{B}}} = kt,$$

and solve for k:

$$k = \frac{1}{t}\ln\frac{c_{\mathrm{B},0}}{c_{\mathrm{B}}} = \frac{1}{1.8\times10^{4}\ \mathrm{s}}\ln\frac{200\ \mathrm{mol\,m^{-3}}}{143\ \mathrm{mol\,m^{-3}}} = \mathbf{1.86\times10^{-5}\ s^{-1}}.$$

2.16.7 Radiocarbon dating

a) Rate coefficient k of the decay process:

We use Equation (16.15),

$$t_{1/2} = \frac{\ln 2}{k},$$

in order to convert the half-life $t_{1/2}$ to the rate coefficient k:

$$k = \frac{\ln 2}{t_{1/2}} = \frac{0.6931}{5,730\ \mathrm{a}} = \mathbf{1.210\times10^{-4}\ a^{-1}}.$$

b) Age t of the artifact:

We start again from Equation (16.12),

$$\ln\frac{c_{\mathrm{B},0}}{c_{\mathrm{B}}} = kt,$$

but solve in this case for t:

$$t = \frac{1}{k}\ln\frac{c_{\mathrm{B},0}}{c_{\mathrm{B}}}.$$

The ratio $c_{\mathrm{B}}/c_{\mathrm{B},0}$ is supposed to be 0.77, i.e. we have:

$$t = \frac{1}{1.210\times10^{-4}\ \mathrm{a}^{-1}}\ln\frac{1}{0.77} \approx \mathbf{2,160\ a}.$$

Consequently, the wood for the beam was cut approximately in the 2nd century BC.

2.16.8 Decomposition of hydrogen iodide

a) Half-life $t_{1/2}$ of the decomposition process:

The decomposition of hydrogen iodide ($2\,HI \rightarrow H_2 + I_2$) is a reaction of the type

$$2\,\mathrm{B} \rightarrow \text{products}\,.$$

Therefore, Equation (16.23) is used to calculate the half-life, i.e. the time it takes to reduce the concentration of the hydrogen iodide to half of its initial value:

$$t_{1/2} = \frac{1}{2kc_{\mathrm{B},0}} = \frac{1}{2\times(1.0\times10^{-5}\,\mathrm{m^3\,mol^{-1}\,s^{-1}})\times 20\,\mathrm{mol\,m^{-3}}} = \mathbf{2500\,s}\,.$$

b) Time t required for the the concentration of HI to decrease to one-eighth of its initial value:

We start from Equation (16.21),

$$\frac{1}{c_\mathrm{B}} = \frac{1}{c_{\mathrm{B},0}} + 2kt\,,$$

and solve for t:

$$t = \frac{1}{2k}\left(\frac{1}{c_\mathrm{B}} - \frac{1}{c_{\mathrm{B},0}}\right).$$

The concentration c_B of hydrogen iodide is supposed to be $c_{\mathrm{B},0}/8 = 2.5\,\mathrm{mol\,m^{-1}}$:

$$t = \frac{1}{2\times(1.0\times10^{-5}\,\mathrm{m^3\,mol^{-1}\,s^{-1}})}\left(\frac{1}{2.5\,\mathrm{mol\,m^{-3}}} - \frac{1}{20\,\mathrm{mol\,m^{-3}}}\right)$$

$$= \mathbf{17500\,s}\ (\approx 4\,\mathrm{h}\,52\,\mathrm{min})\,.$$

2.16.9 Alkaline saponification of esters

Rate coefficient k of alkaline saponification of ethyl acetate:

The alkaline saponification of ethyl acetate is a reaction of the type

$$\mathrm{B} + \mathrm{B'} \rightarrow \text{products}\,.$$

However, since the initial concentrations of the two reactants B ($CH_3COOC_2H_5$) and B′ (OH^-) are identical, and both reaction partners participate equally in the reaction, we can use Equation (16.24) for the calculation:

$$\frac{1}{c_\mathrm{B}} = \frac{1}{c_{\mathrm{B},0}} + kt\,.$$

Solving for k results in:

$$k = \frac{1}{t}\left(\frac{1}{c_{\mathrm{B}}} - \frac{1}{c_{\mathrm{B},0}}\right).$$

$$k = \frac{1}{3600\ \mathrm{s}}\left(\frac{1}{62\times10^{3}\ \mathrm{mol\,m^{-3}}} - \frac{1}{100\times10^{3}\ \mathrm{mol\,m^{-3}}}\right) = \mathbf{1.70\times10^{-6}\ m^{3}\,mol^{-1}\,s^{-1}}.$$

Half-life $t_{1/2}$ of alkaline saponification of ethyl acetate:

The half-life $t_{1/2}$ is calculated according to Equation (16.25),

$$t_{1/2} = \frac{1}{kc_{\mathrm{B},0}} = \frac{1}{(1.7\times10^{-6}\ \mathrm{m^{3}\,mol^{-1}\,s^{-1}})\times(100\ \mathrm{mol\,m^{-3}})} = \mathbf{5900\ s}\ (\approx 1\ \mathrm{h}\ 38\ \mathrm{min}).$$

2.17 Composite Reactions

2.17.1 A "sweet" equilibrium reaction

a) Rate coefficient k_{-1} for the backward reaction:

The ratio of the rate coefficients for the forward and the backward reaction corresponds to the conventional equilibrium constant [Eq. (17.5)]:

$$\overset{\circ}{K}_c = \frac{k_{+1}}{k_{-1}} \quad \Rightarrow$$

$$k_{-1} = \frac{k_{+1}}{\overset{\circ}{K}_c} = \frac{8.75\times10^{-4}\ \mathrm{s}^{-1}}{1.64} = \mathbf{5.34\times10^{-4}\ s^{-1}}.$$

b) Equilibrium number $\overset{\circ}{\mathcal{K}}_c$:

The relationship between conventional equilibrium constant $\overset{\circ}{K}_c$ and equilibrium number $\overset{\circ}{\mathcal{K}}_c$ is given by Equation (6.20):

$$\overset{\circ}{K}_c = \kappa\overset{\circ}{\mathcal{K}}_c, \quad \text{where} \quad \kappa = (c^\ominus)^{\nu_c} \quad \text{with} \quad \nu_c = \nu_\mathrm{B} + \nu_\mathrm{D}.$$

ν_c is the sum of the conversion numbers of those substances, which show a dependence of the chemical potential upon concentration c. ν_c is equal to 0 in the present case because of $\nu_c = (-1)+1$. $\overset{\circ}{K}_c$ and $\overset{\circ}{\mathcal{K}}_c$ are therefore identical.

Basic value $\overset{\circ}{\mathcal{A}}$ of the chemical drive:

The basic value $\overset{\circ}{\mathcal{A}}$ of the chemical drive results from Equation (6.22):

$$\overset{\circ}{\mathcal{A}} = RT\ln\overset{\circ}{\mathcal{K}}_c = 8.314\ \mathrm{G\,K^{-1}}\times310\ \mathrm{K}\times\ln1.64 = 1280\ \mathrm{G} = \mathbf{1.28\ kG}.$$

c*) Reaction time t:

The concentration of β-D-glucose (substance D) is just the same as that of α-D-glucose (substance B) if the initial concentration of α-D-glucose has been reduced to one-half of its initial value, i.e. $c_\mathrm{B}/c_{\mathrm{B},0} = 0.5$. To determine the reaction time, the integrated rate law has to be used [Eq (17.11)]:

$$c_\mathrm{B} = \frac{k_{-1} + k_{+1}\times\mathrm{e}^{-(k_{+1}+k_{-1})t}}{k_{+1}+k_{-1}}c_{\mathrm{B},0}.$$

The equation becomes clearer when we abbreviate the sum $(k_{+1} + k_{-1})$ by k:

$$c_\mathrm{B} = \frac{k_{-1} + k_{+1}\times\mathrm{e}^{-kt}}{k}c_{\mathrm{B},0}.$$

Solving for t results in:

$$t = -\frac{1}{k}\ln\left[\frac{1}{k_{+1}}\times\left(\frac{c_{\mathrm{B}}\times k}{c_{\mathrm{B},0}} - k_{-1}\right)\right].$$

We obtain with $k = (8.75\times10^{-4}\ \mathrm{s}^{-1} + 5.34\times10^{-4}\ \mathrm{s}^{-1}) = 14.09\times10^{-4}\ \mathrm{s}^{-1}$:

$$t = -\frac{1}{14.09\times10^{-4}\ \mathrm{s}^{-1}}\cdot$$

$$\ln\left\{\frac{1}{8.75\times10^{-4}\ \mathrm{s}^{-1}}\times\left[0.5\times(14.09\times10^{-4}\ \mathrm{s}^{-1}) - (5.34\times10^{-4}\ \mathrm{s}^{-1})\right]\right\}$$

$$= 1161\ \mathrm{s} = \mathbf{19\ min\ 21\ s}.$$

2.17.2 Equilibrium reaction

a) Conventional equilibrium constant $\overset{\circ}{K}_c$:

The conventional equilibrium constant results according to Equation (6.21) in:

$$\overset{\circ}{K}_c = \frac{c_{\mathrm{D,eq}}}{c_{\mathrm{B,eq}}}.$$

The equilibrium concentration $c_{\mathrm{D,eq.}}$, however, is $c_{\mathrm{D,eq.}} = c_{\mathrm{B},0} - c_{\mathrm{B,eq.}} = 1.00\ \mathrm{kmol\,m^{-3}} - 0.20\ \mathrm{kmol\,m^{-3}} = 0.80\ \mathrm{kmol\,m^{-3}}$. Hence, we obtain:

$$\overset{\circ}{K}_c = \frac{0.80\times10^3\ \mathrm{mol\,m^{-3}}}{0.20\times10^3\ \mathrm{mol\,m^{-3}}} = \mathbf{4.00}.$$

b*) Rate coefficient k_{-1}:

Again, we start from the integrated rate law [Eq. (17.11)] [see also Solution 2.17.1 c)]:

$$c_{\mathrm{B}} = \frac{k_{-1} + k_{+1}\times\mathrm{e}^{-kt}}{k_{+1} + k_{-1}}c_{\mathrm{B},0}.$$

Inserting of $k_{+1} = k_{-1}\times\overset{\circ}{K}_c$ results in:

$$\frac{c_{\mathrm{B}}}{c_{\mathrm{B},0}} = \frac{k_{-1} + k_{-1}\times\overset{\circ}{K}_c\times\mathrm{e}^{-kt}}{k_{-1} + k_{-1}\times\overset{\circ}{K}_c} = \frac{k_{-1}(1+\overset{\circ}{K}_c\times\mathrm{e}^{-kt})}{k_{-1}(1+\overset{\circ}{K}_c)} = \frac{1+\overset{\circ}{K}_c\times\mathrm{e}^{-kt}}{1+\overset{\circ}{K}_c}.$$

If we know the sum k, we can get the rate coefficient k_{-1} from the relationship

$$k = k_{+1} + k_{-1} = k_{-1}\times\overset{\circ}{K}_c + k_{-1} = k_{-1}(\overset{\circ}{K}_c+1) \quad\Rightarrow\quad k_{-1} = \frac{k}{\overset{\circ}{K}_c+1}.$$

Therefore, we solve the integrated rate law for k,

$$e^{-kt} = \frac{\frac{c_B}{c_{B,0}}(1+\overset{\circ}{K}_c)-1}{\overset{\circ}{K}_c}$$

$$k = -\frac{1}{t}\ln\frac{\frac{c_B}{c_{B,0}}(1+\overset{\circ}{K}_c)-1}{\overset{\circ}{K}_c}$$

$$k = -\frac{1}{600\ \text{s}}\ln\frac{\frac{0.70\times10^3\ \text{mol}\,\text{m}^{-3}}{1.00\times10^3\ \text{mol}\,\text{m}^{-3}}(1+4.00)-1}{4.00} = 7.83\times10^{-4}\ \text{s}^{-1},$$

and obtain

$$k_{-1} = \frac{7.83\times10^{-4}\ \text{s}^{-1}}{4.00+1} = \mathbf{1.57\times10^{-4}\ s^{-1}}.$$

Rate coefficient k_{+1}:

The rate coefficient k_{+1} is the difference between the sum k and the value for k_{-1}:

$$k_{+1} = k - k_{-1} = (7.83\times10^{-4}\ \text{s}^{-1}) - (1.57\times10^{-4}\ \text{s}^{-1}) = \mathbf{6.26\times10^{-4}\ s^{-1}}.$$

2.17.3 Decomposition of acetic acid

a) Reaction time t, after which 90 % of the acetic acid has been consumed:

In first-order reactions, the concentration of the reactant B, in this case acetic acid, decreases exponentially with time. This also applies if two (or more) of such reactions take place in parallel, the rate coefficient k being the sum of the rate coefficients of the individual reactions [Eq. (17.18)]:

$$c_B = c_{B,0}e^{-kt} \quad \text{with} \quad k = k_1 + k_2.$$

Solving for the reaction time t as well as inserting of $k = 3.74\ \text{s}^{-1} + 4.65\ \text{s}^{-1} = 8.39\ \text{s}^{-1}$ and $c_B = (1-0.90)c_{B,0} = 0.10c_{B,0}$ results in:

$$t = -\frac{1}{k}\ln\frac{c_B}{c_{B,0}} = -\frac{1}{8.39\ \text{s}^{-1}}\ln 0.10 = \mathbf{0.27\ s}.$$

b) Concentration ratio c_D:$c_{D'}$ of the reaction products methane and ketene:

The products methane (D) and ketene (D′) compete proportionally to their rate coefficients for the reactant concentration [Eq. (17.23)]:

$$\frac{c_D}{c_{D'}} = \frac{k_1}{k_2} = \frac{3.74\ \text{s}^{-1}}{4.65\ \text{s}^{-1}} = \mathbf{0.80}.$$

The ratio of products is 0.80 and time independent.

c) Maximum yield $\eta_{D',max}$ of ketene:

The fraction of a resulting product is the higher, the greater the corresponding rate coefficient is. Therefore, the maximum yield $\eta_{D',max}$ of ketene results in:

$$\eta_{D',max} = \frac{k_2}{k_1 + k_2} = \frac{4.65\ s^{-1}}{3.74\ s^{-1} + 4.65\ s^{-1}} = 0.55 = \mathbf{55\ \%},$$

related to the initial amount of acetic acid.

2.17.4* Parallel reactions

Sum k of the rate coefficients:

The sum k of the two rate coefficients k_1 and k_2 can be determined by means of Equation (17.18) [see also Solution 2.173 a)]:

$$c_B = c_{B,0}e^{-kt} \quad \Rightarrow \quad k = -\frac{1}{t}\ln\frac{c_B}{c_{B,0}}.$$

$$k = -\frac{1}{2400\ s^{-1}}\ln\frac{0.05\times10^3\ mol\,m^{-3}}{0.50\times10^3\ mol\,m^{-3}} = 9.6\times10^{-4}\ s^{-1}.$$

Rate coefficient k_2:

The change of concentration over time of product D′ due to reaction 2 is described by Equation (17.22):

$$c_{D'} = \frac{k_2}{k}c_{B,0}(1 - e^{-kt}).$$

Solving for the rate coefficient k_2 results in:

$$k_2 = \frac{k \times c_{D'}}{c_{B,0}(1 - e^{-kt})}.$$

$$k_2 = \frac{(9.6\times10^{-4}\ s^{-1})\times(0.10\times10^3\ mol\,m^{-3})}{(0.50\times10^3\ mol\,m^{-3})\times\{1 - \exp[(-9.6\times10^{-4}\ s^{-1})\times2400\ s]\}}$$

$$k_2 = \frac{0.096\ s^{-1}\ mol\,m^{-3}}{450\ mol\,m^{-3}} = \mathbf{2.1\times10^{-4}\ s^{-1}}.$$

Rate coefficient k_1:

Consequently, the rate coefficient k_1 is:

$$k_1 = k - k_2 = (9.6\times10^{-4}\ s^{-1}) - (2.1\times10^{-4}\ s^{-1}) = \mathbf{7.5\times10^{-4}\ s^{-1}}.$$

2.17.5* Consecutive reactions

a) Time t_{max} at which the concentration of I reaches its maximum value:

The following equation is given in the text of the exercise for the change in concentration of the intermediate substance I with time:

$$c_I(t) = \frac{k_1}{k_2 - k_1} c_{B,0}(e^{-k_1 t} - e^{-k_2 t}) .$$

The concentration c_I reaches its maximum value when $dc_I(t)/dt$ is equal to zero. The first derivative of the function $c_I(t)$ is:

$$\frac{dc_I(t)}{dt} = \frac{k_1}{k_2 - k_1} c_{B,0}(-k_1 e^{-k_1 t} + k_2 e^{-k_2 t}) .$$

Here, the rules (A1.7) and (A1.13; chain rule) for calculating derivatives were used. As already mentioned, the derivative is then set equal to zero:

$$\frac{k_1}{k_2 - k_1} c_{B,0}[-k_1 \exp(-k_1 t_{max}) + k_2 \exp(-k_2 t_{max})] = 0$$

$$-k_1 \exp(-k_1 t_{max}) + k_2 \exp(-k_2 t_{max}) = 0$$

$$k_2 \exp(-k_2 t_{max}) = k_1 \exp(-k_1 t_{max})$$

$$\frac{\exp(-k_2 t_{max})}{\exp(-k_1 t_{max})} = \frac{k_1}{k_2} .$$

According to the calculation rule $e^a/e^b = e^{a-b}$ we obtain

$$\exp[(k_1 - k_2)t_{max}] = \frac{k_1}{k_2}$$

and finally by taking the logarithm on both sides:

$$(k_1 - k_2) \times t_{max} = \ln \frac{k_2}{k_1}$$

$$t_{max} = \frac{1}{k_1 - k_2} \ln \frac{k_1}{k_2} .$$

b) Time t_{max} calculated for the given example:

$$t_{max} = \frac{1}{0.006\ s^{-1} - 0.010\ s^{-1}} \ln \frac{0.006\ s^{-1}}{0.010\ s^{-1}} = \mathbf{128\ s} .$$

c) Maximum concentration $c_{\mathrm{I,max}}$ of intermediate I:

$$c_{\mathrm{I,max}} = \frac{0.010\ \mathrm{s}^{-1}}{0.006\ \mathrm{s}^{-1} - 0.010\ \mathrm{s}^{-1}} \times (1.00\times10^{3}\ \mathrm{mol\,m^{-3}})\times$$

$$[\exp(-0.010\ \mathrm{s}^{-1}\times 128\ \mathrm{s}) - \exp(-0.006\ \mathrm{s}^{-1}\times 128\ \mathrm{s})]$$

$$c_{\mathrm{I,max}} = \frac{-1.859\ \mathrm{s^{-1}\,mol\,m^{-3}}}{-0.004\ \mathrm{s}^{-1}} = 465\ \mathrm{mol\,m^{-3}} = \mathbf{0.47\ kmol\,m^{-3}}\,.$$

2.17.6* Construction of $c(t)$ curves for a multistep reaction

a) Equations for the change in concentration $\dot{c} = \mathrm{d}c/\mathrm{d}t$ of the four substances B, I, I′, D:

$$\dot{c}_{\mathrm{B}} = -r_1 = k_1 c_{\mathrm{B}} c_{\mathrm{I}}\,,$$

$$\dot{c}_{\mathrm{I}} = r_1 - r_2 = k_1 c_{\mathrm{B}} c_{\mathrm{I}} - k_2 c_{\mathrm{I}} c_{\mathrm{I'}},$$

$$\dot{c}_{\mathrm{I'}} = r_2 - r_3 = k_2 c_{\mathrm{I}} c_{\mathrm{I'}} - k_3 c_{\mathrm{I'}}\,,$$

$$\dot{c}_{\mathrm{D}} = r_3 = k_3 c_{\mathrm{I'}}\,.$$

b) Construction of the $c(t)$ curves step by step:

To illustrate the procedure, the values in the first line as well as the first values in the second line should be calculated.

First line:

$$c_{\mathrm{I},0} = 50.0\ \mathrm{mol\,m^{-3}}\ \text{(given)}$$

$$c_{\mathrm{I'},0} = 35.0\ \mathrm{mol\,m^{-3}}\ \text{(given)}$$

$$k_1 c_{\mathrm{B}} c_{\mathrm{I}} = (5\times10^{-6}\ \mathrm{mol^{-1}\,m^3\,s^{-1}})\times 1000\ \mathrm{mol\,m^{-3}}\times 50.0\ \mathrm{mol\,m^{-3}} = 0.250\ \mathrm{mol\,m^{-3}\,s^{-1}}$$

$$k_2 c_{\mathrm{I}} c_{\mathrm{I'}} = (100\times10^{-6}\ \mathrm{mol^{-1}\,m^3\,s^{-1}})\times 50.0\ \mathrm{mol\,m^{-3}}\times 35.0\ \mathrm{mol\,m^{-3}} = 0.175\ \mathrm{mol\,m^{-3}\,s^{-1}}$$

$$k_3 c_{\mathrm{I'}} = (5\times10^{-3}\ \mathrm{s^{-1}})\times 35.0\ \mathrm{mol\,m^{-3}} = 0.175\ \mathrm{mol\,m^{-3}\,s^{-1}}$$

$$\dot{c}_{\mathrm{I}}\times\Delta t = (k_1 c_{\mathrm{B}} c_{\mathrm{I}} - k_2 c_{\mathrm{I}} c_{\mathrm{I'}})\times\Delta t$$

$$= (0.250\ \mathrm{mol\,m^{-3}\,s^{-1}} - 0.175\ \mathrm{mol\,m^{-3}\,s^{-1}})\times 100\ \mathrm{s} = 7.5\ \mathrm{mol\,m^{-3}}$$

$$\dot{c}_{\mathrm{I'}}\times\Delta t = (k_2 c_{\mathrm{I}} c_{\mathrm{I'}} - k_3 c_{\mathrm{I'}})\times\Delta t$$

$$= (0.175\ \mathrm{mol\,m^{-3}\,s^{-1}} - 0.175\ \mathrm{mol\,m^{-3}\,s^{-1}})\times 100\ \mathrm{s} = 0\ \mathrm{mol\,m^{-3}}$$

Second line:

$$c_{\mathrm{I}} = c_{\mathrm{I},0} + \dot{c}_{\mathrm{I}}\times\Delta t = 50.0\ \mathrm{mol\,m^{-3}} + 7.5\ \mathrm{mol\,m^{-3}} = 57.5\ \mathrm{mol\,m^{-3}}$$

$$c_{\mathrm{I'}} = c_{\mathrm{I'},0} + \dot{c}_{\mathrm{I'}}\times\Delta t = 35.0\ \mathrm{mol\,m^{-3}} + 0\ \mathrm{mol\,m^{-3}} = 35.0\ \mathrm{mol\,m^{-3}}\,,\ldots$$

$t/$ s	$c_I/$ mol m^{-3}	$c_{I'}/$ mol m^{-3}	$k_1c_Bc_I/$ mol m^{-3} s^{-1}	$k_2c_Ic_{I'}/$ mol m^{-3} s^{-1}	$k_3c_{I'}/$ mol m^{-3} s^{-1}	$\dot{c}_I \times \Delta t/$ mol m^{-3}	$\dot{c}_{I'} \times \Delta t/$ mol m^{-3}
0	**50.0**	**35.0**	**0.250**	**0.175**	**0.175**	**7.5**	**0**
100	**57.5**	**35.0**	**0.288**	**0.201**	**0.175**	**8.7**	**2.6**
200	**66.2**	**37.6**	0.331	0.249	0.188	8.2	6.1
300	74.4	43.7	0.372	0.325	0.219	4.7	10.6
400	79.1	54.3	0.396	0.430	0.272	−3.4	15.8
500	75.7	70.1	0.379	0.531	0.351	−15.2	18.0
600	60.5	88.1	0.303	0.533	0.441	−23.0	9.2
700	37.5	97.3	0.188	0.365	0.487	−17.7	−12.2
800	19.8	85.1	0.099	0.168	0.426	−6.9	−25.8
900	12.9	59.3	0.065	0.076	0.297	−1.1	−22.1
1000	11.8	37.2	–	–	–	–	–

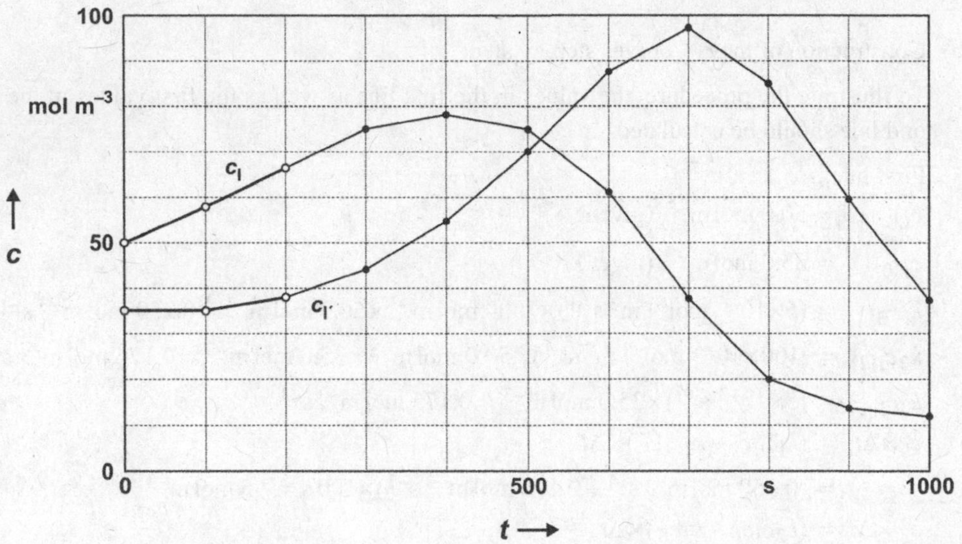

The figure gives a first, but only rough impression from the change in concentration of the intermediate substances I and I′. In order to obtain more accurate results, the time interval Δt would have to be significantly shorter.

2.18 Theory of Rate of Reaction

2.18.1 Decomposition of dinitrogen pentoxide

Rate coeffizient k at a temperature of $\vartheta = 65\ °C$:

We start from the ARRHENIUS equation [Eq. (18.2)]:

$$k(T) = k_\infty \exp\frac{-W_A}{RT}.$$

Inserting the given values results in:

$$k(338\ K) = 4.94\times10^{13}\ s^{-1}\times\exp\frac{-103\times10^3\ J\,mol^{-1}}{8.314\ G\,K^{-1}\times338\,K}$$

$$= 4.94\times10^{13}\ s^{-1}\times\exp\frac{-103\times10^3\ J\,mol^{-1}}{8.314\ J\,mol^{-1}\,K^{-1}\times338\,K} = \mathbf{5.96\times10^{-3}\ s^{-1}}.$$

2.18.2 Kinetics in everyday life

Activation energy W_A:

We start again from the ARRHENIUS equation. In our everyday example, a temperature increase from $T_1 = 281$ K to $T_2 = 303$ K should cause the reaction rate and thus the rate coefficient to increase by 40 times:

$$\frac{k_2}{k_1} = \frac{k_\infty e^{-W_A/RT_2}}{k_\infty e^{-W_A/RT_1}} = \exp\frac{W_A}{R}\left(\frac{1}{T_1}-\frac{1}{T_2}\right) \approx 40.$$

Taking the logarithm and solving for the activation energy W_A in question results in:

$$W_A = \frac{R\times\ln\frac{k_2}{k_1}}{\frac{1}{T_1}-\frac{1}{T_2}} \approx \frac{8.314\ J\,mol^{-1}\,K^{-1}\times\ln 40}{\frac{1}{281\ K}-\frac{1}{303\ K}} \approx 120\times10^3\ J\,mol^{-1} \approx \mathbf{120\ kJ\,mol^{-1}}.$$

2.18.3 Acid-catalyzed hydrolysis of benzylpenicillin

a) Rate coefficient k_1:

Between half-life $t_{1/2}$ and rate coefficient k_1, the following relationship exists for a first-order reaction [Eq. (16.15)]:

$$t_{1/2,1} = \frac{\ln 2}{k_1} \quad\Rightarrow$$

$$k_1 = \frac{\ln 2}{t_{1/2,1}} = \frac{\ln 2}{1098\ s} = 6.31\times10^{-4}\ s^{-1}.$$

Frequency factor k_∞:

The frequency factor k_∞ can be determined by means of the ARRHENIUS equation [Eq. (18.2)]:

$$k_1 = k_\infty \exp\frac{-W_A}{RT_1} \quad \Rightarrow$$

$$k_\infty = \frac{k_1}{\exp\frac{-W_A}{RT_1}} = \frac{6.31\times10^{-4}\ \text{s}^{-1}}{\exp\frac{-87.14\times10^3\ \text{J mol}^{-1}}{8.314\ \text{J mol}^{-1}\ \text{K}^{-1}\times 333\ \text{K}}} = \mathbf{2.95\times10^{10}\ s^{-1}}.$$

b) Rate coefficient k_2:

$$k_2 = k_\infty \exp\frac{-W_A}{RT_2}.$$

$$k_2 = (2.95\times10^{10}\ \text{s}^{-1})\times\exp\frac{-87.14\times10^3\ \text{J mol}^{-1}}{8.314\ \text{J mol}^{-1}\ \text{K}^{-1}\times 303\ \text{K}} = 2.80\times10^{-5}\ \text{s}^{-1}.$$

Half-life $t_{1/2,2}$:

$$t_{1/2,2} = \frac{\ln 2}{k_2} = \frac{\ln 2}{2.80\times10^{-5}\ \text{s}^{-1}} = 24800\ \text{s} = \mathbf{6.9\ h}.$$

As expected, the reaction proceeds much more slowly at the lower temperature.

2.18.4 Rate of the decomposition of dintrogen tetroxide

a) Rate coefficient k:

To determine the rate coefficient k, the ARRHENIUS equation [Eq. (18.2)] is used again:

$$k = k_\infty \exp\frac{-W_A}{RT}.$$

$$k = (2\times10^{11}\ \text{m}^3\ \text{mol}^{-1}\ \text{s}^{-1})\times\exp\frac{-46\times10^3\ \text{J mol}^{-1}}{8.314\ \text{J mol}^{-1}\ \text{K}^{-1}\times 298\ \text{K}} = 1730\ \text{m}^3\ \text{mol}^{-1}\ \text{s}^{-1}.$$

Rate density r:

$$r = k\times c_B\times c_{B'} = 1730\ \text{m}^3\ \text{mol}^{-1}\ \text{s}^{-1}\times 1\ \text{mol m}^{-3}\times 40\ \text{mol m}^{-3} = \mathbf{6.9\times10^4\ mol\ m^{-3}\ s^{-1}}.$$

b) Half-live $t_{1/2}$:

In the present case, we are dealing with a so-called pseudo first-order reaction. The constant concentration $c_{B'}$ of the nitrogen can be combined with the actual rate coefficient k to create a new rate coefficient k':

$$k' = k\times c_{B'}.$$

The half-life $t_{1/2}$ can then be calculated according to Equation (16.15):

$$t_{1/2} = \frac{\ln 2}{k'} = \frac{\ln 2}{k \times c_{B'}} = \frac{\ln 2}{1730\ \text{m}^3\ \text{mol}^{-1}\ \text{s}^{-1} \times 40\ \text{mol m}^{-3}} = 10.0 \times 10^{-6}\ \text{s} = \mathbf{10\ \mu s}.$$

2.18.5 Hydrogen iodide equilibrium

Rate coefficient k_{+1} for the forward reaction at different temperatures:

From the ARRHENIUS equation [Eq. (18.2)],

$$k(T) = k_\infty \exp\frac{-W_A}{RT},$$

we obtain the velocity coefficient k_{+1} at the two temperatures $T_1 = 629$ K and $T_2 = 700$ K:

$$k_{+1}(T_1) = (3.90 \times 10^8\ \text{m}^3\ \text{mol}^{-1}\ \text{s}^{-1}) \times \exp\frac{-165.7 \times 10^3\ \text{J mol}^{-1}}{8.314\ \text{J mol}^{-1}\ \text{K}^{-1} \times 629\ \text{K}}$$

$$= 6.76 \times 10^{-6}\ \text{m}^3\ \text{mol}^{-1}\ \text{s}^{-1}.$$

$$k_{+1}(T_2) = (3.90 \times 10^8\ \text{m}^3\ \text{mol}^{-1}\ \text{s}^{-1}) \times \exp\frac{-165.7 \times 10^3\ \text{J mol}^{-1}}{8.314\ \text{J mol}^{-1}\ \text{K}^{-1} \times 700\ \text{K}}$$

$$= 1.68 \times 10^{-4}\ \text{m}^3\ \text{mol}^{-1}\ \text{s}^{-1}.$$

Rate coefficient k_{-1} for the backward reaction at different temperatures:

The rate coefficient k_{-1} for the backward reaction can be derived from the relationship (17.5) for an equilibrium reaction,

$$\overset{\circ}{K}_c = \frac{k_{+1}}{k_{-1}}.$$

Solving for k_{-1} yields:

$$k_{-1} = \frac{k_{+1}}{\overset{\circ}{K}_c}.$$

Inserting the values calculated in part a) results in:

$$k_{-1}(T_1) = \frac{6.76 \times 10^{-6}\ \text{m}^3\ \text{mol}^{-1}\ \text{s}^{-1}}{76.4} = 8.85 \times 10^{-8}\ \text{m}^3\ \text{mol}^{-1}\ \text{s}^{-1}.$$

$$k_{-1}(T_2) = \frac{1.68 \times 10^{-4}\ \text{m}^3\ \text{mol}^{-1}\ \text{s}^{-1}}{52.0} = 3.23 \times 10^{-6}\ \text{m}^3\ \text{mol}^{-1}\ \text{s}^{-1}.$$

Activation energy $W_{A,-1}$ for the backward reaction:

The equation from solution 2.18.2 can be used to determine the activation energy $W_{A,-1}$ for the backward reaction:

$$W_{A,-1} = \frac{R \times \ln \frac{k_{-1}(T_2)}{k_{-1}(T_1)}}{\frac{1}{T_1} - \frac{1}{T_2}} .$$

$$W_{A,-1} = \frac{8.314\ \mathrm{J\,mol^{-1}\,K^{-1}} \times \ln \frac{3.23 \times 10^{-6}\ \mathrm{m^3\,mol^{-1}\,s^{-1}}}{8.85 \times 10^{-8}\ \mathrm{m^3\,mol^{-1}\,s^{-1}}}}{\frac{1}{629\ \mathrm{K}} - \frac{1}{700\ \mathrm{K}}}$$

$$= \frac{29.91\ \mathrm{J\,mol^{-1}\,K^{-1}}}{1.613 \times 10^{-4}\ \mathrm{K^{-1}}} = \mathbf{185.4\ kJ\,mol\,s^{-1}} .$$

Frequeny factor $k_{\infty,-1}$ for the backward reaction:

The ARRHENIUS equation is solved for $k_{\infty,-1}$ and the corresponding values for a particular temperature, for example T_1, are inserted:

$$k_{\infty,-1} = \frac{k_{-1}(T_1)}{\exp \frac{-W_A}{RT_1}} = \frac{8.85 \times 10^{-8}\ \mathrm{m^3\,mol^{-1}\,s^{-1}}}{\exp \frac{-185.4 \times 10^3\ \mathrm{J\,mol^{-1}}}{8.314\ \mathrm{J\,mol^{-1}\,K^{-1}} \times 629\ \mathrm{K}}} = \mathbf{2.21 \times 10^8\ m^3\,mol^{-1}\,s^{-1}} .$$

2.18.6 Collision theory

a) Fraction q_1 of all particles having a sufficient kinetic energy at $T_1 = 300$ K:

The fraction q_1 of all particles with a kinetic energy of at least W_{min} at a temperature T_1 = 300 K results according to Equation (18.6) in:

$$q_1 \approx \exp \frac{-W_{min}}{RT_1} .$$

Inserting the given values yields:

$$q_1 \approx \exp \frac{-60 \times 10^3\ \mathrm{J\,mol^{-1}}}{8.314\ \mathrm{J\,mol^{-1}\,s^{-1}} \times 300\ \mathrm{K}} = \mathbf{3.6 \times 10^{-11}} .$$

b) Fraction q_2 of all particles having a sufficient kinetic energy at $T_2 = 400$ K:

The fraction q_2 of particles with sufficient kinetic energy resulting after an increase in temperature by 100 K is

$$q_2 \quad \approx \exp\frac{-60\times10^3\ \mathrm{J\,mol^{-1}}}{8.314\ \mathrm{J\,mol^{-1}\,s^{-1}}\times400\ \mathrm{K}} = \mathbf{1.5\times10^{-8}}.$$

As expected, the fraction of reactive particles increases strongly with increasing temperature (to about 500 times).

2.18.7* Nitrogen oxides in air

a) Calculation of a chemical potential $\overset{\circ}{\mu}$ at elevated temperature using the example of the gas NO:

To describe the temperature dependence of the chemical potential, we can choose a linear approach as the simplest possibility [cf. Eq. (5.2)]:

$$\overset{\circ}{\mu}(T_1) \quad = \overset{\circ}{\mu}(T_0) + \alpha(T_1 - T_0).$$

Insertion yields for the concrete example:

$$\overset{\circ}{\mu}(398\ \mathrm{K}) = 87.6\times10^3\ \mathrm{G} + (-211\ \mathrm{G\,K^{-1}})\times(398\ \mathrm{K} - 298\ \mathrm{K}) = 66.5\times10^3\ \mathrm{G} = \mathbf{66.5\ kG}.$$

The remaining values are calculated in the same way.

In summary, we obtain:

	$2\,NO$	$+\ O_2$	$\rightarrow \ddagger$	$\rightarrow 2\,NO_2$
$\overset{\circ}{\mu}(298\ \mathrm{K})/\mathrm{kG}$	2×87.6	**0**	+240	**2×52.3**
	175.2			**104.6**
$\alpha/\mathrm{G\,K^{-1}}$	2×(−211)	**−205**	−356	**2×(−240)**
$\overset{\circ}{\mu}(398\ \mathrm{K})/\mathrm{kG}$	**2×66.5**	**−20.5**	**+204.4**	**2×28.3**
	112.5			**56.6**

In the adjacent diagram, the potentials for 298 K are shown as black bars and those for 398 K as gray bars.

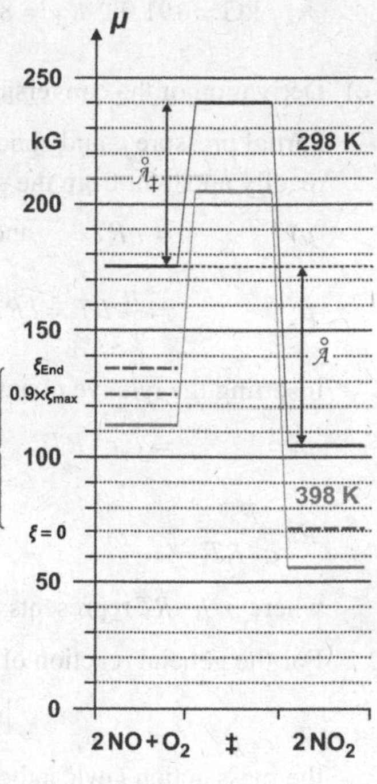

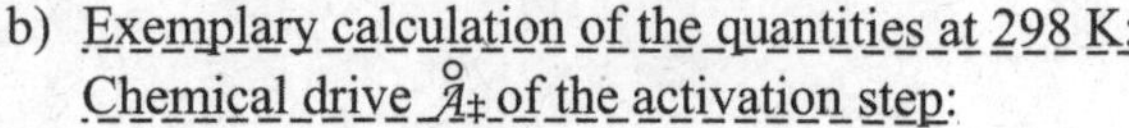

b) Exemplary calculation of the quantities at 298 K:
Chemical drive $\overset{\circ}{\mathcal{A}}_\ddagger$ of the activation step:

The drive $\mathcal{A}$ of any transformation is, as we have often used in our calculations [cf. Eq. (4.3)]:

$$\mathcal{A} \quad = \sum_{\text{initial}} |\nu_i|\mu_i - \sum_{\text{final}} \nu_j\mu_j\,.$$

Thus, we obtain for the basic value of the chemical drive of the activation step:

$$\overset{\circ}{\mathcal{A}}_\ddagger \quad = [2\,\overset{\circ}{\mu}(\mathrm{NO}) + \overset{\circ}{\mu}(\mathrm{O_2})] - \overset{\circ}{\mu}(\ddagger) = [175.2 - 240]\ \mathrm{kG} = \mathbf{-64.8\ kG}.$$

Equilibrium number $\overset{\circ}{\mathcal{K}}_{\ddagger}$ for the activation step:

The equilibrium number $\overset{\circ}{\mathcal{K}}_{\ddagger}$ can be calculated from the drive $\overset{\circ}{\mathcal{A}}_{\ddagger}$ by means of Equation (6.18):

$$\overset{\circ}{\mathcal{K}}_{\ddagger} = \exp\frac{\overset{\circ}{\mathcal{A}}_{\ddagger}}{RT} = \exp\frac{-64.8\times10^{3}\ \text{G}}{8.314\ \text{G}\,\text{K}^{-1}\times298\ \text{K}} = \mathbf{4.4\times10^{-12}}.$$

Chemical drive $\overset{\circ}{\mathcal{A}}$ for the overall process:

$$\overset{\circ}{\mathcal{A}} = [2\,\overset{\circ}{\mu}(\text{NO})+\overset{\circ}{\mu}(\text{O}_2)]-2\,\overset{\circ}{\mu}(\text{NO}_2) = [175.2-104.6]\ \text{kG} = \mathbf{+70.6\ kG}.$$

Equilibrium number $\overset{\circ}{\mathcal{K}}$ for the overall process:

$$\overset{\circ}{\mathcal{K}} = \exp\frac{\overset{\circ}{\mathcal{A}}}{RT} = \exp\frac{70.6\times10^{3}\ \text{G}}{8.314\ \text{G}\,\text{K}^{-1}\times298\ \text{K}} = \mathbf{2.4\times10^{12}}.$$

The corresponding values at a temperature of 398 K can be calculated analogously. In summary, we obtain:

$$\overset{\circ}{\mathcal{A}}_{\ddagger}/\text{kG} = -64.8;\quad \overset{\circ}{\mathcal{K}}_{\ddagger} = 4.4\times10^{-12};\quad \overset{\circ}{\mathcal{A}}/\text{kG} = +70.6;\quad \overset{\circ}{\mathcal{K}} = 2.4\times10^{12}\qquad (298\ \text{K}),$$

$$\overset{\circ}{\mathcal{A}}_{\ddagger}/\text{kG} = -91.9;\quad \overset{\circ}{\mathcal{K}}_{\ddagger} = 8.7\times10^{-13};\quad \overset{\circ}{\mathcal{A}}/\text{kG} = +55.9;\quad \overset{\circ}{\mathcal{K}} = 2.2\times10^{7}\qquad (398\ \text{K}).$$

c) Derivation of the conversion factor:

Partial pressure p and concentration c are proportional to each other for thin gases. This is readily apparent from the general gas law:

$$pV = nRT \quad \text{and therefore}$$

$$p = \frac{n}{V}RT = cRT.$$

Inserting the relative quantities $p_r = p/p^{\ominus}$ and $c_r = c/c^{\ominus}$ results in:

$$p_r p^{\ominus} = c_r c^{\ominus} RT \quad \text{and}$$

$$p_r \frac{p^{\ominus}}{c^{\ominus}RT} = c_r,$$

where $p^{\ominus}/c^{\ominus}RT$ represents a fixed factor.

For the general reaction of thin gases,

$$|\nu_B|\text{B}+|\nu_{B'}|\text{B}'+\ldots \rightleftarrows \nu_D\text{D}+\nu_{D'}\text{D}'+\ldots,$$

the mass action law can be expressed in the following form [cf. Eq. (6.29)]:

$$\frac{p_r(\text{D})^{\nu_D}\times p_r(\text{D}')^{\nu_{D'}}\times\ldots}{p_r(\text{B})^{|\nu_B|}\times p_r(\text{B}')^{|\nu_{B'}|}\times\ldots} = \overset{\circ}{\mathcal{K}}_p.$$

Multiplication by the factor $(p^\ominus/c^\ominus RT)^\nu$ with $\nu = \nu_B + \nu_{B'} + ... + \nu_D + \nu_{D'} + ...$ yields:

$$\frac{c_r(D)^{\nu_D} \times c_r(D')^{\nu_{D'}} \times ...}{c_r(B)^{|\nu_B|} \times c_r(B')^{|\nu_{B'}|} \times ...} = \left(\frac{p^\ominus}{c^\ominus RT}\right)^\nu \times \mathring{\mathcal{K}}_p = \mathring{\mathcal{K}}_c .$$

Equilibrium number $\mathring{\mathcal{K}}_{\ddagger c}$ at 298 K:

The equilibrium number $\mathring{\mathcal{K}}_{\ddagger c}$ is calculated as follows

$$\mathring{\mathcal{K}}_{\ddagger c} = \left(\frac{p^\ominus}{c^\ominus RT}\right)^\nu \times \mathring{\mathcal{K}}_\ddagger ,$$

where $\nu = \nu(NO) + \nu(O_2) + \nu(\ddagger) = (-2) + (-1) + 1 = -2.$

$$\mathring{\mathcal{K}}_{\ddagger c} = \left[\frac{100 \times 10^3 \text{ Pa}}{(1 \times 10^3 \text{ mol m}^{-3}) \times 8.314 \text{ J mol}^{-1} \text{K}^{-1} \times 298 \text{ K}}\right]^{-2} \times (4.4 \times 10^{-12})$$

$$= \mathbf{2.7 \times 10^{-9}} .$$

Conversion of units:

$$\frac{\text{Pa}}{\text{mol m}^{-3} \text{ J mol}^{-1} \text{K}^{-1} \text{ K}} = \frac{\text{N m}^{-2}}{\text{m}^{-3} \text{ N m}} = 1 .$$

Rate coefficient k at 298 K:

The rate coefficient k is then given by:

$$k = \kappa_\ddagger \times \frac{k_B T}{h} \times \mathring{\mathcal{K}}_{\ddagger c} .$$

With $\kappa_\ddagger = (c^\ominus)^\nu = (c^\ominus)^{-2}$ we obtain:

$$k = (1 \times 10^3 \text{ mol m}^{-3})^{-2} \times \frac{(1.38 \times 10^{-23} \text{ J K}^{-1}) \times 298 \text{ K}}{6.626 \times 10^{-34} \text{ J s}} \times (2.7 \times 10^{-9})$$

$$k = \mathbf{1.68 \times 10^{-2} \text{ m}^6 \text{ mol}^{-2} \text{ s}^{-1}} .$$

The corresponding values at a temperature of 398 K can be calculated in the same way. In summary, we have:

$\mathring{\mathcal{K}}_{\ddagger c} = 2.7 \times 10^{-9}$; $k/\text{m}^6 \text{ mol}^{-2} \text{ s}^{-1} = 1.68 \times 10^{-2}$ (298 K),

$\mathring{\mathcal{K}}_{\ddagger c} = 9.5 \times 10^{-10}$; $k/\text{m}^6 \text{ mol}^{-2} \text{ s}^{-1} = 7.87 \times 10^{-3}$ (398 K).

d) Chemical potential $\mu_0(NO)$ of NO in the air at the beginning ($\xi = 0$):

At the beginning, the air should have a NO content of $x_0(NO) = 0.001$ at 298 K. In order to take into account the dependence of the chemical potential on the content of the substance in question, the mass action equation 3 [Eq. (13.1)] is used in the present case:

$$\mu_0(NO) = \mathring{\mu}(NO) + RT \ln x_0(NO) .$$

As mentioned, $\overset{\circ}{\mu}\,(=\overset{\bullet}{\mu})$ is a special basic value, namely the chemical potential of the pure substance.

$$\mu_0(\text{NO}) = (87.6\times10^3\ \text{G}) + 8.314\ \text{G K}^{-1}\times 298\ \text{K}\times\ln 0.001 = 70500\ \text{G} = 70.5\ \text{kG}\,.$$

Chemical potential $\mu_0(O_2)$ of O_2 in the air at the beginning ($\xi = 0$):

Air has an O_2 content of $x_0(O_2) = 0.21$. The chemical potential of oxygen in the air thus results in:

$$\mu_0(\text{O}_2) = \overset{\circ}{\mu}(\text{O}_2) + RT\ln x(\text{O}_2)\,.$$

$$\mu_0(\text{O}_2) = 0\ \text{G} + 8.314\ \text{G K}^{-1}\times 298\ \text{K}\times\ln 0.21 = -3900\ \text{G} = -3.9\ \text{kG}\,.$$

Due to the large excess of oxygen, its chemical potential does not change during the reaction [$\mu(O_2) = \mu_0(O_2)$].

Potential $\mu_0 = \mu_0(2\,NO + O_2)$ in the air at the beginning ($\xi = 0$):

$$\mu_0 = 2\mu_0(\text{NO}) + \mu_0(\text{O}_2)\,.$$

$$\mu_0 = 2\times(70.5\times10^3\ \text{G}) + (-3.9\times10^3\ \text{G}) = 137.1\times10^3\ \text{G} = \mathbf{137.1\ kG}\,.$$

This potential (as well as the following two μ_1 and μ_2) is shown as a dashed dark gray bar in the diagram in part a).

Chemical potential $\mu_1(NO)$ of NO in the air after oxidation of 90 % of the NO:

If 90 % of the nitrogen monoxide is oxidized, a mole fraction $x_1(NO) = 0.0001$ remains in the air. Consequently, the chemical potential $\mu_1(NO)$ results in:

$$\mu_1(\text{NO}) = \overset{\circ}{\mu}(\text{NO}) + RT\ln x_1(\text{NO})\,.$$

$$\mu_1(\text{NO}) = (87.6\times10^3\ \text{G}) + 8.314\ \text{G K}^{-1}\times 298\ \text{K}\times\ln 0.0001 = 64800\ \text{G} = 64.8\ \text{kG}\,.$$

Potential $\mu_1 = \mu_1(2\,NO + O_2)$ in the air at $\xi = 0.9\times\xi_{max}$:

$$\mu_1 = 2\mu_1(\text{NO}) + \mu_0(\text{O}_2)\,.$$

$$\mu_1 = 2\times(64.8\times10^3\ \text{G}) + (-3.9\times10^3\ \text{G}) = 125.7\times10^3\ \text{G} = \mathbf{125.7\ kG}\,.$$

Chemical potential $\mu(NO_2)$ of NO_2 at the end (ξ_{final}):

Each nitrogen monoxide molecule produces a nitrogen dioxide molecule, i.e. at the end, NO_2 has a mole fraction of $x(NO_2) = 0.001$in the air. Therefore, the chemical potential of NO_2 is:

$$\mu(\text{NO}_2) = \overset{\circ}{\mu}(\text{NO}_2) + RT\ln x(\text{NO}_2)\,.$$

$$\mu(\text{NO}_2) = (52.3\times10^3\ \text{G}) + 8.314\ \text{G K}^{-1}\times 298\ \text{K}\times\ln 0{,}001 = 35200\ \text{G} = 35.2\ \text{kG}\,.$$

Potential $\mu_2 = \mu_2(2\ NO_2)$ at the end (ξ_{final}):

$$\mu_2 = 2\mu(NO_2) = 2\times(35.2\times10^3\ G) = 70.4\times10^3\ G = \mathbf{70.4\ kG}.$$

e) Concentration $c_0(O_2)$ of oxygen in the air:

The concentration $c_0(O_2)$ of oxygen can be calculated using the general gas law [see Solution 2.18.7 c)]:

$$p_0(O_2) = c_0(O_2)\times R\times T \quad \Rightarrow$$

$$c_0(O_2) = \frac{p_0(O_2)}{RT}.$$

Partial pressure $p_0(O_2)$ of oxygen in the air:

$$p_0(O_2) = x_0(O_2)\times p^{\ominus} = 0.21\times(100\times10^3\ Pa) = 21\times10^3\ Pa.$$

$$c_0(O_2) = \frac{21\times10^3\ Pa}{8.314\ J\,mol^{-1}\,K^{-1}\times298\ K} = 8.48\ mol\,m^{-3}.$$

Due to the large excess of oxygen, its concentration does not change during the reaction [$c(O_2) = c_0(O_2)$].

Initial concentration $c_0(NO)$ of nitrogen monoxide in the air:

The calculation is carried out in the same way as the previous one:

$$c_0(NO) = \frac{p_0(NO)}{RT}.$$

Initial partial pressure $p_0(NO)$ of nitrogen monoxide in the air:

$$p_0(NO) = x_0(NO)\times p^{\ominus} = 0.001\times(100\times10^3\ Pa) = 100\ Pa.$$

$$c_0(NO) = \frac{100\ Pa}{8.314\ J\,mol^{-1}\,K^{-1}\times298\ K} = 0.0404\ mol\,m^{-3}.$$

Half-life $t_{1/2}$ of the NO oxidation:

The reaction is supposed to be a pseudo second-order reaction, i.e. the half-life $t_{1/2}$ can be calculated according to Equation (16.23):

$$t_{1/2} = \frac{1}{2k'\times c_0(NO)} = \frac{1}{2k\times c(O_2)\times c_0(NO)}.$$

$$t_{1/2} = \frac{1}{2\times(1.68\times10^{-2}\ mol^{-2}\,m^6\,s^{-1})\times8.48\ mol\,m^{-3}\times0.0404\ mol\,m^{-3}} = \mathbf{87\ s}.$$

f) Time t until the concentration of NO has dropped to 1/1000 of its initial value:

We start from Equation (16.21),

$$\frac{1}{c(\mathrm{NO})} = \frac{1}{c_0(\mathrm{NO})} + 2k't\,,$$

and solve for t:

$$t = \frac{1}{2k'}\left(\frac{1}{c(\mathrm{NO})} - \frac{1}{c_0(\mathrm{NO})}\right) = \frac{1}{2k \times c(\mathrm{O_2})}\left(\frac{1}{c(\mathrm{NO})} - \frac{1}{c_0(\mathrm{NO})}\right).$$

The concentration $c(\mathrm{NO})$ should be $c_0(\mathrm{NO})/1000 = 4.04 \times 10^{-5}\ \mathrm{mol\,m^{-3}}$:

$$t = \frac{1}{2 \times (1.68 \times 10^{-2}\ \mathrm{mol^{-2}\,m^6\,s^{-1}}) \times 8.48\ \mathrm{mol\,m^{-3}}} \cdot \left(\frac{1}{4.04 \times 10^{-5}\ \mathrm{mol\,m^{-3}}} - \frac{1}{4.04 \times 10^{-2}\ \mathrm{mol\,m^{-3}}}\right)$$

$$\approx 86800\ \mathrm{s} \approx \mathbf{1\ d}\,.$$

The concentration of NO has dropped to 1/1000 of its initial value only after about one day.

2.19 Catalysis

2.19.1 Catalytic decomposition of hydrogen peroxide by iodide

Ratio of the rate coefficients of catalyzed and uncatalyzed reaction:

We start from the ARRHENIUS equation [Eq. (18.2)]. The ratio of the rate coefficients of catalyzed reaction (k_{cat}) and uncatalyzed reaction (k_{uncat}) at 298 K then results in:

$$\frac{k_{cat}}{k_{uncat}} = \frac{k_{\infty} e^{-W_{A,cat}/RT}}{k_{\infty} e^{-W_{A,uncat}/RT}} = \exp\frac{-W_{A,cat}+W_{A,uncat}}{RT} .$$

$$\frac{k_{cat}}{k_{uncat}} = \exp\frac{(-59\times10^3\ \mathrm{J\,mol^{-1}})+(76\times10^3\ \mathrm{J\,mol^{-1}})}{8.314\ \mathrm{J\,mol^{-1}\,K^{-1}}\times298\ \mathrm{K}} = \mathbf{955} .$$

The reaction is thus accelerated almost by a factor of 1000.

2.19.2 Catalytic decomposition of acetylcholine

Initial rate density r_0:

The initial rate density r_0 for any initial concentration $c_{S,0}$ of substrate can be determined using Equation (19.11):

$$r_0 = k_2 \times \frac{c_{E,0} \times c_{S,0}}{K_M + c_{S,0}} .$$

$$r_0 = (1.4\times10^4\ \mathrm{s^{-1}})\times\frac{(5\times10^{-6}\ \mathrm{mol\,m^{-3}})\times10\ \mathrm{mol\,m^{-3}}}{(9\times10^{-2}\ \mathrm{mol\,m^{-3}})+10\ \mathrm{mol\,m^{-3}}} = \mathbf{0.069\ mol\,m^{-3}\,s^{-1}} .$$

Maximum initial rate density $r_{0,max}$:

The maximum initial rate density $r_{0,max}$ corresponds to the product of turnover number k_2 and the enzyme concentration $c_{E,0}$ [Eq. (19.13)]:

$$r_{0,max} = k_2 \times c_{E,0} = (1.4\times10^4\ \mathrm{s^{-1}})\times(5\times10^{-6}\ \mathrm{mol\,m^{-3}}) = \mathbf{0.070\ mol\,m^{-3}\,s^{-1}} .$$

2.19.3 Application of MICHAELIS-MENTEN kinetics

a) Dissociation constant $\mathring{K}_{diss}$ of the enzyme-substrate complex:

The mechanism proposed by Leonor MICHAELIS and Maud Leonora MENTEN assumes that enzyme E and substrate S rapidly and reversibly form an enzyme-substrate complex ES:

$$\mathrm{E}+\mathrm{S} \underset{k_{-1}}{\overset{k_1}{\rightleftarrows}} \mathrm{ES} \xrightarrow{k_2} \mathrm{E}+\mathrm{P} .$$

The dissociation constant for the complex, i.e. the equilibrium constant for its decomposition into enzyme and substrate, can then be calculated according to Equation (17.5):

$$\overset{\circ}{K}_{\text{diss}} = \frac{k_{-1}}{k_{+1}} = \frac{4\times10^4\ \text{s}^{-1}}{1.0\times10^5\ \text{m}^3\ \text{mol}^{-1}\ \text{s}^{-1}} = \mathbf{0.4\ mol\ m^{-3}}\,.$$

b) Michaelis constant K_M:

The Michaelis constant K_M is calculated using Equation (19.6):

$$K_M = \frac{k_{-1}+k_2}{k_{+1}} = \frac{(4\times10^4\ \text{s}^{-1})+(8\times10^5\ \text{s}^{-1})}{1.0\times10^5\ \text{m}^3\ \text{mol}^{-1}\ \text{s}^{-1}} = \mathbf{8.4\ mol\ m^{-3}}\,.$$

Catalytic efficiency k_2/K_M:

The catalytic efficiency can be described by the quotient k_2/K_M:

$$\frac{k_2}{K_M} = \frac{8\times10^5\ \text{s}^{-1}}{8.4\ \text{mol}\,\text{m}^{-3}} = \mathbf{9.5\times10^4\ m^3\ mol^{-1}\ s^{-1}}\,.$$

c) K_M can only be used as a measure for the substrate affinity of the enzyme when k_2 is much smaller than k_{-1}. This condition is by no means fulfilled in the present case.

2.19.4 Hydration of carbon dioxide by carbonic anhydrase

a) Determination of K_M and $r_{0,\max}$:

The so-called Lineweaver-Burke plot can be used to determine the characteristic quantities K_M and $r_{0,\max}$ for an enzyme in question. In order to do this, we must find the reciprocal of the Michaelis-Menten equation. After transforming, we obtain [Eq. (19.15)]:

$$\frac{1}{r_0} = \frac{K_M}{r_{0,\max}}\times\frac{1}{c_{S,0}}+\frac{1}{r_{0,\max}}\,.$$

If we now plot $1/r_0$ versus $1/c_{S,0}$, we obtain a straight line (if a fixed total concentration $c_{E,0}$ of enzyme is used), from whose extrapolated intersection point with the ordinate the value of $r_{0,\max}$ can be determined. The slope $K_M/r_{0,\max}$ can then be used to calculate K_M.

$1/c_{S,0}$ / (m^3 mol^{-1})	$1/r_0$ / (m^3 s mol^{-1})
0.8	36.0
0.4	20.1
0.2	12.3
0.05	6.5
0.025	5.3

Alternatively, the intersection point with the abscissa can also be used, since it corresponds to $-1/K_M$.

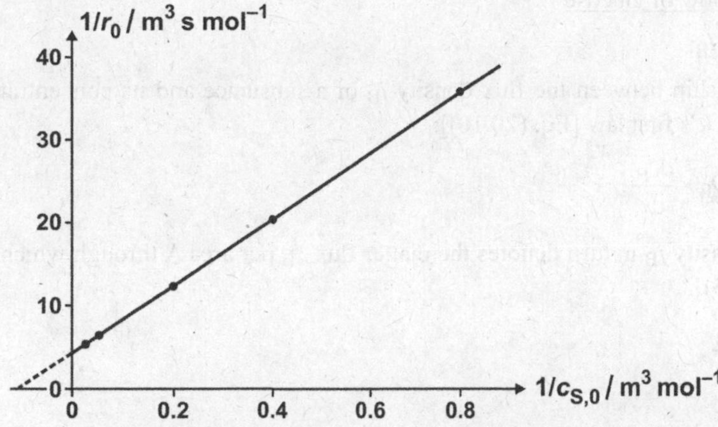

From the diagram or by linear regression we obtain an ordinate intercept b of $4.4\ \mathrm{m^3\,s\,mol^{-1}}$ and a slope m of 39.5 s.

The maximum initial rate density $r_{0,\max}$ then results in:

$$b = \frac{1}{r_{0,\max}} \quad \Rightarrow$$

$$r_{0,\max} = \frac{1}{b} = \frac{1}{4.4\ \mathrm{m^3\,s\,mol^{-1}}} = \mathbf{0.23\ mol\,m^{-3}\,s^{-1}}.$$

The MICHAELIS constant K_M is obtained from the slope m of the straight line according to

$$m = \frac{K_M}{r_{0,\max}} \quad \Rightarrow$$

$$K_M = m \times r_{0,\max} = 39.5\ \mathrm{s} \times 0{,}24\ \mathrm{mol\,m^{-3}\,s^{-1}} = \mathbf{9.1\ mol\,m^{-3}}.$$

b) Turnover number k_2:

The turnover number k_2 can be determined by means of the relationship (19.13):

$$r_{0,\max} = k_2 \times c_{E,0} \quad \Rightarrow$$

$$k_2 = \frac{r_{0,\max}}{c_{E,0}} = \frac{0.23\ \mathrm{mol\,m^{-3}\,s^{-1}}}{2.8\times10^{-6}\ \mathrm{mol\,m^{-3}}} = 8.2\times10^4\ \mathrm{s^{-1}}.$$

Catalytic efficiency k_2/K_M:

$$\frac{k_2}{K_M} = \frac{8.2\times10^4\ \mathrm{s^{-1}}}{9.1\ \mathrm{mol\,m^{-3}}} = \mathbf{9.0\times10^3\ m^3\,mol^{-1}\,s^{-1}}.$$

2.20 Transport Phenomena

2.20.1 Flow of glucose

Matter flux J_B:

The relationship between the flux density j_B of a substance and its concentration gradient is given by FICK's first law [Eq. (20.10)]:

$$j_B = -D_B \times \frac{dc_B}{dx}.$$

The flux density j_B in turn denotes the matter flux J_B per area A through which the flow passes [Eq. (20.5)],

$$j_B = \frac{J_B}{A},$$

i.e. we have:

$$\frac{J_B}{A} = -D_B \times \frac{dc_B}{dx}.$$

Solving for the matter flux J_B in question yields:

$$J_B = -A \times D_B \times \frac{dc_B}{dx}.$$

Since the concentration gradient should be linear, we can write:

$$J_B = -A \times D_B \times \frac{\Delta c_B}{l} = -A \times D_B \times \frac{c_{B,2} - c_{B,1}}{l}.$$

Therefore, we obtain for the flux J_B of glucose:

$$J_B = -(5 \times 10^{-4}\ \text{m}^2) \times (0.67 \times 10^{-9}\ \text{m}^2\,\text{s}^{-1}) \times \frac{0.2\ \text{mol}\,\text{m}^{-3} - 1.0\ \text{mol}\,\text{m}^{-3}}{1 \times 10^{-3}\ \text{m}}$$

$$= \mathbf{2.7 \times 10^{-10}\ mol\ s^{-1}}.$$

2.20.2 Diffusion of carbon in iron

a) Diffusion coeffizient D_B of carbon in iron at 750 °C:

One starts again from FICK's first law,

$$j_B = -D_B \times \frac{dc_B}{dx} = -D_B \times \frac{\Delta c_B}{d},$$

and solves for D_B:

$$D_B = -j_B \times \frac{d}{\Delta c_B} = -j_B \times \frac{d}{c_{B,2} - c_{B,1}}.$$

$$D_B = -(28.3\times10^{-6}\ \text{mol m}^{-2}\ \text{s}^{-1})\times\frac{3\times10^{-3}\ \text{m}}{185\ \text{mol m}^{-3} - 1850\ \text{mol m}^{-3}} = \mathbf{5.1\times10^{-11}\ m^2\ s^{-1}}.$$

b) Activation energy W_A for the diffusion process:

In order to determine the activation energy W_A, the equation from Solution 2.18.2 can be used in a modified form:

$$W_A = \frac{R\times\ln\dfrac{D_B(T_2)}{D_B(T_1)}}{\dfrac{1}{T_1}-\dfrac{1}{T_2}}.$$

$$W_A = \frac{8.314\ \text{J mol}^{-1}\ \text{K}^{-1}\times\ln\dfrac{1.7\times10^{-10}\ \text{m}^2\ \text{s}^{-1}}{5.1\times10^{-11}\ \text{m}^2\ \text{s}^{-1}}}{\dfrac{1}{1023\ \text{K}}-\dfrac{1}{1173\ \text{K}}}$$

$$= \frac{10.0\ \text{J mol}^{-1}\ \text{K}^{-1}}{1.25\times10^{-4}\ \text{K}^{-1}} = 80000\ \text{J} = \mathbf{80{,}0\ kJ\ mol\ s^{-1}}.$$

"Frequeny factor" $D_{B,0}$:

The ARRHENIUS approach is solved for $D_{B,0}$ and the corresponding values for a specific temperature, e.g. T_1, are used:

$$D_{B,0} = \frac{D_B(T_1)}{\exp\dfrac{-W_A}{RT_1}} = \frac{5.1\times10^{-11}\ \text{m}^2\ \text{s}^{-1}}{\exp\dfrac{-80\times10^3\ \text{J mol}^{-1}}{8.314\ \text{J K}^{-1}\times1023\ \text{K}}} = \mathbf{6.2\times10^{-7}\ m^2\ s^{-1}}.$$

2.20.3* Diffusion of sucrose

a) Concentration $c_{B,0}$ of sucrose (uniform distribution in the glass of tea):

$$c_{B,0} = \frac{n_{B,0}}{V_S}.$$

Amount $n_{B,0}$ of sucrose:

With the molar mass $M_B = 342.0\times10^{-3}$ kg mol^{-1} for sucrose ($C_{12}H_{22}O_{11}$), one obtains:

$$n_{B,0} = \frac{m_{B,0}}{M_B} = \frac{10\times10^{-3}\ \text{kg}}{342.0\times10^{-3}\ \text{kg mol}^{-1}} = 0.029\ \text{mol}.$$

Volume V_S of the solution:

$$V_S = l \times A = l \times \pi r^2 = l \times \pi (d/2)^2 .$$

$$V_S = (7\times10^{-2}\ \text{m})\times 3.142\times[(5\times10^{-2}\ \text{m})/2]^2 = 137\times10^{-6}\ \text{m}^3 .$$

$$c_{B,0} = \frac{0.029\ \text{mol}}{137\times10^{-6}\ \text{m}^3} = \mathbf{212\ mol\ m^{-3}} .$$

Concentration $c_{B,1}$ at the bottom:

Since there should exist a linear concentration gradient and the concentration $c_{B,2}$ at the surface should be 0, the concentration $c_{B,1}$ at the bottom results in:

$$c_{B,1} = 2\times c_{B,0} = 2\times 212\ \text{mol m}^{-3} = \mathbf{424\ mol\ m^{-3}} .$$

b) Flux density j_B:

The flux density j_B results from FICK's first law:

$$j_B = -D_B \times \frac{\text{d}c_B}{\text{d}z} = -D_B \times \frac{\Delta c_B}{l} = -D_B \times \frac{c_{B,2} - c_{B,1}}{l} .$$

$$j_B = -(0.5\times10^{-9}\ \text{m}^2\,\text{s}^{-1})\times \frac{0 - 424\ \text{mol m}^{-3}}{7\times10^{-2}\ \text{m}} = \mathbf{3.0\times10^{-6}\ mol\ m^{-2}\ s^{-1}} .$$

Migration velocity $v_{B,1}$:

The migration velocity $v_{B,1}$ can be calculated by means of Equation (20.5):

$$j_B = c_{B,1} \times v_{B,1} \quad \Rightarrow$$

$$v_{B,1} = \frac{j_B}{c_{B,1}} = \frac{3.0\times10^{-6}\ \text{mol m}^{-2}\,\text{s}^{-1}}{424\ \text{mol m}^{-3}} = \mathbf{7.1\times10^{-9}\ m\ s^{-1}} .$$

c) Duration Δt of concentration equalization:

The duration Δt needed to approximately equalize the concentration in the glass of tea can be determined using Equation (20.13):

$$\Delta t = \frac{l^2}{8D_B} = \frac{(7\times10^{-2}\ \text{m})^2}{8\times(0.5\times10^{-9}\ \text{m}^2\,\text{s}^{-1})} = \mathbf{1225000\ s} \approx 14\ \text{Tage} .$$

2.20.4 Oil as lubricant

Friction force F_f:

The friction force F_f to be overcome is determined by NEWTON's law of friction [Eq. (20.15)],

$$F_{\mathrm{f}} = -\eta \times A \times \frac{\Delta v_x}{\Delta z},$$

where, in our case, we simply have $\Delta v_x/\Delta z = v_0/d$:

$$F_{\mathrm{f}} = -\eta \times A \times \frac{v_0}{d}.$$

$$F_{\mathrm{f}} = -0.1\,\mathrm{Pa\,s} \times 0.18\,\mathrm{m}^2 \times \frac{0.3\,\mathrm{m\,s^{-1}}}{0.1\times10^{-3}\,\mathrm{m}} = -0.1\,\mathrm{N\,m^{-2}\,s} \times 0.18\,\mathrm{m}^2 \times \frac{0.3\,\mathrm{m\,s^{-1}}}{0.1\times10^{-3}\,\mathrm{m}} = -54\,\mathrm{N}.$$

Thus, a force $F = -F_{\mathrm{f}} =$ **54 N** must act on the block to overcome the friction force opposing the movement.

2.20.5 Molecular radius and volume

a) Molecular radius r_{B}:

Equation (20.20) is used to estimate the molecular radius:

$$D_{\mathrm{B}} = \frac{k_{\mathrm{B}}T}{6\pi \times \eta \times r_{\mathrm{B}}},$$

where k_{B} is the BOLTZMANN constant. Solving for r_{B} results in:

$$r_{\mathrm{B}} = \frac{k_{\mathrm{B}}T}{6\pi \times \eta \times D_{\mathrm{B}}}.$$

Myoglobin:

$$r_{\mathrm{myo}} = \frac{k_{\mathrm{B}}T}{6\pi \times \eta \times D_{\mathrm{myo}}}.$$

$$r_{\mathrm{myo}} = \frac{(1.38\times10^{-23}\,\mathrm{J\,K^{-1}})\times 293\,\mathrm{K}}{6\times3.142\times(1.002\times10^{-3}\,\mathrm{Pa\,s})\times(0.113\times10^{-9}\,\mathrm{m^2\,s^{-1}})} = \mathbf{1.9\times10^{-9}\,m}.$$

Conversion of units:

$$\frac{\mathrm{J\,K^{-1}\,K}}{\mathrm{Pa\,s\,m^2\,s^{-1}}} = \frac{\mathrm{N\,m}}{\mathrm{N\,m^{-2}\,m^2}} = \mathrm{m}.$$

Hemoglobin:

$$r_{\mathrm{hem}} = \frac{k_{\mathrm{B}}T}{6\pi \times \eta \times D_{\mathrm{hem}}}.$$

$$r_{\mathrm{hem}} = \frac{(1.38\times10^{-23}\,\mathrm{J\,K^{-1}})\times 293\,\mathrm{K}}{6\times3.142\times(1.002\times10^{-3}\,\mathrm{Pa\,s})\times(0.069\times10^{-9}\,\mathrm{m^2\,s^{-1}})} = \mathbf{3.1\times10^{-9}\,m}.$$

Molecular volume V_B:

Since myoglobin and hemoglobin molecules are roughly spherical, the molecular volume is:

$$V_B = \frac{4}{3}\pi r_B^3 .$$

Myoglobin:

$$V_{myo} = \frac{4}{3}\pi r_{myo}^3 = \frac{4}{3}\times 3.142\times(1.9\times10^{-9}\ \text{m})^3 = \mathbf{2.9\times10^{-26}\ m^3} .$$

Hemoglobin:

$$V_{hem} = \frac{4}{3}\pi r_{hem}^3 = \frac{4}{3}\times 3.142\times(3.1\times10^{-9}\ \text{m})^3 = \mathbf{12.5\times10^{-26}\ m^3} .$$

b) The volume ratio of hemoglobin to myoglobin molecule is

$$\frac{12.5\times10^{-26}\ \text{m}^3}{2.9\times10^{-26}\ \text{m}^3} = 4.3 .$$

Since hemoglobin consists of four subunits that are similar in structure to the monomeric myoglobin, the volume ratio obtained makes sense.

2.20.6 Sinking of pollutant particles in air

a) Volume V_p of a spherical pollutant particles:

$$V_p = \frac{4}{3}\pi r_p^3 = \frac{4}{3}\times 3.142\times(8\times10^{-6}\ \text{m})^3 = 2.14\times10^{-15}\ \text{m}^3 .$$

Mass m_p of a pollutant particles:

$$\rho_p = \frac{m_p}{V_p} \quad \Rightarrow$$

$$m_p = \rho_p \times V_P = (2500\ \text{kg m}^{-3})\times(2.14\times10^{-15}\ \text{m}^3) = 5.35\times10^{-12}\ \text{kg} .$$

Sinking velocity v_p of the particles:

At constant sinking velocity, the gravitational force F_g,

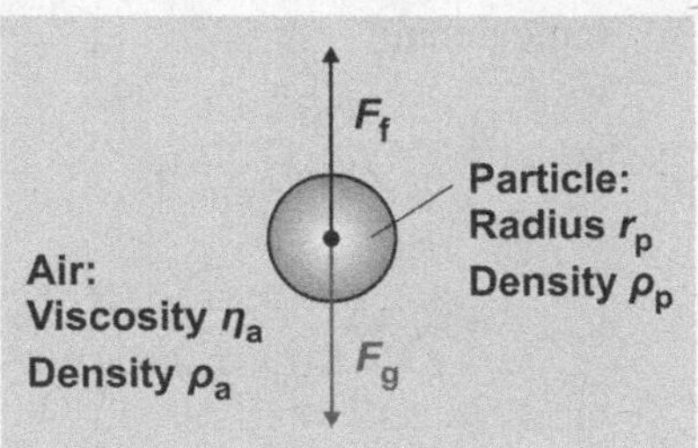

$$F_g = m_p \times g ,$$

exactly balances STOKES' friction force F_f [Eq. (20.18)],

$$F_f = 6\pi \times \eta_a \times r_p \times v_p .$$

We therefore have:

$$F_g = -F_f \Rightarrow$$
$$m_p \times g = -6\pi \times \eta_a \times r_p \times v_p .$$

Solving for v_p yields:

$$v_p = -\frac{m_p \times g}{6\pi \times \eta_a \times r_p} = -\frac{(5.35\times10^{-12}\ \text{kg})\times 9{,}8\ \text{m}\,\text{s}^{-2}}{6\times3.142\times(18.5\times10^{-6}\ \text{Pa}\,\text{s})\times(8\times10^{-6}\ \text{m})} = \mathbf{-0.019\ m\,s^{-1}} .$$

Conversion of units:

$$\frac{\text{kg}\,\text{m}\,\text{s}^{-2}}{\text{Pa}\,\text{s}\,\text{m}} = \frac{\text{N}}{\text{N}\,\text{m}^{-2}\,\text{s}\,\text{m}} = \frac{\text{m}}{\text{s}} .$$

b) Duration Δt of sinking:

For a uniform motion, we obtain the following equation:

$$v = \frac{\Delta h}{\Delta t} \Rightarrow$$

$$\Delta t = \frac{\Delta h}{v} = \frac{-50\ \text{m}}{-0.019\ \text{m}\,\text{s}^{-1}} = \mathbf{2630\ s} \approx 44\ \text{min} .$$

c) Buoyant force F_b:

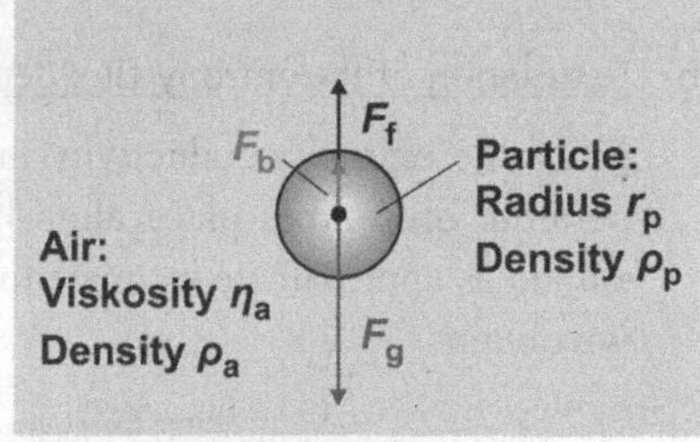

If the buoyancy force is taken into account, the force equilibrium has to be formulated as follows:

$$F_g = -F_f - F_b \Rightarrow$$
$$F_g + F_b = -F_f .$$

The gravitational force is reduced by the buoyancy force, a finding, which can be expressed by the following relationship:

$$F_g + F_b = (\rho_p \times V_p - \rho_a \times V_p) \times g = (\rho_p - \rho_a) \times V_p \times g .$$

Since the density of the particle ($\rho_p = 2500\ \text{kg}\,\text{m}^{-3}$) is more than two thousand times higher than the density of the air ($\rho_a = 1.2\ \text{kg}\,\text{m}^{-3}$), the buoyancy force is so small that it plays no role here.

2.20.7 Entropy conduction through a copper plate

Density j_S of the entropy flux:

The density j_S of the entropy flux is directly proportional to the temperature gradient $\mathrm{d}T/\mathrm{d}x$ [FOURIER's law; Eq. (20.22)],

$$j_S = -\sigma_S \times \frac{dT}{dx}.$$

σ_S is the entropy conductivity of the material. Since there should exist a linear concentration gradient, we can write:

$$j_S = -\sigma_S \times \frac{\Delta T}{d} = -\sigma_S \times \frac{T_2 - T_1}{d}.$$

For the temperature range in question (average temperature $T = 298$ K) we can use the entropy conductivity for copper from Table 20.5 in the textbook "Physical Chemistry from a Different Angle":

$$j_S = -1.3\,\text{Ct}\,\text{K}^{-1}\,\text{s}^{-1}\,\text{m}^{-1} \times \frac{273\,\text{K} - 323\,\text{K}}{20\times10^{-3}\,\text{m}} = \mathbf{3250\,Ct\,s^{-1}\,m^{-2}}.$$

2.20.8* Entropy loss in buildings

a) Entropy flux density $j_{S,\text{B}}$ for the non-insulated wall:

The entropy flux density $j_{S,\text{B}}$ for the non-insulated wall can be calculated using FOURIER's law [Eq. (20.22)]:

$$j_{S,\text{B}} = -\sigma_{S,\text{B}} \times \frac{\Delta T_{\text{total}}}{d_\text{B}} = -0.003\,\text{Ct}\,\text{K}^{-1}\,\text{s}^{-1}\,\text{m}^{-1} \times \frac{-20\,\text{K}}{12\times10^{-3}\,\text{m}} = \mathbf{5.0\,Ct\,s^{-1}\,m^{-1}}.$$

b) Estimation of the entropy flux density $j_{S,\text{BP},1}$ for the insulated wall:

Since the entropy conductivity in polystyrene is less than $\frac{1}{20}$ of that in bricks, the temperature drop takes place almost completely in the polystyrene; therefore, it can be assumed as approximate estimate that the wall consists only of the polystyrene boards. Thus, we obtain:

$$j_{S,\text{BP},1} \approx j_{S,\text{P}} = -\sigma_{S,\text{P}} \times \frac{\Delta T_{\text{total}}}{d_\text{P}}.$$

$$j_{S,\text{BP},1} \approx -0.00013\,\text{Ct}\,\text{K}^{-1}\,\text{s}^{-1}\,\text{m}^{-1} \times \frac{-20\,\text{K}}{5\times10^{-3}\,\text{m}} = \mathbf{0.52\,Ct\,s^{-1}\,m^{-1}}.$$

c) Temperature profile through the insulated wall:

The entropy flux flowing through the polystyrene ($J_{S,\text{P}}$) should be equal to the entropy flux flowing through the bricks ($J_{S,\text{B}}$), i.e. we have

$$J_{S,\text{P}} = J_{S,\text{B}} \quad \text{and therefore}$$

$$-\sigma_{S,\text{P}} \times \frac{\Delta T_\text{P}}{d_\text{P}} = -\sigma_{S,\text{B}} \times \frac{\Delta T_\text{B}}{d_\text{B}}.$$

Solving for ΔT_P, the temperature change through the polystyrene boards, results in:

$$\Delta T_P = \frac{\sigma_{S,B} \times d_P}{\sigma_{S,P} \times d_B} \times \Delta T_B = \frac{0.003\ \text{Ct}\,\text{K}^{-1}\,\text{s}^{-1}\,\text{m}^{-1} \times (5\times10^{-3}\ \text{m})}{0.00013\ \text{Ct}\,\text{K}^{-1}\,\text{s}^{-1}\,\text{m}^{-1} \times (12\times10^{-3}\ \text{m})} \times \Delta T_B = 9.62 \times \Delta T_B\,.$$

The temperature difference between inside and outside corresponds to the sum of the temperature changes through the brick wall and through the polystyrene boards:

$$\Delta T_{\text{total}} = \Delta T_B + \Delta T_P = \Delta T_B + 9.62 \times \Delta T_B = 10.62 \times \Delta T_B\,.$$

The temperature change ΔT_B through the brick wall is therefore:

$$\Delta T_B = \frac{\Delta T_{\text{total}}}{10.62} = \frac{-20\ \text{K}}{10.62} = -1.88\ \text{K}\,.$$

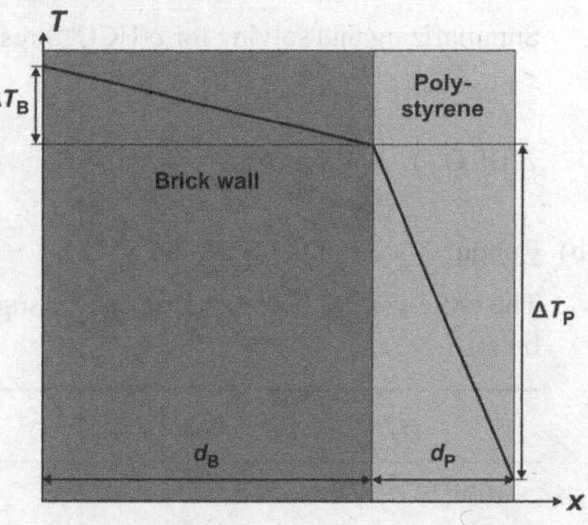

Accordingly, the temperature change ΔT_P through the polystyrene boards results in -18.12 K (see the adjoining figure).

Entropy flux density $j_{S,BP,2}$:

We can calculate the entropy flux density $j_{S,BP,2}$ by inserting the corresponding data into FOURIER's law for one of the two layers (this is possible since $J_{S,P} = J_{S,B}$):

$$j_{S,BP,2} = -\sigma_{S,B} \times \frac{\Delta T_B}{d_B} = -0.003\ \text{Ct}\,\text{K}^{-1}\,\text{s}^{-1}\,\text{m}^{-1} \times \frac{-1.88\ \text{K}}{12\times10^{-3}\ \text{m}} = \mathbf{0.47\ Ct\,s^{-1}\,m^{-1}}\,.$$

This value is about 10 % lower than the estimated value from b), so that the approximation is still acceptable. A comparison with the value for the non-insulated wall shows that the entropy loss could be reduced by the insulation to approximately $\frac{1}{10}$ of the original value.

2.21 Electrolyte Solutions

2.21.1 Conductivity of tap water

a) Concentration $c(HCO_3^-)$ of the hydrogencarbonate ions:

In tap water, as in any electrolyte solution, the electroneutrality rule [Eq. (21.1)] is valid:

$\sum_i z_i c_i = 0$, summed over all ion types i, i.e.

$$[(+2)\times 1.0+(+2)\times 0.2+1\times 0.5+(-2)\times 0.3+(-1)\times 0.5]\ \text{mol m}^{-3}+(-1)\times c(HCO_3^-) = 0.$$

Summarizing and solving for $c(HCO_3^-)$ results in:

$$1.8\ \text{mol m}^{-3} - c(HCO_3^-) = 0$$

$$c(HCO_3^-) = \mathbf{1.8\ mol\ m^{-3}}.$$

b) Conductivity σ of tap water at 25 °C:

The table given in the Exercise can be supplemented by means of Table 21.2 in the textbook:

	Ca^{2+}	Mg^{2+}	Na^+	SO_4^{2-}	Cl^-	HCO_3^-
$c\,/\,\text{mol m}^{-3}$	1.0	0.2	0.5	0.3	0.5	1.8
$u\,/\,10^{-8}\ \text{m}^2\ \text{V}^{-1}\ \text{s}^{-1}$	6.2	5.5	5.2	–8.3	–7.9	–4.6

To calculate the conductivity σ of tap water, the "four-factor formula" [Eq. (21.26)] is used:

$$\sigma = \sum_i z_i \mathcal{F} \times c_i \times u_i = \mathcal{F} \times \sum_i z_i \times c_i \times u_i .$$

$$\sigma = 96485\ \text{C mol}^{-1} \times [(+2)\times 1.0\times 6.2+(+2)\times 0.2\times 5.5+(+1)\times 0.5\times 5.2+ (-2)\times 0.3\times(-8.3)+(-1)\times 0.5\times(-7.9)+(-1)\times 1.8\times(-4.6)]\times 10^{-8}\ \text{mol m}^{-1}\ \text{V}^{-1}\ \text{s}^{-1}$$

$$\sigma = 96485\ \text{A s mol}^{-1}\times(34.41\times 10^{-8}\ \text{mol m}^{-1}\ \text{V}^{-1}\ \text{s}^{-1})$$

$$= 0.033\ \text{A V}^{-1}\ \text{m}^{-1} = \mathbf{0.033\ S\ m^{-1}}.$$

c)* Conductivity σ of tap water at 100 °C:

The specific conductivity changes by 2 % when the temperature increases by 1 degree, i.e. we have:

$$\sigma(373\ \text{K}) = \sigma(298\ \text{K})\times(1.02)^{373-298} = \sigma(298\ \text{K})\times(1.02)^{75} = 0.033\ \text{S m}^{-1}\times 4.4$$

$$= \mathbf{0.15\ S\ m^{-1}}.$$

2.21.2 Migration of Zn^{2+} ions

a) Electric mobility u of Zn^{2+} ions:

The electric mobility u of Zn^{2+} ions can be calculated by means of Equation (21.11):

$$v = u \times E \quad \Rightarrow$$

$$u = \frac{v}{E} = \frac{2.74 \times 10^{-5}\ \mathrm{m\,s^{-1}}}{500\ \mathrm{V\,m^{-1}}} = 5.5 \times 10^{-8}\ \mathrm{m^2\,V^{-1}\,s^{-1}}.$$

Radius r of the hydrated Zn^{2+} ion:

The radius r of the hydrated Zn^{2+} ion can be estimated using Equation (21.14):

$$u = \frac{z \times e_0}{6\pi \times \eta \times r} \quad \Rightarrow$$

$$r = \frac{z \times e_0}{6\pi \times \eta \times u}.$$

$$r = \frac{(+2) \times (1.60 \times 10^{-19}\ \mathrm{C})}{6 \times 3.142 \times (0.890 \times 10^{-3}\ \mathrm{Pa\,s}) \times (5.5 \times 10^{-8}\ \mathrm{m^2\,V^{-1}\,s^{-1}})} = \mathbf{3.5 \times 10^{-10}\ m}.$$

Conversion of units:

$$\frac{\mathrm{C}}{\mathrm{Pa\,s\,m^2\,V^{-1}\,s^{-1}}} = \frac{\mathrm{A\,s\,V}}{\mathrm{N\,m^{-2}\,m^2}} = \frac{\mathrm{N\,m}}{\mathrm{N}} = \mathrm{m}.$$

The following relationships exist between the units of the electric energy $W = U \times I \times \Delta t$: 1 V A s = 1 J = 1 N m.

b) Diffusion coefficient D of the hydrated Zn^{2+} ion:

To estimate the diffusion coefficient D of the hydrated Zn^{2+} ion, we can have recourse to the EINSTEIN-SMOLUCHOWSKI equation [Eq. (20.11)],

$$D = \omega RT,$$

and the defining equation of the electric mobility u,

$$u = \omega z \mathcal{F}.$$

The combination of both relationships results in:

$$D = \frac{u}{z \times \mathcal{F}} \times RT = \frac{5.5 \times 10^{-8}\ \mathrm{m^2\,V^{-1}\,s^{-1}}}{(+2) \times 96485\ \mathrm{C\,mol^{-1}}} \times 8.314\ \mathrm{G\,K^{-1}} \times 298\ \mathrm{K} = \mathbf{7.1 \times 10^{-10}\ m^2\,s^{-1}}.$$

Conversion of units:

$$\frac{\mathrm{m^2\,V^{-1}\,s^{-1}}}{\mathrm{C\,mol^{-1}}}\ \mathrm{G\,K^{-1}\,K} = \frac{\mathrm{m^2\,s^{-1}}}{\mathrm{V\,A\,s\,mol^{-1}}}\ \mathrm{J\,mol^{-1}} = \mathrm{m^2\,s^{-1}}.$$

2.21.3 Dissociation of formic acid in aqueous solution

a) Determination of the cell constant Z:

We start from the following equation

$$R_C = \rho_C \times Z = \frac{1}{\sigma_C} \times Z \qquad \text{[cf. Eq. (21.25)]}$$

and hence obtain

$$Z = R_C \times \sigma_C = 40.0\ \Omega \times 1.288\ \mathrm{S\,m^{-1}} = 40.0\ \mathrm{S^{-1}} \times 1.288\ \mathrm{S\,m^{-1}} = \mathbf{51.52\ m^{-1}}.$$

b) Conductivity σ of the solution of formic acid:

Again, we have:

$$R = \frac{1}{\sigma} \times Z \quad \Rightarrow$$

$$\sigma = \frac{1}{R} \times Z = \frac{1}{1026.3\ \Omega} \times 51.52\ \mathrm{m^{-1}} = \mathbf{0.050200\ S\,m^{-1}}.$$

Molar conductivity Λ of the solution of formic acid:

The following relationship exists between conductivity σ and molar conductivity Λ [Eq. (21.28)]:

$$\Lambda = \frac{\sigma}{c} = \frac{0.050200\ \mathrm{S\,m^{-1}}}{10\ \mathrm{mol\,m^{-3}}} = 5.020 \times 10^{-3}\ \mathrm{S\,m^2\,mol^{-1}} = \mathbf{5.020\ mS\,m^2\,mol^{-1}}.$$

c) Degree of dissociation α:

The degree of dissociation α is calculated using Equation (21.42):

$$\alpha = \frac{\Lambda}{\Lambda^{0}} = \frac{5.020 \times 10^{-3}\ \mathrm{S\,m^2\,mol^{-1}}}{40.43 \times 10^{-3}\ \mathrm{S\,m^2\,mol^{-1}}} = \mathbf{0.124}.$$

Equilibrium constant $\overset{\ominus}{K}_c$:

Here, the dissociation of formic acid in water is considered,

$$\mathrm{HCOOH|w} \rightleftarrows \mathrm{H^+|w} + \mathrm{HCOO^-|w}.$$

Using OSTWALD's dilution law [Eq. (21.41)], we can calculate the equilibrium constant in question:

$$\overset{\ominus}{K}_c = \frac{\alpha^2}{(1-\alpha)} \times c = \frac{(0.124)^2}{(1-0.124)} \times 10\ \mathrm{mol\,m^{-3}} = \mathbf{0.176\ mol\,m^{-3}}.$$

Equilibrium number $\mathcal{K}_c^\ominus$:

To convert the conventional equilibrium constant $K_c^\ominus$ into the equilibrium number $\mathcal{K}_c^\ominus$, Equation (6.20) is used,

$$K_c^\ominus = \kappa \mathcal{K}_c^\ominus , \quad \text{where} \quad \kappa = (c^\ominus)^{\nu_c} \quad \text{with} \quad \nu_c = \nu_B + \nu_{B'} + ... + \nu_D + \nu_{D'} + ...$$

In the present case, we have $\nu_c = -1 + 1 + 1 = +1$ and therefore $\kappa = (1 \text{ kmol m}^{-3})^1 = 1 \text{ kmol m}^{-3}$. Solving the equation for the equilibrium number $\mathcal{K}_c^\ominus$ in question results in:

$$\mathcal{K}_c^\ominus = \frac{K_c^\ominus}{\kappa} = \frac{0.176 \text{ mol m}^{-3}}{1 \times 10^3 \text{ mol m}^{-3}} = 1.76 \times 10^{-4} .$$

d) Standard value $\mu_p^\ominus$ of the proton potential of the acid-base pair formic acid/formate:

The dissociation of formic acid in water can, in BRØNSTED's sense, also be interpreted as proton transfer from formic acid to water:

$$\text{HCOOH|w} + \text{H}_2\text{O|w} \rightleftarrows \text{H}_3\text{O}^+\text{|w} + \text{HCOO}^-\text{|w} .$$

The proton potential $\mu_p^\ominus(\text{HCOOH/HCOO}^-)$ can then be calculated according to Equation (7.8):

$$\mu_p^\ominus(\text{HCOOH/HCOO}^-) - \mu_p^\ominus(\text{H}_3\text{O}^+/\text{H}_2\text{O}) = RT \ln \frac{c_r(\text{HCOO}^-) \times c_r(\text{H}_3\text{O}^+)}{c_r(\text{HCOOH})} .$$

The proton potential $\mu_p^\ominus(\text{H}_3\text{O}^+/\text{H}_2\text{O})$ is equal to 0 and the argument of the logarithm corresponds to the equilibrium number $\mathcal{K}_c^\ominus$. Thus, we obtain for the standard value of the proton potential:

$$\mu_p^\ominus(\text{HCOOH/HCOO}^-) = RT \ln \mathcal{K}_c^\ominus .$$

$$\mu_p^\ominus(\text{HCOOH/HCOO}^-) = 8.314 \text{ G K}^{-1} \times 298 \text{ K} \times \ln(1.76 \times 10^{-4}) = -21400 \text{ G} = \mathbf{-21.4\ kG} .$$

The pair formic acid/formate is therefore, as expected, a weak acid-base pair (see also Table 7.2).

Proton potential μ_p in the formic acid solution:

For the proton potential of a weak acid-base pair such as formic acid/formate dissolved in pure water, the following relationship is valid [cf. Eq. (7.6)]:

$$\mu_p = \tfrac{1}{2} \times \left\{ \mu_p^\ominus(\text{HCOOH/HCOO}^-) + \mu_p^\ominus(\text{H}_3\text{O}^+/\text{H}_2\text{O}) + RT \ln[c(\text{HCOOH})/c^\ominus] \right\} .$$

The degree of dissociation α of the present formic acid was 0.124, i.e. the dissociated fraction of 12.4 % is considerably higher than 5 %. The approximation used in Section 7.2 that the undissociated fraction $c(\text{Ad})$ can be equated to the initial concentration c_0 is no longer permissible here. Instead, the concentration $c(\text{HCOOH}) = (1 - \alpha) \times c_0 = (1 - 0.124) \times 0.010 \text{ kmol m}^{-3} = 0.00876 \text{ kmol m}^{-3}$ has to be used:

$$\mu_p = \tfrac{1}{2}\times[-21.4\times10^3\ \text{G} + 0\ \text{G} + 8.314\ \text{G}\,\text{K}^{-1}\times298\ \text{K}\times\ln(0.00876\ \text{kmol}\,\text{m}^{-3}/1\ \text{kmol}\,\text{m}^{-3})]$$

$$= -16600\ \text{G} = \mathbf{-16.6\ kG}\,.$$

2.21.4* Solubility product of lead sulfate

Conductivity σ of the saturated solution of $PbSO_4$:

Also in this case we have:

$$\sigma = \frac{1}{R}\times Z = \frac{1}{10340\ \Omega}\times51.52\ \text{m}^{-1} = 49.83\times10^{-4}\ \text{S}\,\text{m}^{-1}\,.$$

Saturation concentration c_{sd}:

We start from Equation (21.28):

$$\Lambda = \frac{\sigma}{c}\,.$$

Since $PbSO_4$ belongs to the sparingly soluble salts, its saturated solution is so diluted that the molar conductivity Λ of the solution can be equated with the limiting molar conductivity Λ^{0}. The latter, in turn, is the sum of the contributions of cation and anion, $\Lambda^{0} = \Lambda^{0}(Pb^{2+}) + \Lambda^{0}(SO_4^{2-})$ [cf. Eq. (21.35)]. In addition, the conductivity σ of the solution has to be corrected for the conductivity σ_W of the water. Therefore, we obtain:

$$\Lambda^{0}(Pb^{2+}) + \Lambda^{0}(SO_4^{2-}) = \frac{\sigma-\sigma_W}{c_{sd}} \quad\Rightarrow$$

$$c_{sd} = \frac{\sigma-\sigma_W}{\Lambda^{0}(Pb^{2+}) + \Lambda^{0}(SO_4^{2-})}\,.$$

$$c_{sd} = \frac{(49.83\times10^{-4}\ \text{S}\,\text{m}^{-1}) - (1.80\times10^{-4}\ \text{S}\,\text{m}^{-1})}{(142\times10^{-4}\ \text{S}\,\text{m}^2\,\text{mol}^{-1}) + (160\times10^{-4}\ \text{S}\,\text{m}^2\,\text{mol}^{-1})} = 0.1590\ \text{mol}\,\text{m}^{-3}\,.$$

Solubility product $\mathcal{K}_{sd}^{\ominus}$ of $PbSO_4$:

The solubility product $\mathcal{K}_{sd}^{\ominus}$ of $PbSO_4$ is calculated according to Section 6.6:

$$\mathcal{K}_{sd}^{\ominus} = c_r(Pb^{2+})\times c_r(SO_4^{2-}) = (c_{sd}/c^{\ominus})^2\,.$$

$$\mathcal{K}_{sd}^{\ominus} = [(0.1590\times10^{-3}\ \text{kmol}\,\text{m}^{-3})/1\ \text{kmol}\,\text{m}^{-3}]^2 = \mathbf{2.53\times10^{-8}}\,.$$

2.21.5 Limiting molar conductivity of silver bromide

Limiting molar conductivity Λ^{Θ} of AgBr:

The law of independent migration of ions [Eq. (21.31)] can be used to determine the limiting molar conductivity of AgBr. By appropriate combination of the given limiting molar conductivities,

$$\Lambda^{\Theta}(\text{AgBr}) = \Lambda^{\Theta}(\text{Ag}^+)+\Lambda^{\Theta}(\text{Br}^-)+\Lambda^{\Theta}(\text{Na}^+)-\Lambda^{\Theta}(\text{Na}^+)+\Lambda^{\Theta}(\text{NO}_3^-)-\Lambda^{\Theta}(\text{NO}_3^-),$$

the value in question can be calculated:

$$\Lambda^{\Theta}(\text{AgBr}) = \Lambda^{\Theta}(\text{AgNO}_3)+\Lambda^{\Theta}(\text{NaBr})-\Lambda^{\Theta}(\text{NaNO}_3).$$

$$\Lambda^{\Theta}(\text{AgBr}) = [13.33+12.82-12.15]\times 10^{-3}\ \text{S}\,\text{m}^2\,\text{mol}^{-1} = \mathbf{14.00\times 10^{-3}\ S\,m^2\,mol^{-1}}.$$

2.21.6 Experimental determination of the limiting molar conductivity

a) In order to check whether an electrolyte is strong or weak, KOHLRAUSCH's square root law [Eq. (21.38)],

$$\Lambda = \Lambda^{\Theta} - b\sqrt{c},$$

which is valid for sufficiently diluted strong electrolytes, can be used. In the case of a strong electrolyte, one should obtain a linear relation if Λ is plotted as a function of $\sqrt{c}$ at concentrations that are not too high. First of all, the conductivity σ,

$$\sigma = \frac{1}{R}\times Z,$$ and from this the molar conductivity Λ,

$$\Lambda = \frac{\sigma}{c},$$ has to be determined for each concentration.

$\sqrt{c}\,/\,\text{mmol}^{1/2}\,\text{m}^{-3/2}$	7.07	4.47	3.16	2.24	1.00	0.71
$\sigma\,/\,\text{S}\,\text{m}^{-1}$	0.3846	0.1625	0.08376	0.04286	0.00885	0.00446
$\Lambda\,/\,\text{mS}\,\text{m}^2\,\text{mol}^{-1}$	7.69	8.13	8.38	8.57	8.85	8.92

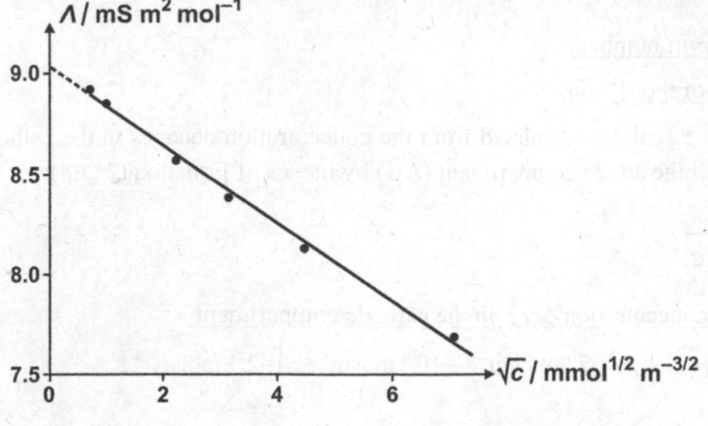

In fact, a linear relationship is obtained, i.e., it is a strong electrolyte.

b) Limiting molar conductivity $\Lambda^{\ominus}$:

The limiting molar conductivity $\Lambda^{\ominus}$ of the electrolyte can be determined from extrapolation to the vertical intercept of the straight line. From the diagram or by linear regression we obtain an intercept of 9.02 mS m^2 mol^{-1}. This corresponds to the limiting molar conductivity $\Lambda^{\ominus}$.

2.21.7 Migration of permanganate ions in the electric field

Limiting molar conductivity $\Lambda^{\ominus}$ of the MnO_4^- ions:

The molar ionic conductivity of any ion is given by

$$\Lambda_i = z_i \times u_i \times \mathcal{F} . \qquad \text{Eq. (21.29)}$$

Taking into account the relationship

$$v = u \times \frac{U}{l}, \qquad \text{Eq. (21.16)}$$

one obtains

$$\Lambda_i = z_i \times v_i \times \frac{l}{U} \times \mathcal{F} .$$

Thus, the molar conductivity Λ_- of the MnO_4^- ions results in:

$$\Lambda_- = (-1) \times (-8.5 \times 10^{-6} \text{ m s}^{-1}) \times \frac{0.15 \text{ m}}{20 \text{ V}} \times 96485 \text{ C mol}^{-1} = 61.5 \times 10^{-4} \text{ S m}^2 \text{ mol}^{-1} .$$

Transport number $t_-^{\ominus}$ of the MnO_4^- ion:

The transport number $t_-^{\ominus}$ of the MnO_4^- ion is given in analogy to Equation (21.55) by:

$$t_-^{\ominus} = \frac{\nu_- \Lambda_-^{\ominus}}{\Lambda^{\ominus}} = \frac{\nu_- \Lambda_-^{\ominus}}{\Lambda_+^{\ominus} + \Lambda_-^{\ominus}} = \frac{1 \times (61.5 \times 10^{-4} \text{ S m}^2 \text{ mol}^{-1})}{(73.5 + 61.5) \times 10^{-4} \text{ S m}^2 \text{ mol}^{-1}} = \mathbf{0.46} .$$

2.21.8 HITTORF transport numbers

a) Transport number t_+ of the H^+ ion:

The transport number t_+ can be calculated from the concentration changes in the cathode compartment (CC) and the anode compartment (AC) by means of Equation (21.68):

$$t_+ = \frac{\Delta c_{AC}}{\Delta c_{CC} + \Delta c_{AC}} .$$

Change in concentration Δc_{CC} in the cathode compartment:

$$\Delta c_{CC} = c_{CC} - c_0 = 97.9 \text{ mol m}^{-3} - 100 \text{ mol m}^{-3} = -2.1 \text{ mol m}^{-3} .$$

Change in concentration Δc_{AC} in the anode compartment:

$$\Delta c_{AC} = c_{AC} - c_0 = 90.4\ \text{mol m}^{-3} - 100\ \text{mol m}^{-3} = -9.6\ \text{mol m}^{-3}.$$

$$t_+ = \frac{-9.6\ \text{mol m}^{-3}}{(-2.1\ \text{mol m}^{-3}) + (-9.6\ \text{mol m}^{-3})} = \mathbf{0.821}.$$

Transport number t_- of the Cl^- ion:

The transport number t_- can be obtained analogously [Eq. (21.67)]:

$$t_- = \frac{\Delta c_{CC}}{\Delta c_{CC} + \Delta c_{AC}} = \frac{-2.1\ \text{mol m}^{-3}}{(-2.1\ \text{mol m}^{-3}) + (-9.6\ \text{mol m}^{-3})} = \mathbf{0.179}.$$

b) Ionic conductivity $\Lambda_+^{\emptyset}$ of the H^+ ion at infinite dilution:

The ionic conductivity $\Lambda_+^{\emptyset}$ of the cation can be determined from the transport number t_+ using Equation (21.54) or (21.55), respectively. For simplicity's sake, we assume that the transport number t_+ determined in part a) corresponds approximately to $t_+^{\emptyset}$.

$$t_+^{\emptyset} = \frac{\nu_+ \Lambda_+^{\emptyset}}{\Lambda^{\emptyset}} \quad \Rightarrow$$

$$\Lambda_+^{\emptyset} = \frac{t_+^{\emptyset} \times \Lambda^{\emptyset}}{\nu_+} = \frac{0.821 \times (426.0 \times 10^{-4}\ \text{S m}^2\ \text{mol}^{-1})}{1} = \mathbf{349.7 \times 10^{-4}\ S\ m^{-2}\ mol^{-1}}.$$

Ionic conductivity $\Lambda_-^{\emptyset}$ of the Cl^- ion at infinite dilution:

The ionic conductivity $\Lambda_-^{\emptyset}$ of the anion can be determined analogously:

$$\Lambda_-^{\emptyset} = \frac{t_-^{\emptyset} \times \Lambda^{\emptyset}}{\nu_-} = \frac{0.179 \times (426.0 \times 10^{-4}\ \text{S m}^{-2}\ \text{mol}^{-1})}{1} = \mathbf{76.3 \times 10^{-4}\ S\ m^{-2}\ mol^{-1}}.$$

2.21.9* HITTORF transport numbers for advanced students

Total charge Q being transported:

When a current I is allowed to flow through a cell for a period of time Δt, the following charge is transported in total:

$$Q = I \times \Delta t = (80 \times 10^{-3}\ \text{A}) \times 3600\ \text{s} = 288\ \text{C}.$$

Mass $m_{CC,0}$ of KOH in the cathode compartment before the electrolysis:

The mass $m_{CC,0}$ of KOH can be easily calculated from the corresponding mass fraction w [see Eq. (1.8)]:

$$w_0(\text{KOH}) = \frac{m_{CC,0}(\text{KOH})}{m_{\text{total}}} \quad \Rightarrow$$

$$m_{CC,0}(KOH) = w_0(KOH) \times m_{total} = 0.002 \times (100.00 \times 10^{-3}\ kg) = 0.200 \times 10^{-3}\ kg\ .$$

Mass m_{CC} of KOH in the cathode compartment after the electrolysis:

If one converts the analytical value after the electrolysis from a mass of 25.00×10^{-3} kg of the solution to a mass of 100.00×10^{-3} kg, one obtains

$$m_{CC}(KOH) = (0.0615 \times 10^{-3}\ kg) \times 4 = 0.246 \times 10^{-3}\ kg\ .$$

Change Δm_{CC} in mass of KOH in the cathode compartment:

The change Δm_{CC} of KOH in the cathode compartment results in:

$$\Delta m_{CC}(KOH) = m_{CC}(KOH) - m_{CC,0}(KOH)\ .$$

$$\Delta m_{CC}(KOH) = (0.246 \times 10^{-3}\ kg) - (0.200 \times 10^{-3}\ kg) = 0.046 \times 10^{-3}\ kg\ .$$

Change Δn_{CC}^{+} in amount of the K^+ ions in the cathode compartment:

The change Δn_{CC}^{+} in amount of the K^+ ions in the cathode compartment corresponds to the change Δn_{CC} of amount of KOH:

$$\Delta n_{CC}^{+} = \Delta n_{CC}(KOH) = \frac{\Delta m_{CC}(KOH)}{M(KOH)} = \frac{0.046 \times 10^{-3}\ kg}{56.1 \times 10^{-3}\ kg\,mol^{-1}} = 8.2 \times 10^{-4}\ mol\ .$$

Fraction Q_+ of the charge transported by the K^+ ions:

Since the K^+ ions are not discharged at the cathode, only the increase of their amount in the cathode compartment by migration from the anode compartment must be taken into account, i.e. we obtain:

$$\Delta n_{CC}^{+} = \frac{Q_+}{z_+ \mathcal{F}} \quad \Rightarrow$$

$$Q_+ = z_+ \mathcal{F} \times \Delta n_{CC}^{+} = (+1) \times 96495\ C\,mol^{-1} \times (8.2 \times 10^{-4}\ mol) = 79.1\ C\ .$$

Transport number t_+ of the K^+ ion:

The transport number t_+ of the K^+ ion describes the fraction of the total current I and therefore also the total charge Q, which is transported by this ionic species:

$$t_+ = \frac{I_+}{I} = \frac{Q_+}{Q} = \frac{79.1\ C}{288\ C} = \mathbf{0.275}\ .$$

Transport number t_- of the OH^- ion:

The transport number t_- of the OH^- ion can be calculated using Equation (21.52):

$$t_+ + t_- = 1 \quad \Rightarrow$$

$$t_- = 1 - t_+ = 1 - 0.275 = \mathbf{0.725}\ .$$

2.22 Electrode Reactions and Galvani Potential Differences

2.22.1 Electron potential

Conversion formula: $NO|g + 2\,H_2O|l \rightarrow NO_3^-|w + 4\,H^+|w + 3\,e^-$

$\mu^\ominus$/kG: +87.6 $2\times(-237.1)$ -108.7 4×0

Electron potential $\mu_e^\ominus$ of the redox pair:

The standard value of the electron potential of a redox pair Rd/Ox results based on Equation (22.18) in:

$$\mu_e^\ominus(\text{Rd/Ox}) := \frac{1}{\nu_e}\left[\mu_{\text{Rd}}^\ominus - \mu_{\text{Ox}}^\ominus\right].$$

In the present case, both the reducing agent and the oxidizing agent represent a combination of substances, i.e. we have:

$$\mu_{\text{Rd}}^\ominus = \nu_{\text{Rd}'}\mu_{\text{Rd}'}^\ominus + \nu_{\text{Rd}''}\mu_{\text{Rd}''}^\ominus = \{[1\times87.6]+[2\times(-237.1)]\}\ \text{kG} = \mathbf{-386.6\ kG}.$$

$$\mu_{\text{Ox}}^\ominus = \nu_{\text{Ox}'}\mu_{\text{Ox}'}^\ominus + \nu_{\text{Ox}''}\mu_{\text{Ox}''}^\ominus = \{[1\times(-108.7)+[4\times0]\}\ \text{kG} = \mathbf{-108.7\ kG}.$$

Thus, we obtain for the standard value of the electron potential

$$\mu_e^\ominus(\text{Rd/Ox}) = \frac{1}{3}\left[(-386.6\ \text{kG})-(-108.7\ \text{kG})\right] = \mathbf{-92.6\ kG}.$$

2.22.2 Galvani potential difference and difference of chemical potentials in electrochemical equilibrium

Difference of the chemical potentials of the zinc ions in metal and in solution:

The Galvani potential difference $U_{\text{S}\rightarrow\text{Me}}$ is generally calculated according to Equation (22.12):

$$U_{\text{S}\rightarrow\text{Me}} = -\frac{\mathcal{A}_\text{J}(\text{J|d} \rightarrow \text{J|m})}{z_\text{J}\mathcal{F}}.$$

Correspondingly, we can formulate for our example:

$$U_{\text{S}\rightarrow\text{Me}} = -\frac{\mathcal{A}(\text{Zn}^{2+}|\text{w} \rightarrow \text{Zn}^{2+}|\text{m})}{2\mathcal{F}},$$

where

$$\mathcal{A} = \mu(\text{Zn}^{2+}|\text{w}) - \mu(\text{Zn}^{2+}|\text{m}).$$

Solving for the difference in question results in:

$$\mu(\text{Zn}^{2+}|\text{m}) - \mu(\text{Zn}^{2+}|\text{w}) = +2\mathcal{F}\times U_{\text{S}\rightarrow\text{Me}}.$$

$$\mu(\text{Zn}^{2+}|\text{m}) - \mu(\text{Zn}^{2+}|\text{w}) = 2\times96485\ \text{C mol}^{-1}\times(0.3\ \text{V}) = 58000\ \text{G} = \mathbf{58\ kG}.$$

Conversion of units:

$\mathrm{C\,mol^{-1}\,V = A\,s\,mol^{-1}\,V = J\,mol^{-1} = G}$.

Due to the formula $W_{el} = U \times I \times \Delta t$ for the electric energy, we have 1 J = 1 V A s, as already discussed in solution 2.21.2.

2.22.3 Galvani potential difference of redox electrodes

Conversion formula: $2\,Cr^{3+}|w + 7\,H_2O|l \rightarrow Cr_2O_7^{2-}\,|w + 14\,H^+|w + 6\,e^-$

NERNST's equation:

Since we have a composite redox pair Rd → Ox + ν_ee, in which Rd stands for the combination of substances $\nu_{Rd'}$Rd′+ $\nu_{Rd''}$Rd″ + ... and Ox for the combination $\nu_{Ox'}$Ox′ + $\nu_{Ox''}$Ox″ + ... , we have to use the generalized form of NERNST's equation [Eq. (22.24)]:

$$\Delta\varphi = \Delta\overset{\circ}{\varphi} + \frac{RT}{\nu_e \mathcal{F}} \ln \frac{c_r(\mathrm{Ox'})^{\nu_{Ox'}} \times c_r(\mathrm{Ox''})^{\nu_{Ox''}} \times ...}{c_r(\mathrm{Rd'})^{\nu_{Rd'}} \times c_r(\mathrm{Rd''})^{\nu_{Rd''}} \times ...}.$$

In the case of our example, we obtain:

$$\Delta\varphi = \Delta\overset{\circ}{\varphi}(\mathrm{Cr^{3+}/Cr_2O_7^{2-}}) + \frac{RT}{6\mathcal{F}} \ln \frac{c_r(\mathrm{Cr_2O_7^{2-}})}{c_r(\mathrm{Cr^{3+}})^2 \times c_r(\mathrm{H^+})^{14}}.$$

The liquid water as a pure liquid does not appear in the argument of the logarithm.

2.22.4 Concentration dependence of the Galvani potential difference of metal-metal ion electrodes

Change $\Delta(\Delta\varphi)$ of the Galvani potential difference with a change of concentration:

The concentration dependence of the Galvani potential difference of a metal-metal ion electrode, as represented by the Cu/Cu^{2+} electrode, is described by Equation (22.16):

$$\Delta\varphi = \Delta\overset{\circ}{\varphi} + \frac{RT}{z_J \mathcal{F}} \ln \frac{c_J(\mathrm{S})}{c^\ominus}.$$

If the ion concentration is increased by a factor of 5, we obtain for our example the following change of the Galvani potential difference:

$$\Delta(\Delta\varphi) = \left[\Delta\overset{\circ}{\varphi}(\mathrm{Cu/Cu^{2+}}) + \frac{RT}{2\mathcal{F}} \ln \frac{5c(\mathrm{Cu^{2+}})}{c^\ominus}\right] - \left[\Delta\overset{\circ}{\varphi}(\mathrm{Cu/Cu^{2+}}) + \frac{RT}{2\mathcal{F}} \ln \frac{c(\mathrm{Cu^{2+}})}{c^\ominus}\right]$$

$$= \frac{RT}{2\mathcal{F}} \ln \frac{5c(\mathrm{Cu^{2+}}) \times c^\ominus}{c^\ominus \times c(\mathrm{Cu^{2+}})} = \frac{RT}{2\mathcal{F}} \ln 5 = \frac{8.314\,\mathrm{G\,K^{-1}} \times 298\,\mathrm{K}}{2 \times 96485\,\mathrm{C\,mol^{-1}}} \ln 5 = \mathbf{0.0207\,V}.$$

Conversion of units:

$$\frac{\text{G K}^{-1}\,\text{K}}{\text{C mol}^{-1}} = \frac{\text{J mol}^{-1}}{\text{A s mol}^{-1}} = \frac{\text{V A s}}{\text{A s}} = \text{V}\,.$$

2.22.5 Diffusion voltage

a) Diffusion voltage U_{diff} in the presence of NaCl solutions:

The diffusion voltage U_{Diff} in the steady state for a 1-1 electrolyte (such as NaCl) results according to Equation (22.30) in

$$U_{\text{diff}} = \frac{t_+ - t_-}{\mathcal{F}} RT \ln \frac{c(\text{II})}{c(\text{I})}\,.$$

Transport number t_+:

The following relationship [Eq. (21.53)] exists between the transport number t_+ and the electric mobilities u_+ and u_-:

$$t_+ = \frac{u_+}{u_+ - u_-} = \frac{5.2\times10^{-8}\ \text{m}^2\,\text{V}^{-1}\,\text{s}^{-1}}{[5.2-(-7.9)]\times10^{-8}\ \text{m}^2\,\text{V}^{-1}\,\text{s}^{-1}} = 0.40\,.$$

Transport number t_-:

From Equation (21.52) immediately follows:

$$t_- = 1 - t_+ = 1 - 0.40 = 0.60\,.$$

$$U_{\text{Diff}} = \frac{0.40-0.60}{96485\ \text{C mol}^{-1}} \times 8.314\ \text{G K}^{-1} \times 298\ \text{K} \times \ln\frac{100\ \text{mol m}^{-3}}{300\ \text{mol m}^{-3}}$$

$$= 0.0056\ \text{V} = \mathbf{5.6\ mV}.$$

b) Diffusion voltage U_{diff} in the presence of KCl solutions:

In this case, one obtains for the diffusion voltage U_{diff} with

$$t_+ = \frac{7.6\times10^{-8}\ \text{m}^2\,\text{V}^{-1}\,\text{s}^{-1}}{[7.6-(-7.9)]\times10^{-8}\ \text{m}^2\,\text{V}^{-1}\,\text{s}^{-1}} = 0.49 \quad \text{and} \quad t_- = 1 - 0.49 = 0.51$$

the following value:

$$U_{\text{diff}} = \frac{0.49-0.51}{96485\ \text{C mol}^{-1}} \times 8.314\ \text{G K}^{-1} \times 298\ \text{K} \times \ln\frac{100\ \text{mol m}^{-3}}{300\ \text{mol m}^{-3}}$$

$$= 0.00056\ \text{V} = \mathbf{0.56\ mV}.$$

By changing the cation, the diffusion voltage has decreased to about one tenth of the original value.

2.22.6 Membrane voltage

a) Because of the electroneutrality rule we have:
External solution (I): $c_{K^+}(I) = c_{Cl^-}(I)$,
Internal solution (II): $5c_{Prot^{5+}}(II) + c_{K^+}(II) = c_{Cl^-}(II)$.

b) Internal Cl^- concentration $c_{Cl^-}(II)$:

After establishing of the equilibrium, the DONNAN equation [Eq. (22.35)] is applicable:

$$c_{K^+}(I) \times c_{Cl^-}(I) = c_{K^+}(II) \times c_{Cl^-}(II).$$

Solving for $c_{Cl^-}(II)$ results in:

$$c_{Cl^-}(II) = \frac{c_{K^+}(I)}{c_{K^+}(II)} \times c_{Cl^-}(I) = 1.08 \times 150\ \text{mol}\,\text{m}^{-3} = \mathbf{162\ mol\,m^{-3}}.$$

c) Membrane voltage U_{mem}:

The membrane voltage (DONNAN voltage) U_{mem} is given by Equation (22.32):

$$U_{mem} = \frac{RT}{\mathcal{F}} \ln\left(\frac{c_{K^+}(II)}{c_{K^+}(I)}\right) = \frac{8.314\ \text{G}\,\text{K}^{-1} \times 298\ \text{K}}{96485\ \text{C}\,\text{mol}^{-1}} \ln\frac{1}{1.08} = -0.0020\ \text{V} = \mathbf{-2.0\ mV}.$$

1.22.7* Membrane voltage for advanced students

a) Final concentrations of the ions on both sides of the membrane:

If one refers to the initial concentration of the protein with a charge z on one side of the membrane as $c_{0,II}$, then the concentration of the associated ion (in our example Na^+) is $z \times c_{0,II}$. On the other side of the membrane there is a NaCl solution with a concentration $c_{0,I}$. As a result, Na^+ and Cl^- ions begin to flow into the protein solution. Until equilibrium is established, the concentration of the external NaCl solution is reduced by c_x, i.e. its final concentration is $c_{0,I} - c_x$. The concentration of Na^+ ions in the internal solution has increased to $z \times c_{0,II} + c_x$, that of the (previously not present) Cl^- ions to c_x. Inserting these expressions into the DONNAN equation [Eq. (22.35)] (and assuming equal volumes on both sides of the membrane) yields:

$$(c_{0,I} - c_x) \times (c_{0,I} - c_x) = (z \times c_{0,II} + c_x) \times c_x, \text{ i.e.,}$$

$$c_{0,I}^2 - 2c_{0,I}c_x + c_x^2 = zc_{0,II}c_x + c_x^2.$$

Solving for c_x results via the intermediate step

$$(zc_{0,II} + 2c_{0,I})c_x = c_{0,I}^2 \quad \text{finally in}$$

$$c_x = \frac{c_{0,I}^2}{zc_{0,II} + 2c_{0,I}}.$$

If we now insert the given values, we obtain

$$c_x = \frac{(50\ \text{mol m}^{-3})^2}{(+3)\times 3\ \text{mol m}^{-3} + 2\times 50\ \text{mol m}^{-3}} = 22.9\ \text{mol m}^{-3}$$

and thus for the concentrations in the external solution:

$$c_{\text{Na}^+}(\text{I}) = c_{\text{Cl}^-}(\text{I}) = c_{0,\text{I}} - c_x = 50\ \text{mol m}^{-3} - 22.9\ \text{mol m}^{-3} = \mathbf{27.1\ mol\ m^{-3}}.$$

The concentrations in the internal solution, on the other hand, result in:

$$c_{\text{Prot}^{z-}}(\text{II}) = c_{0,\text{II}} = \mathbf{3\ mol\ m^{-3}},$$

$$c_{\text{Na}^+}(\text{II}) = z\times c_{0,\text{II}} + c_x = (+3)\times 3\ \text{mol m}^{-3} + 22.9\ \text{mol m}^{-3} = \mathbf{31.9\ mol\ m^{-3}},$$

$$c_{\text{Cl}^-}(\text{II}) = c_x = \mathbf{22.9\ mol\ m^{-3}}.$$

b) Membrane voltage U_{mem}:
The membrane voltage U_{mem} results according to Equation (22.32) in:

$$U_{\text{mem}} = \frac{RT}{\mathcal{F}}\ln\left(\frac{c_{\text{Na}^+}(\text{II})}{c_{\text{Na}^+}(\text{I})}\right).$$

$$U_{\text{mem}} = \frac{8.314\ \text{G K}^{-1}\times 298\ \text{K}}{96485\ \text{C mol}^{-1}}\ln\frac{31.9\ \text{mol m}^{-3}}{27.1\ \text{mol m}^{-3}} = 0.0042\ \text{V} = \mathbf{4.2\ mV}.$$

2.23 Redox Potentials and Galvanic Cells

2.23.1 NERNST's equation of a half-cell

a) Conversion formula: $\mathrm{Ag|s + Cl^-|w \rightarrow AgCl|s + e^-}$

$\mu^\ominus/\mathrm{kG}$: 0 −131.2 −109.8

Redox potential $E^\ominus$ of the half-cell in question:

For a composite redox pair Rd → Ox + ν_ee, in which Rd represents the combination of substances $\nu_{\mathrm{Rd'}}\mathrm{Rd'} + \nu_{\mathrm{Rd''}}\mathrm{Rd''} + ...$ and Ox the combination $\nu_{\mathrm{Ox'}}\mathrm{Ox'} + \nu_{\mathrm{Ox''}}\mathrm{Ox''} + ...$, the standard value $E^\ominus$ of the redox potential can be calculated based on Equation (23.6):

$$E^\ominus(\mathrm{Rd/Ox}) = -\frac{\left[\nu_{\mathrm{Rd'}}\mu^\ominus(\mathrm{Rd'}) + \nu_{\mathrm{Rd''}}\mu^\ominus(\mathrm{Rd''}) + ... - \nu_{\mathrm{Ox'}}\mu^\ominus(\mathrm{Ox'}) - ...\right] - \nu_e\mu_e^\ominus(\mathrm{H_2/H^+})}{\nu_e F}.$$

[Unfortunately, an error had crept into Equation (23.6) in the textbook "Physical Chemistry from a Different Angle" and the factor ν_e in the numerator had been forgotten].

The term $\mu_e^\ominus(\mathrm{H_2/H^+})$ is equal to zero and we therefore obtain:

$$E^\ominus(\mathrm{Rd/Ox}) = -\frac{\nu_{\mathrm{Rd'}}\mu^\ominus(\mathrm{Rd'}) + \nu_{\mathrm{Rd''}}\mu^\ominus(\mathrm{Rd''}) + ... - \nu_{\mathrm{Ox'}}\mu^\ominus(\mathrm{Ox'}) - \nu_{\mathrm{Ox'}}\mu^\ominus(\mathrm{Ox''}) - ...}{\nu_e F}.$$

Inserting the values of the concrete example results in:

$$E^\ominus(\mathrm{Ag{+}Cl^-/Ag}) = -\frac{\{[0+(-131.2)]-[-109.8]\}\times 10^3\ \mathrm{G}}{1\times 96485\ \mathrm{C\,mol^{-1}}} = \mathbf{+0.222\ V}.$$

NERNST's equation of the half-cell in question:

For a composite redox pair, NERNST's equation can be formulated as follows:

$$E = \overset{\circ}{E}(\mathrm{Rd/Ox}) + \frac{E_{\mathrm{N}}(T)}{\nu_e}\times\lg\frac{c_r(\mathrm{Ox'})^{\nu_{\mathrm{Ox'}}}\times c_r(\mathrm{Ox''})^{\nu_{\mathrm{Ox''}}}\times ...}{c_r(\mathrm{Rd'})^{\nu_{\mathrm{Rd'}}}\times c_r(\mathrm{Rd''})^{\nu_{\mathrm{Rd''}}}\times ...},$$

where c_r (= $c/c^\ominus$) represents the relative concentration.

At the standard temperature $T^\ominus = 298$ K, $E_{\mathrm{N}}^\ominus$ is 0.059 V and we can write:

$$E = E^\ominus(\mathrm{Rd/Ox}) + \frac{0.059\ \mathrm{V}}{\nu_e}\times\lg\frac{c_r(\mathrm{Ox'})^{\nu_{\mathrm{Ox'}}}\times c_r(\mathrm{Ox''})^{\nu_{\mathrm{Ox''}}}\times ...}{c_r(\mathrm{Rd'})^{\nu_{\mathrm{Rd'}}}\times c_r(\mathrm{Rd''})^{\nu_{\mathrm{Rd''}}}\times ...}.$$

Thus, we obtain for the silver-silver chloride electrode

$$E = +0.222\ \mathrm{V} + \frac{0.059\ \mathrm{V}}{1}\times\lg\frac{1}{c_r(\mathrm{Cl^-})},$$

because the mass action term for pure solid substances such as Ag and AgCl is omitted.

b) Conversion formula: $Mn^{2+}|w + 4\ H_2O|l \rightarrow MnO_4^-|w + 8\ H^+|w + 5\ e^-$

$\mu^\ominus$/kG: −228.1 4×(−237.1) −447.2 8×0

Redox potential $E^\ominus$ of the half-cell in question:

$$E^\ominus(Mn^{2+}/MnO_4^-) = -\frac{\{[(-228.1)+4\times(-237.1)]-[(-447.2)+8\times 0]\}\times 10^3\ G}{5\times 96485\ C\,mol^{-1}}$$

$$= \mathbf{+1.512\ V}\,.$$

NERNST's equation of the half-cell in question:

$$E = +1.512\ V + \frac{0.059\ V}{5}\times \lg\frac{c_r(MnO_4^-)\times c_r(H^+)^8}{c_r(Mn^{2+})}\,.$$

The liquid water as pure liquid does not appear in the argument of the logarithm.

c) Conversion formula: $HCOOH|w \rightarrow CO_2|g + 2\ H^+|w + 2\ e^-$

$\mu^\ominus$/kG: −372.4 −394.4 2×0

Redox potential $E^\ominus$ of the half-cell in question:

$$E^\ominus(HCOOH/CO_2) = -\frac{\{[-372.4]-[(-394.4)+2\times 0]\}\times 10^3\ G}{2\times 96485\ C\,mol^{-1}} = \mathbf{-0.114\ V}\,.$$

NERNST's equation of the half-cell in question:

$$E = -0.114\ V + \frac{0.059\ V}{2}\times \lg\frac{p_r(CO_2)\times c_r(H^+)^2}{c_r(HCOOH)}\,.$$

In the case of carbon dioxide gas, its relative pressure $p_r\ (= p/p^\ominus)$ is used.

d) Conversion formula: $HCOO^-|w + 3\ OH^-|w \rightarrow CO_3^{2-}|w + 2\ H_2O|l + 2\ e^-$

$\mu^\ominus$/kG: −351.0 3×(−157.2) −527,8 2×(−237.1)

Redox potential $E^\ominus$ of the half-cell in question:

$$E^\ominus(HCOO^-/CO_3^{2-}) = -\frac{\{[(-351.0)+3\times(-157.2)]-[(-527.8)+2\times(-237.1)]\}\times 10^3\ G}{2\times 96485\ C\,mol^{-1}}$$

$$= \mathbf{-0.930\ V}\,.$$

NERNST's equation of the half-cell in question:

$$E = -0.930\ V + \frac{0.059\ V}{2}\times \lg\frac{c_r(CO_3^{2-})}{c_r(HCOO^-)\times c_r(OH^-)^3}\,.$$

2.23.2 Concentration dependence of redox potentials

a) Conversion formula: $Fe^{2+}|w \rightarrow Fe^{3+}|w + e^-$

Redox potential E of the half-cell in question:

The redox potential E at standard temperature $T^\ominus$ = 298 K can be calculated based on Equation (23.8):

$$E = E^\ominus(\text{Rd/Ox}) + \frac{0.059\ \text{V}}{\nu_e} \times \lg \frac{c(\text{Ox})}{c(\text{Rd})}.$$

For the redox pair of bivalent and trivalent iron ions from our example, we thus obtain:

$$E = E^\ominus(\text{Fe}^{2+}/\text{Fe}^{3+}) + \frac{0.059\ \text{V}}{1} \times \lg \frac{c(\text{Fe}^{3+})}{c(\text{Fe}^{2+})}.$$

The standard value of the redox potential can be found in Table 23.1 in the textbook "Physical Chemistry from a Different Angle": $E^\ominus(Fe^{2+}/Fe^{3+}) = +0.771$ V.

$$E = +0.771\ \text{V} + 0.059\ \text{V} \times \lg \frac{0.010\ \text{kmol m}^{-3}}{0.005\ \text{kmol m}^{-3}} = \mathbf{+0.789\ V}.$$

b) Conversion formula: $Cl^-|w + 4\ H_2O|l \rightarrow ClO_4^-|w + 8\ H^+|w + 8\ e^-$

Redox potential E of the half-cell in question:

For the calculation of the redox potential, NERNST'S equation for composite redox pairs already used in Solution 2.22.1 can be applied:

$$E = E^\ominus(\text{Rd/Ox}) + \frac{0.059\ \text{V}}{\nu_e} \times \lg \frac{c_r(\text{Ox}')^{\nu_{\text{Ox}'}} \times c_r(\text{Ox}'')^{\nu_{\text{Ox}''}} \times ...}{c_r(\text{Rd}')^{\nu_{\text{Rd}'}} \times c_r(\text{Rd}'')^{\nu_{\text{Rd}''}} \times ...}.$$

For the example given, we obtain:

$$E = E^\ominus(\text{Cl}^-/\text{ClO}_4^-) + \frac{0.059\ \text{V}}{8} \times \lg \frac{[c(\text{ClO}_4^-)/c^\ominus] \times [c(\text{H}^+)/c^\ominus]^8}{[c(\text{Cl}^-)/c^\ominus]}.$$

Hydrogen ion concentration $c(H^+)$ in the solution:

The hydrogen ion concentration can be determined according to the following relationship (see Section 7.3 in the textbook "Physical Chemistry from a Different Angle"):

$$c(\text{H}^+)/c^\ominus = 10^{-\text{pH}} \Rightarrow$$

$$c(\text{H}^+) = 10^{-\text{pH}} \times c^\ominus = 10^{-3.0} \times 1\ \text{kmol m}^{-3} = 0.001\ \text{kmol m}^{-3}.$$

$$E = +1.389\ \text{V} + \frac{0.059\ \text{V}}{8} \times \lg \frac{[0.020\ \text{kmol m}^{-3}/1\ \text{kmol m}^{-3}] \times [0.001\ \text{kmol m}^{-3}/1\ \text{kmol m}^{-3}]^8}{[0.005\ \text{kmol m}^{-3}/1\ \text{kmol m}^{-3}]}$$

$$= \mathbf{+1.216\ V}.$$

c) Conversion formula: $ClO_3^-|w + 2\ OH^-|w \rightarrow ClO_4^-|w + H_2O|l + 2\ e^-$

Redox potential E of the half-cell in question:

$$E = E^{\ominus}(ClO_3^-/ClO_4^-) + \frac{0.059\ \text{V}}{2} \times \lg \frac{[c(ClO_4^-)/c^{\ominus}]}{[c(ClO_3^-)/c^{\ominus}] \times [c(OH^-)/c^{\ominus}]^2}.$$

$$E = +0.36\ \text{V} + \frac{0.059\ \text{V}}{2} \times \lg \frac{[0.020\ \text{kmol m}^{-3}/1\ \text{kmol m}^{-3}]}{[0.010\ \text{kmol m}^{-3}/1\ \text{kmol m}^{-3}] \times [0.010\ \text{kmol m}^{-3}/1\ \text{kmol m}^{-3}]^2}$$

$$= \mathbf{+0.49\ V}.$$

2.23.3 Redox potential of gas electrodes

Conversion formula: $4\ OH^-|w \rightarrow O_2|g + 2\ H_2O|l + 4\ e^-$

Redox potential E of the oxygen electrode:

The redox potential can be calculated according to

$$E = E^{\ominus}(OH^-/O_2) + \frac{0.059\ \text{V}}{4} \times \lg \frac{[p(O_2)/p^{\ominus}]}{[c(OH^-)/c^{\ominus}]^4}.$$

Hydrogen ion concentration $c(H^+)$ in the solution:

The hydrogen ion concentration results in [see Solution 2.23.2 b)]:

$$c(H^+) = 10^{-pH} \times c^{\ominus} = 10^{-9.0} \times 1\ \text{kmol m}^{-3} = 1.0 \times 10^{-9}\ \text{kmol m}^{-3}.$$

Hydroxide ion concentration $c(OH^-)$ in the solution:

$$K_w = c(H^+) \times c(OH^-) \quad \Rightarrow$$

$$c(OH^-) = \frac{K_w}{c(H^+)} = \frac{1.0 \times 10^{-14}\ \text{kmol}^2\ \text{m}^{-6}}{1.0 \times 10^{-9}\ \text{kmol m}^{-3}} = 1.0 \times 10^{-5}\ \text{kmol m}^{-3}.$$

$$E = +0.401\ \text{V} + \frac{0.059\ \text{V}}{4} \times \lg \frac{[25\ \text{kPa}/100\ \text{kPa})]}{[1.0 \times 10^{-5}\ \text{kmol m}^{-3}/1\ \text{kmol m}^{-3}]^4}$$

$$= \mathbf{+0.687\ V}.$$

2.23.4 Calculation of redox potentials from redox potentials

Conversion formulas: (1) Cr $\rightarrow Cr^{2+} + 2\,e^-$

(2) Cr $\rightarrow Cr^{3+} + 3\,e^-$

(3) $Cr^{2+} \rightarrow Cr^{3+} + e^-$

Redox potential $E^\ominus$ of the half-cell in question:

For a simple redox pair Rd → Ox + ν_ee, the standard value $E^\ominus$ of the redox potential can be calculated based on Equation (23.6):

$$E^\ominus = -\frac{\mu^\ominus(\mathrm{Rd}) - \mu^\ominus(\mathrm{Ox})}{\nu_e \mathcal{F}}.$$

For the drive $\mathcal{A}^\ominus$ of the half-cell reaction, one obtains:

$$\mathcal{A}^\ominus = \mu^\ominus(\mathrm{Rd}) - \mu^\ominus(\mathrm{Ox}) - \nu_e \mu^\ominus(\mathrm{e}^-).$$

The term $\mu^\ominus(\mathrm{Rd}) - \mu^\ominus(\mathrm{Ox})$ can be replaced by the drive $\mathcal{A}^\ominus$, since $\mu^\ominus(\mathrm{e}^-) = 0$, i.e. we have

$$E^\ominus = -\frac{\mathcal{A}^\ominus}{\nu_e \mathcal{F}}.$$

As we know from Chapter 4 in the textbook "Physical Chemistry from a Different Angle," the drives of reactions can be added or subtracted. However, this generally does not apply to the redox potentials, since different factors ν_e can occur. Thus, we first have to calculate the drives of the half-reactions from the redox potentials.

$$\mathcal{A}_1^\ominus = -\nu_{e,1}\mathcal{F} \times E_1^\ominus.$$

$$\mathcal{A}_2^\ominus = -\nu_{e,2}\mathcal{F} \times E_2^\ominus.$$

Due to (3) = – (1) + (2), the drive $\mathcal{A}_3^\ominus$ results in:

$$\mathcal{A}_3^\ominus = -\mathcal{A}_1^\ominus + \mathcal{A}_2^\ominus,$$

i.e., we have

$$-\nu_{e,3}\mathcal{F} \times E_3^\ominus = \nu_{e,1}\mathcal{F} \times E_1^\ominus - \nu_{e,2}\mathcal{F} \times E_2^\ominus.$$

Solving for the redox potential $E_3^\ominus$ results in:

$$E_3^\ominus = \frac{\nu_{e,2}E_2^\ominus - \nu_{e,1}E_1^\ominus}{\nu_{e,3}} = \frac{3\times(-0.744\ \mathrm{V}) - 2\times(-0.913\ \mathrm{V})}{1} = \mathbf{-0.406\ V}.$$

2.23.5 Solubility product of silver iodide

Solubility product $\mathcal{K}_{sd}^\ominus$ of silver iodide:

The NERNST equation for the silver-silver ion electrode is:

$$E_1 \quad = E^{\ominus}(\text{Ag/Ag}^+) + \frac{RT^{\ominus}}{\mathcal{F}} \times \ln c_r(\text{Ag}^+).$$

The Ag^+ ion concentration in the present case is to be determined by the solubility product of the sparingly soluble salt AgI:

$$\mathcal{K}_{sd}^{\ominus} \quad = c_r(\text{Ag}^+) \times c_r(\text{I}^-) \quad \Rightarrow$$

$$c_r(\text{Ag}^+) = \frac{\mathcal{K}_{sd}^{\ominus}}{c_r(\text{I}^-)}.$$

Inserting this expression into the equation above results in:

$$E_1 \quad = E^{\ominus}(\text{Ag/Ag}^+) + \frac{RT^{\ominus}}{\mathcal{F}} \times \ln \frac{\mathcal{K}_{sd}^{\ominus}}{c_r(\text{I}^-)}.$$

For the NERNST equation of the silver-silver iodide electrode one obtains [cf. Eq. (22.29)]:

$$E_2 \quad = E^{\ominus}(\text{Ag} + \text{I}^-/\text{AgI}) + \frac{RT^{\ominus}}{\mathcal{F}} \times \ln \frac{1}{c_r(\text{I}^-)}.$$

Comparing this relationship with the equation above, and considering that E_1 and E_2 describe the same experimental situations and have therefore to be equal, it follows that

$$E^{\ominus}(\text{Ag} + \text{I}^-/\text{AgI}) \quad = E^{\ominus}(\text{Ag/Ag}^+) + \frac{RT^{\ominus}}{\mathcal{F}} \times \ln \mathcal{K}_{sd}^{\ominus}.$$

$c_r(I^-)$ cancelled out on both sides. Solving for $\mathcal{K}_{sd}^{\ominus}$ yields:

$$\mathcal{K}_{sd}^{\ominus} \quad = \exp \frac{[E^{\ominus}(\text{Ag} + \text{I}^-/\text{AgI}) - E^{\ominus}(\text{Ag/Ag}^+)] \times \mathcal{F}}{RT^{\ominus}}.$$

$$\mathcal{K}_{sd}^{\ominus} \quad = \exp \frac{[-0.1522\ \text{V} - 0.7996\ \text{V}] \times 96485\ \text{C mol}^{-1}}{8.314\ \text{G K}^{-1} \times 298.15\ \text{K}} = \mathbf{8.14 \times 10^{-17}}.$$

Saturation concentration c_{sd}:

The saturation concentration c_{sd} of the sparingly soluble silver iodide in aqueous solution can be obtained from the solubility product,

$$\mathcal{K}_{sd}^{\ominus} \quad = c_r(\text{Ag}^+) \times c_r(\text{I}^-),$$

by taking into account that

$$c_{sd} \quad = c(\text{Ag}^+) = c(\text{I}^-)$$

(see Section 6.6 of the textbook "Physical Chemistry of a Different Angle"). Inserting this relationship into the equation for the solubility product

$$\mathcal{K}_{sd}^{\ominus} \quad = (c_{sd}/c^{\ominus}) \times (c_{sd}/c^{\ominus}) = (c_{sd}/c^{\ominus})^2,$$

and solving for c_{sd} results in:

$$c_{sd} = \sqrt{K_{sd}^{\ominus}}c^{\ominus} = \sqrt{8.14\times10^{-17}}\times1\ \text{kmol m}^{-3} = \mathbf{9.02\times10^{-9}\ kmol\ m^{-3}}.$$

2.23.6 Concentration cell

Difference ΔE of the redox potentials for the galvanic cell in question:

The redox potential of a silver-silver ion electrode is described by the following NERNST equation:

$$E = E^{\ominus}(\text{Ag/Ag}^+) + \frac{RT^{\ominus}}{\mathcal{F}}\times\ln c_r(\text{Ag}^+).$$

Thus, we obtain for the difference ΔE of the redox potentials:

$$\Delta E = E_2 - E_1$$

$$\Delta E = \left[E^{\ominus}(\text{Ag/Ag}^+) + \frac{RT^{\ominus}}{\mathcal{F}}\times\ln c_2(\text{Ag}^+)/c^{\ominus}\right] - \left[E^{\ominus}(\text{Ag/Ag}^+) + \frac{RT^{\ominus}}{\mathcal{F}}\times\ln c_1(\text{Ag}^+)/c^{\ominus}\right]$$

The term $E^{\ominus}(\text{Ag/Ag}^+)$ cancels out, i.e. it remains

$$\Delta E = \frac{RT^{\ominus}}{\mathcal{F}}\times\ln\frac{c_2(\text{Ag}^+)}{c_1(\text{Ag}^+)} \quad \text{and therefore}$$

$$\Delta E = \frac{8.314\ \text{G K}^{-1}\times298\ \text{K}}{96485\ \text{C mol}^{-1}}\times\ln\frac{0.0100\ \text{kmol m}^{-3}}{0.0005\ \text{kmol m}^{-3}} = \mathbf{+0.077\ V}.$$

Such galvanic cells, which differ only in the concentration of the electrolyte, are referred to as *concentration cells*.

2.23.7 Galvanic cell

a) Half-reactions and total reaction of the cell:

Oxidation: $Cr^{2+} \rightarrow Cr^{3+} + e^-$ (left, in the cell diagram),
Reduction: $Fe^{3+} + e^- \rightarrow Fe^{2+}$ (right, in the cell diagram),

Total reaction: $Cr^{2+} + Fe^{3+} \rightarrow Cr^{3+} + Fe^{2+}$.

b) Concentration dependence of the cell voltage:

The NERNST equation for the general reaction

$$\text{Rd} + \text{Ox}^* \rightarrow \text{Rd}^* + \text{Ox}$$

is [see Eq. (23.16)]:

$$\Delta E = \Delta\mathring{E} + \frac{RT}{\nu_e\mathcal{F}}\times\ln\frac{c(\text{Ox}^*)\times c(\text{Rd})}{c(\text{Rd}^*)\times c(\text{Ox})}$$

with $\Delta\overset{\circ}{E} = \overset{\circ}{E}(\text{Ox*/Rd*}) - \overset{\circ}{E}(\text{Ox/Rd})$ as the basic value of the potential difference ΔE for the considered cell. For the concentration dependence of ΔE at 25 °C, we obtain in the present example:

$$\Delta E \quad = \Delta E^{\ominus} + \frac{RT^{\ominus}}{\mathcal{F}} \times \ln \frac{c_r(\text{Fe}^{3+}) \times c_r(\text{Cr}^{2+})}{c_r(\text{Fe}^{2+}) \times c_r(\text{Cr}^{3+})} \, .$$

c) Standard value $\Delta E^{\ominus}$ of the potential difference:

The standard values $E^{\ominus}$ of the redox potentials are given in Table 23.1 in the textbook "Physical Chemistry from a Different Angle":

$$E^{\ominus}(\text{Cr}^{2+}/\text{Cr}^{3+}) = -0.407 \text{ V},$$

$$E^{\ominus}(\text{Fe}^{2+}/\text{Fe}^{3+}) = +0.771 \text{ V}.$$

Correspondingly, we obtain for the standard value $\Delta E^{\ominus}$ of the potential difference [cf. Eq. (23.11)]:

$$\Delta E^{\ominus} \quad = E^{\ominus}(\text{Fe}^{2+}/\text{Fe}^{3+}) - E^{\ominus}(\text{Cr}^{2+}/\text{Cr}^{3+}) = (+0.771 \text{ V}) - (-0.407 \text{ V}) = \mathbf{+1.178 \text{ V}} \, .$$

d) Chemical drive $\mathcal{A}^{\ominus}$ of the cell reaction:

The following relationship exists between the chemical drive $\mathcal{A}^{\ominus}$ of the cell reaction and the potential difference $\Delta E^{\ominus}$ of the cell [see Eq. (23.15)]:

$$\Delta E^{\ominus} \quad = \frac{\mathcal{A}^{\ominus}}{\nu_e \mathcal{F}} \quad \Rightarrow$$

$$\mathcal{A}^{\ominus} \quad = \nu_e \mathcal{F} \times \Delta E^{\ominus} \, .$$

$$\mathcal{A}^{\ominus} \quad = 1 \times 96485 \text{ C mol}^{-1} \times (+1.178 \text{ V}) = +113.7 \times 10^3 \text{ G} = \mathbf{+113.7 \text{ kG}} \, .$$

The chemical drive is positive, i.e. the cell reaction takes place spontaneously in the direction indicated by the conversion formula. If the reaction proceeds spontaneously, the electrode with the higher redox potential (here $\text{Pt}|\text{Fe}^{2+},\text{Fe}^{3+}$) becomes the cathode and the electrode with the lower redox potential (here $\text{Pt}|\text{Cr}^{2+},\text{Cr}^{3+}$) the anode.

1.23.8* Potentiometric redox titration

a) Redox potential E as function of τ before the equivalence point:

Before the equivalence point, the redox pair $\text{Fe}^{2+}/\text{Fe}^{3+}$ has to be considered. For the dependence of the redox potential E of this pair on the degree of titration τ, the following equation was given in the text of the exercise:

$$E \quad = E^{\ominus} + \frac{0{,}059 \text{ V}}{\nu_e} \times \lg \frac{\tau}{1-\tau} \, .$$

For the redox potential $E^\ominus$ of the pair Fe^{2+}/Fe^{3+} under standard conditions, we find the value +0.771 V in Table 23.1 of the textbook "Physical Chemistry from a Different Angle." Inserting results in:

$$E = +0.771\,\text{V} + 0.059\,\text{V} \times \lg \frac{\tau}{1-\tau}.$$

At the beginning ($\tau = 0$), there are theoretically only Fe^{2+} ions in the solution and no Fe^{3+} ions. Accordingly, the value of E would mathematically be equal to $-\infty$.

If the Ce^{4+} solution is gradually added to the Fe^{2+} solution, more and more Fe^{3+} ions are formed, which are present in the solution together with the remaining Fe^{2+} ions; the Ce^{4+} ions, however, are almost completely converted into Ce^{3+} ions. For $\tau = 1/101$, we obtain by inserting this value into the above equation:

$$E = +0.771\,\text{V} + 0.059\,\text{V} \times \lg \frac{1/101}{1 - 1/101} = +0.771\,\text{V} + 0.059\,\text{V} \times \lg \frac{1/101}{100/101}$$

$$= +0.771\,\text{V} + 0.059\,\text{V} \times \lg \frac{1}{100} = +0.771\,\text{V} + 0.059\,\text{V} \times (-2) = +0.653\,\text{V}.$$

The remaining values can be calculated analogously.

τ	$1/101$	$1/11$	$1/2$	$10/11$	$100/101$
$\lg[\tau/(1-\tau)]$	−2	−1	0	+1	+2
E/V	+0.653	+0.712	+0.771	+0.830	+0.889

b) Conversion formula and basic stoichiometric equation:

Conversion formula: $Fe^{2+}|w + Ce^{4+}|w \rightarrow Fe^{3+}|w + Ce^{3+}|w$

Basic equation: $\Delta\xi = \dfrac{\Delta n_{Fe^{2+}}}{\nu_{Fe^{2+}}} = \dfrac{\Delta n_{Ce^{4+}}}{\nu_{Ce^{4+}}}$ Eq. (1.15)

Amount $n_{Fe^{2+},0}$ of Fe^{2+} ions in the initial solution:

The amount $n_{Fe^{2+},0}$ of Fe^{2+} ions in the initial solution can be calculated by means of equation (1.9):

$$c_{Fe^{2+},0} = \frac{n_{Fe^{2+},0}}{V_0} \quad \Rightarrow$$

$$n_{Fe^{2+},0} = c_{Fe^{2+},0} \times V_0 = (0.010 \times 10^3\,\text{mol}\,\text{m}^{-3}) \times (100 \times 10^{-6}\,\text{m}^3) = 0.001\,\text{mol}.$$

Volume $V_{T,ep}$ of Ce^{4+} solution required to reach the equivalence point:

Until the equivalence point, the total amount $n_{Fe^{2+},0}$ of Fe^{2+} ions originally present in the solution was consumed. Therefore, the change $\Delta n_{Fe^{2+}}$ of amount of Fe^{2+} ions results in $\Delta n_{Fe^{2+}} = 0 - n_{Fe^{2+},0} = -0.001$ mol.

The corresponding change $\Delta n_{Ce^{4+}}$ of amount of Ce^{4+} ions can be calculated using the basic stoichiometric equation,

$$\Delta n_{Ce^{4+}} = \frac{\nu_{Ce^{4+}} \times \Delta n_{Fe^{2+}}}{\nu_{Fe^{2+}}} = \frac{(-1)\times(-0.001\,\text{mol})}{(-1)} = -0.001\,\text{mol},$$

i.e., because of $\Delta n_{Ce^{4+}} = 0 - n_{Ce^{4+}}$ an amount $n_{Ce^{4+}} = -\Delta n_{Ce^{4+}} = +0.001$ mol of Ce^{4+} ions was consumed. This amount corresponds to a volume of

$$V_{T,eq} = \frac{n_{Ce^{4+}}}{c_{Ce^{4+}}} = \frac{0.001\,\text{mol}}{1000\,\text{mol}\,\text{m}^{-3}} = 1.000\times10^{-6}\,\text{m}^3 = \mathbf{1.000\,mL}.$$

Redox potential E_{ep} at the equivalence point ($\tau = 1$):

At the equivalence point, theoretically only Fe^{3+} and Ce^{3+} ions should be present. However, the two underlying redox reactions

$$Fe^{2+}|w \rightarrow Fe^{3+}|w + e^-$$

$$Ce^{3+}|w \rightarrow Ce^{4+}|w + e^-$$

are to a certain, albeit very small, extent reversible. Therefore, both redox pairs are present at the equivalence point. For the redox potential E_{ep}, we therefore obtain

$$E_{ep} = E^{\ominus}(Fe^{2+}/Fe^{3+}) + \frac{0.059\,\text{V}}{1} \times \lg\frac{c_{Fe^{3+}}}{c_{Fe^{2+}}},$$

but also

$$E_{ep} = E^{\ominus}(Ce^{3+}/Ce^{4+}) + \frac{0.059\,\text{V}}{1} \times \lg\frac{c_{Ce^{4+}}}{c_{Ce^{3+}}}.$$

Addition of both equations results in:

$$2E_{ep} = E^{\ominus}(Fe^{2+}/Fe^{3+}) + E^{\ominus}(Ce^{3+}/Ce^{4+}) + 0.059\,\text{V} \times \lg\frac{c_{Fe^{3+}} \times c_{Ce^{4+}}}{c_{Fe^{2+}} \times c_{Ce^{3+}}}.$$

Since $c_{Fe^{3+}} = c_{Ce^{3+}}$ as well as $c_{Fe^{2+}} = c_{Ce^{4+}}$ at the equivalence point, we have

$$2E_{ep} = E^{\ominus}(Fe^{2+}/Fe^{3+}) + E^{\ominus}(Ce^{3+}/Ce^{4+}) + 0.059\,\text{V} \times \lg 1$$

and finally

$$E_{ep} = \frac{E^{\ominus}(Fe^{2+}/Fe^{3+}) + E^{\ominus}(Ce^{3+}/Ce^{4+})}{2}.$$

$$E_{ep} = \frac{+0.771\ \text{V} + (+1.70\ \text{V})}{2} = \mathbf{+1.234\ V}.$$

c) Redox potential E as function of τ after the equivalence point:

After the equivalence point, the redox pair Ce^{3+}/Ce^{4+} is decisive for the redox potential. For the dependence of the redox potential E of this pair on the degree of titration τ, the following equation was given in the text of the exercise:

$$E = E^{\ominus} + \frac{0.059\ \text{V}}{v_e} \times \ln(\tau - 1).$$

Inserting results in:

$$E = +1.70\ \text{V} + 0.059\ \text{V} \times \lg(\tau - 1).$$

For $\tau = {}^{101}\!/_{100}$, we obtain:

$$E = +1.70\ \text{V} + 0.059\ \text{V} \times \lg({}^{101}\!/_{100} - 1) = +1.70\ \text{V} + 0.059\ \text{V} \times \lg({}^{101}\!/_{100} - {}^{100}\!/_{100})$$

$$= +1.70\ \text{V} + 0.059\ \text{V} \times \lg\frac{1}{100} = +1.70\ \text{V} + 0.059\ \text{V} \times (-2) = +1.582\ \text{V}.$$

The remaining values can be calculated analogously.

τ	${}^{101}\!/_{100}$	${}^{11}\!/_{10}$	2
$\lg(\tau - \tau)$	−2	−1	0
E/V	+1.582	+1.641	+1.70

d) Redox potential E as function of the added volume of Ce^{4+} solution:

Since it is known that 1.000 mL of Ce^{4+} solution is required to reach the equivalence point, the degree of titration τ can easily be converted into the volume V (in ml). We receive the table below (which summarizes all previous results) and the corresponding titration curve.

What is striking is the great similarity of this titration curve with the curve for the titration of the base of a weakly basic pair with the acid of a strongly basic pair [see Fig. 7.6 b) in the textbook "Physical Chemistry from a Different Angle"]. This is due to the fact that the redox titration is based on the same principles as the acid-base titration. Both the redox reactions and the acid-base reactions are transfer reactions, except that in one case an electron exchange takes place and in the other a proton exchange.

V / mL	E / V
0.010	+0.653
0.091	+0.712
0.500	+0.771
0.909	+0.830
0.990	+0.899
1.000	+1.234
1.010	+1.583
1.100	+1.641
2.000	+1.70

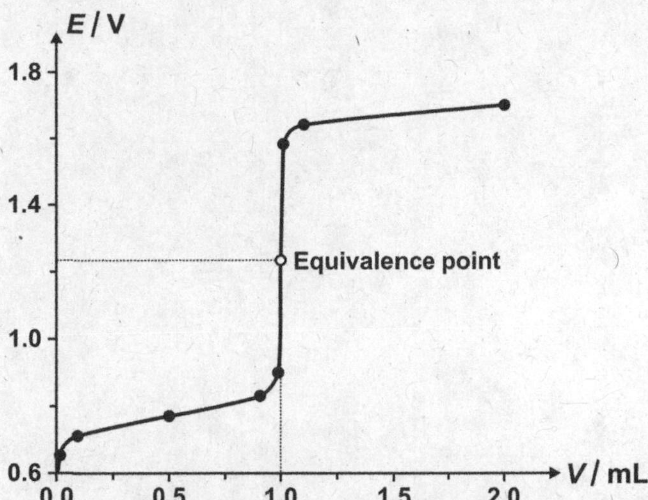

Correction to: Physical Chemistry from a Different Angle Workbook

Correction to:
G. Job and R. Rüffler, *Physical Chemistry from a Different Angle Workbook,*
https://doi.org/10.1007/978-3-030-28491-6

Due to a Publisher error in the original version of the book, the following correction has been incorporated.
The book authors' "Georg Job and Regina Rüffler" affiliation has been updated:

From: "Job-Stiftung, University of Hamburg, Hamburg, Germany",
To: "Job Foundation, Hamburg, Germany".

The updated version of the book can be found at
https://doi.org/10.1007/978-3-030-28491-6

G. Job and R. Rüffler, *Physical Chemistry from a Different Angle Workbook*,
https://doi.org/10.1007/978-3-030-28491-6_3

Printed in the United States
By Bookmasters